P9-APW-819

# Forgotten Books

---

# FREE BOOKS

## *www.forgottenbooks.org*

---

*Truth may seem, but cannot be:*
*Beauty brag, but 'tis not she;*
*Truth and beauty buried be.*

*To this urn let those repair*
*That are either true or fair;*
*For these dead birds sigh a prayer.*

**Bacon**

# MUTUAL AID

## Some Press Opinions

**The Athenæum**: Prince Kropotkin has written a most suggestive and stimulating study, showing not only immense range of reading and observation, and the power of marshalling facts in an orderly manner, but also a most attractive and generous personality, demonstrating unconquerable faith in the innate goodness of humanity and its golden future.

**The Review of Reviews**: There are few more delightful books to read than "Mutual Aid as a Factor in Evolution." It is a good, healthy, cheerful, delightful book, which does one good to read.

**Nature**: The book is undeniably readable throughout. The author has a creed which he preaches with all the fervour of genuine conviction. He is anxious to make converts, but his zeal never leads him to forget fairness and courtesy. Those who disagree with him may learn much by studying the book.

**The Speaker**: Prince Kropotkin has done an important service to sound thinking on questions of sociology.

**The British Weekly**: The array of facts from all departments of life which he here collects is overwhelming. To most readers the facts adduced from mediæval history and from modern conditions will be new. It will be for scientific readers to adjust the burden borne by each of the factors in evolution, but that mutual aid is an essential and important factor no one can doubt who reads this book. Though so full of facts, it is never tedious and never dry, but carries the reader with it.

**The Clarion**: If I described this new work of Prince Kropotkin's as the most important English book published for several years, I should not be guilty of excess. If any work of like value has appeared within the last ten years, I do not remember it. Read it! Read it!

LONDON: WILLIAM HEINEMANN

# MUTUAL AID

## A FACTOR OF EVOLUTION

BY

## P. KROPOTKIN

*Revised and Cheaper Edition*

LONDON
WILLIAM HEINEMANN

First Edition,       price 7s. 6d., 1902
Second Impression,   „       „    1903
Revised and Cheaper Edition,   1904

# CONTENTS

PAGE

INTRODUCTION . . . . . . . . vii

## CHAPTER I
### MUTUAL AID AMONG ANIMALS

Struggle for existence.—Mutual Aid—a law of Nature and chief factor of progressive evolution.—Invertebrates.—Ants and bees.— Birds : Hunting and fishing associations.—Sociability.—Mutual protection among small birds.—Cranes ; parrots . . . . 1

## CHAPTER II
### MUTUAL AID AMONG ANIMALS (*continued*)

Migrations of birds.—Breeding associations.—Autumn societies.— Mammals : small number of unsociable species.—Hunting associations of wolves, lions, etc.—Societies of rodents ; of ruminants ; of monkeys.—Mutual Aid in the struggle for life.—Darwin's arguments to prove the struggle for life within the species.—Natural checks to over-multiplication.—Supposed extermination of intermediate links.— Elimination of competition in Nature . . . . . 32

## CHAPTER III
### MUTUAL AID AMONG SAVAGES

Supposed war of each against all.—Tribal origin of human society. —Late appearance of the separate family.—Bushmen, Hottentots.— Australians, Papuas.—Eskimos, Aleoutes.—Features of savage life difficult to understand for the European.—The Dayak's conception of justice.—Common law . . . . . . 76

## CHAPTER IV
### MUTUAL AID AMONG THE BARBARIANS

The great migrations.—New organization rendered necessary.— The village community.—Communal work.—Judicial procedure.— Inter-tribal law.—Illustrations from the life of our contemporaries.— Buryates.—Kabyles.—Caucasian mountaineers.—African stems . 115

## CHAPTER V
### MUTUAL AID IN THE MEDIÆVAL CITY

Growth of authority in Barbarian society.—Serfdom in the villages. —Revolt of the fortified towns : their liberation ; their charts.— The guild.—Double origin of the free mediæval city.—Self-jurisdiction, self-administration.—Honourable position of labour.—Trade by the guild and by the city . . . . . . 153

PAGE

## CHAPTER VI

### MUTUAL AID IN THE MEDIÆVAL CITY (*continued*)

Likeness and diversity among the mediæval cities.—The craft-guilds: State-attributes in each of them.—Attitude of the city towards the peasants; attempts to free them.—The lords.—Results achieved by the mediæval city: in arts, in learning.—Causes of decay . . . . . . . . . 187

## CHAPTER VII

### MUTUAL AID AMONGST OURSELVES

Popular revolts at the beginning of the State-period.—Mutual Aid institutions of the present time.—The village community: its struggles for resisting its abolition by the State.—Habits derived from the village-community life, retained in our modern villages.—Switzerland, France, Germany, Russia . . . . . 223

## CHAPTER VIII

### MUTUAL AID AMONGST OURSELVES (*continued*)

Labour-unions grown after the destruction of the guilds by the State.—Their struggles.—Mutual Aid in strikes.—Co-operation.—Free associations for various purposes.—Self-sacrifice.—Countless societies for combined action under all possible aspects.—Mutual Aid in slum-life.—Personal aid . . . . . . 262

CONCLUSION . . . . . . . . . 293

## APPENDIX

I. SWARMS OF BUTTERFLIES, DRAGON-FLIES, ETC . . 301

II. THE ANTS . . . . . . . . 302

III. NESTING ASSOCIATIONS . . . . . . 304

IV. SOCIABILITY OF ANIMALS . . . . . 306

V. CHECKS TO OVER-MULTIPLICATION . . . 307

VI. ADAPTATIONS TO AVOID COMPETITION . . . 310

VII. THE ORIGIN OF THE FAMILY . . . . 313

VIII. DESTRUCTION OF PRIVATE PROPERTY ON THE GRAVE . 320

IX. THE "UNDIVIDED FAMILY" . . . . 320

X. THE ORIGIN OF THE GUILDS . . . . 321

XI. THE MARKET AND THE MEDIÆVAL CITY . . 325

XII. MUTUAL-AID ARRANGEMENTS IN THE VILLAGES OF NETHERLAND AT THE PRESENT DAY . . . 327

INDEX . . . . . . . 329

# INTRODUCTION

Two aspects of animal life impressed me most during the journeys which I made in my youth in Eastern Siberia and Northern Manchuria. One of them was the extreme severity of the struggle for existence which most species of animals have to carry on against an inclement Nature; the enormous destruction of life which periodically results from natural agencies; and the consequent paucity of life over the vast territory which fell under my observation. And the other was, that even in those few spots where animal life teemed in abundance, I failed to find—although I was eagerly looking for it—that bitter struggle for the means of existence, *among animals belonging to the same species,* which was considered by most Darwinists (though not always by Darwin himself) as the dominant characteristic of struggle for life, and the main factor of evolution.

The terrible snow-storms which sweep over the northern portion of Eurasia in the later part of the winter, and the glazed frost that often follows them; the frosts and the snow-storms which return every year in the second half of May, when the trees are already in full blossom and insect life swarms everywhere; the early frosts and, occasionally, the heavy snowfalls in July and August, which suddenly destroy myriads of insects, as well as the second broods of the birds in the prairies; the torrential rains, due to the monsoons, which fall in more temperate regions in August and September—resulting in inundations on a scale which is only known in America and in Eastern Asia, and swamping, on the plateaus, areas as wide as European States; and finally, the heavy snowfalls, early in October, which eventually render a territory as large as France and Germany, absolutely impracticable for ruminants, and destroy them by the thousand—these were the conditions under which I saw animal life struggling in Northern Asia. They made me realize at an early date the overwhelming importance in Nature of what Darwin described as "the natural checks to over-multiplication," in comparison to the struggle between individuals of the same species for the means of subsistence, which may go on here and there, to some limited extent, but never attains the importance of the former. Paucity of life, under-population—not over-population—being the distinctive feature of that immense part of the globe which we name Northern Asia, I conceived since then serious doubts—which subsequent study has only confirmed —as to the reality of that fearful competition for food and life within each species, which was an article of faith with most Darwinists, and, consequently, as to the dominant part which this sort of competition was supposed to play in the evolution of new species.

On the other hand, wherever I saw animal life in abundance, as, for instance, on the lakes where scores of species and millions of individuals came together to rear their progeny; in the colonies of rodents; in the migrations of birds which took place at that time on a truly American scale along the Usuri; and especially in a migration of fallow-deer which I witnessed on the Amur, and during which scores of

thousands of these intelligent animals came together
from an immense territory, flying before the coming
deep snow, in order to cross the Amur where it is
narrowest—in all these scenes of animal life which
passed before my eyes, I saw Mutual Aid and Mutual
Support carried on to an extent which made me
suspect in it a feature of the greatest importance for
the maintenance of life, the preservation of each
species, and its further evolution.

And finally, I saw among the semi-wild cattle and
horses in Transbaikalia, among the wild ruminants
everywhere, the squirrels, and so on, that when
animals have to struggle against scarcity of food, in
consequence of one of the above-mentioned causes,
the whole of that portion of the species which is
affected by the calamity, comes out of the ordeal so
much impoverished in vigour and health, that *no
progressive evolution of the species can be based upon
such periods of keen competition.*

Consequently, when my attention was drawn, later
on, to the relations between Darwinism and Sociology,
I could agree with none of the works and pamphlets
that had been written upon this important subject.
They all endeavoured to prove that Man, owing to
his higher intelligence and knowledge, *may* mitigate
the harshness of the struggle for life between men ;
but they all recognized at the same time that
the struggle for the means of existence, of every
animal against all its congeners, and of every man
against all other men, was "a law of Nature." This
view, however, I could not accept, because I was
persuaded that to admit a pitiless inner war for life
within each species, and to see in that war a condition
of progress, was to admit something which not only

had not yet been proved, but also lacked confirmation from direct observation.

On the contrary, a lecture "On the Law of Mutual Aid," which was delivered at a Russian Congress of Naturalists, in January 1880, by the well-known zoologist, Professor Kessler, the then Dean of the St. Petersburg University, struck me as throwing a new light on the whole subject. Kessler's idea was, that besides the *law of Mutual Struggle* there is in Nature *the law of Mutual Aid*, which, for the success of the struggle for life, and especially for the progressive evolution of the species, is far more important than the law of mutual contest. This suggestion— which was, in reality, nothing but a further development of the ideas expressed by Darwin himself in *The Descent of Man*—seemed to me so correct and of so great an importance, that since I became acquainted with it (in 1883) I began to collect materials for further developing the idea, which Kessler had only cursorily sketched in his lecture, but had not lived to develop. He died in 1881.

In one point only I could not entirely endorse Kessler's views. Kessler alluded to "parental feeling" and care for progeny (see below, Chapter I.) as to the source of mutual inclinations in animals. However, to determine how far these two feelings have really been at work in the evolution of sociable instincts, and how far other instincts have been at work in the same direction, seems to me a quite distinct and a very wide question, which we hardly can discuss yet. It will be only after we have well established the facts of mutual aid in different classes of animals, and their importance for evolution, that we shall be able to study what belongs in the evolution of sociable feelings, to parental feelings, and what to sociability proper—the latter having evidently its origin at the earliest stages of the evolution of the animal world, perhaps even at the "colony-stages." I consequently directed my chief attention to establishing first of all, the importance of the Mutual Aid factor of evolution, leaving to ulterior research the task of discovering the *origin* of the Mutual Aid instinct in Nature.

The importance of the Mutual Aid factor—"if its generality could only be demonstrated"—did not escape the naturalist's genius so manifest in Goethe. When Eckermann told once to Goethe—it was in 1827—that two little wren-fledglings, which had run away from him, were found by him next day in a nest of robin redbreasts (*Rothkehlchen*), which fed the little ones, together with their own youngsters, Goethe grew quite excited about this fact. He saw in it a confirmation of his pantheistic views, and said :—" If it be true that this feeding of a stranger goes through all Nature as something having the character of a general law—then many an enigma would be solved." He returned to this matter on the next day, and most earnestly entreated Eckermann (who was, as is known, a zoologist) to make a special study of the subject, adding that he would surely come "to quite invaluable treasuries of results" (*Gespräche*, edition of 1848, vol. iii. pp. 219, 221). Unfortunately, this study was never made, although it is very possible that Brehm, who has accumulated in his works such rich materials relative to mutual aid among animals, might have been inspired by Goethe's remark.

Several works of importance were published in the years 1872–1886, dealing with the intelligence and the mental life of animals (they are mentioned in a footnote in Chapter I. of this book), and three of them dealt more especially with the subject under consideration; namely, *Les Sociétés animales*, by Espinas (Paris, 1877); *La Lutte pour l'existence et l'association pour la lutte*, a lecture by J. L. Lanessan (April 1881); and Louis Büchner's book, *Liebe und Liebes-Leben in der Thierwelt*, of which the first edition appeared in 1882 or 1883, and a second, much enlarged, in 1885. But excellent though each of these works is, they leave ample room for a work in which Mutual Aid would be considered, not only as an argument in favour of a pre-human origin of moral instincts, but also as a law of Nature and a factor of evolution. Espinas devoted his main attention to such animal societies (ants, bees) as are established upon a physiological division of labour, and though his work is full of admirable hints in all possible directions, it was written at a time when the evolution of human societies could not yet be treated with the knowledge we now possess. Lanessan's lecture has more the character of a brilliantly-laid-out general plan of a work, in which mutual support would be dealt with, beginning with rocks in the sea, and then passing in review the world of plants, of animals and men. As to Büchner's work, suggestive though it is and rich in facts, I could not agree with its leading idea. The book begins with a hymn to Love, and nearly all its illustrations are intended to prove the existence of love and sympathy among animals. However, to reduce animal sociability to *love* and *sympathy* means to reduce its generality and its importance, just as human ethics based upon love and personal sympathy only have contributed to narrow the comprehension of the moral feeling as a whole.

It is not love to my neighbour—whom I often do not know at all—which induces me to seize a pail of water and to rush towards his house when I see it on fire; it is a far wider, even though more vague feeling or instinct of human solidarity and sociability which moves me. So it is also with animals. It is not love, and not even sympathy (understood in its proper sense) which induces a herd of ruminants or of horses to form a ring in order to resist an attack of wolves; not love which induces wolves to form a pack for hunting; not love which induces kittens or lambs to play, or a dozen of species of young birds to spend their days together in the autumn; and it is neither love nor personal sympathy which induces many thousand fallow-deer scattered over a territory as large as France to form into a score of separate herds, all marching towards a given spot, in order to cross there a river. It is a feeling infinitely wider than love or personal sympathy—an instinct that has been slowly developed among animals and men in the course of an extremely long evolution, and which has taught animals and men alike the force they can borrow from the practice of mutual aid and support, and the joys they can find in social life.

The importance of this distinction will be easily appreciated by the student of animal psychology, and the more so by the student of human ethics. Love, sympathy and self-sacrifice certainly play an immense part in the progressive development of our moral feelings. But it is not love and not even sympathy upon which Society is based in mankind. It is the conscience—be it only at the stage of an instinct—of human solidarity. It is the unconscious recognition of the force that is borrowed by each man from the practice of mutual aid ; of the close dependency of every one's happiness upon the happiness of all ; and of the sense of justice, or equity, which brings the individual to consider the rights of every other individual as equal to his own. Upon this broad and necessary foundation the still higher moral feelings are developed. But this subject lies outside the scope of the present work, and I shall only indicate here a lecture, " Justice and Morality," which I delivered in reply to Huxley's *Ethics*, and in which the subject has been treated at some length.

Consequently I thought that a book, written on *Mutual Aid as a Law of Nature and a factor of evolution*, might fill an important gap. When Huxley issued, in 1888, his " Struggle-for-life " manifesto (*Struggle for Existence and its Bearing upon Man*), which to my appreciation was a very incorrect representation of the facts of Nature, as one sees them in the bush and in the forest, I communicated with the editor of the *Nineteenth Century*, asking him whether he would give the hospitality of his review to an elaborate reply to the views of one of the most prominent Darwinists ; and Mr. James Knowles received the proposal with fullest sympathy. I also spoke of it to W. Bates. " Yes, certainly ; *that* is true Darwinism," was his reply. " It is horrible what ' they ' have made of Darwin. Write these articles, and when they are printed, I will write to you a letter which you may publish." Unfortunately, it took me nearly seven years to write these articles, and when the last was published, Bates was no longer living.

After having discussed the importance of mutual aid in various classes of animals, I was evidently bound to discuss the importance of the same factor in the evolution of Man. This was the more necessary as there are a number of evolutionists who may not refuse to admit the importance of mutual aid among animals, but who, like Herbert Spencer, will refuse to admit it for Man. For primitive Man—they maintain—war of each against all was *the* law of life. In how far this assertion, which has been too willingly repeated, without sufficient criticism, since the times of Hobbes, is supported by what we know about the early phases of human development, is discussed in the chapters given to the Savages and the Barbarians.

The number and importance of mutual-aid institutions which were developed by the creative genius of the savage and half-savage masses, during the earliest clan-period of mankind and still more during the next village-community period, and the immense influence which these early institutions have exercised upon the subsequent development of mankind, down to the present times, induced me to extend my researches to the later, historical periods as well ; especially, to study that most interesting period—the free mediæval city-republics, of which the universality and influence upon our modern civilization have not yet been duly appreciated. And finally, I have tried to indicate in brief the immense importance which the mutual-support instincts, inherited by mankind from its extremely long evolution, play even now in our modern society, which is supposed to rest upon the principle: " every one for himself, and the State for all," but which it never has succeeded, nor will succeed in realizing.

It may be objected to this book that both animals and men are represented in it under too favourable an aspect ; that their sociable qualities are insisted upon, while their anti-social and self-asserting instincts are hardly touched upon. This was, however, unavoidable. We have heard so much lately of the " harsh, pitiless struggle for life," which was said to be carried on by every animal against all other animals, every "savage" against all other "savages," and every civilized man against all his co-citizens—and these assertions have so much become an article of faith—that it was necessary, first of all, to oppose to them a wide series of facts showing animal and human life under a quite different aspect. It was necessary to indicate the overwhelming importance which sociable habits play in Nature and in the progressive evolution of both the animal species and human beings : to prove that they secure to animals a better protection from their enemies, very often facilities for getting food (winter provisions, migrations, etc.), longevity, and therefore a greater facility for the development of intellectual faculties ; and that they have given to men, in addition to the same advantages, the possibility of working out those institutions which have enabled mankind to survive in its hard struggle against Nature, and to progress, notwithstanding all the vicissitudes of its history. It is a book on the law of Mutual Aid, viewed at as one of the chief factors of evolution—not on *all* factors of evolution and their respective values ; and this first book had to be written, before the latter could become possible.

I should certainly be the last to underrate the part which the self-assertion of the individual has played in the evolution of mankind. However, this subject requires, I believe, a much deeper treatment than the one it has hitherto received. In the history of mankind, individual self-assertion has often been, and continually is, something quite different from, and far larger and deeper than, the petty, unintelligent narrow-mindedness, which, with a large class of writers, goes for "individualism" and "self-assertion." Nor have history-making individuals been limited to those whom historians have represented as heroes. My intention, consequently, is, if circumstances permit it, to discuss separately the part taken by the self-assertion of the individual in the progressive evolution of mankind. I can only make in this place the following general remark :—When the Mutual Aid institutions—the tribe, the village community, the guilds, the mediæval city—began, in the course of history, to lose their primitive character, to be invaded by parasitic growths, and thus to become hindrances to progress, the revolt of individuals against these institutions took always two different aspects. Part of those who rose up strove to purify the old institutions, or to work out a higher form of commonwealth, based upon the same Mutual Aid principles; they tried, for instance, to introduce the principle of "compensation," instead of the *lex talionis*, and later on, the pardon of offences, or a still higher ideal of equality before the human conscience, *in lieu* of "compensation," according to class-value. But at the very same time, another portion of the same individual rebels endeavoured to break down the protective institutions of mutual support, with no other intention but to increase their own wealth and their own powers. In this three-cornered contest, between the two classes of revolted individuals and the supporters of what existed, lies the real tragedy of history. But to delineate that contest, and honestly to study the part played in the evolution of mankind by each one of these three forces, would require at least as many years as it took me to write this book.

Of works dealing with nearly the same subject which have been published since the publication of my articles on Mutual Aid among Animals, I must mention *The Lowell Lectures on the Ascent of Man*, by Henry Drummond (London, 1894), and *The Origin and Growth of the Moral Instinct*, by A. Sutherland (London, 1898). Both are constructed chiefly on the lines taken in Büchner's *Love*, and in the second work the parental and familial feeling as the sole influence at work in the development of the moral feelings has been dealt with at some length. A third work dealing with man and written on similar lines is *The Principles of Sociology*, by Prof. F. A. Giddings, the first edition of which was published in 1896 at New York and London, and the leading ideas of which were sketched by the author in a pamphlet in 1894. I must leave, however, to literary critics the task of discussing the points of contact, resemblance, or divergence between these works and mine.

The different chapters of this book were published first in the *Nineteenth Century* ("Mutual Aid among Animals," in September and November 1890; "Mutual Aid among Savages," in April 1891; "Mutual Aid among the Barbarians," in January 1892; "Mutual Aid in the Mediæval City," in August and September 1894; and "Mutual Aid amongst Modern Men," in January and June 1896). In bringing them out in a book form my first intention was to embody in an Appendix the mass of materials, as well as the discussion of several secondary points, which had to be omitted in the review articles. It appeared, however, that the Appendix would double the size of the book, and I was compelled to abandon, or, at least, to postpone its publication. The present

Appendix includes the discussion of only a few points which have been the matter of scientific controversy during the last few years; and into the text I have introduced only such matter as could be introduced without altering the structure of the work.

I am glad of this opportunity for expressing to the editor of the *Nineteenth Century*, Mr. James Knowles, my very best thanks, both for the kind hospitality which he offered to these papers in his review, as soon as he knew their general idea, and the permission he kindly gave me to reprint them.

*Bromley, Kent,*
*1902.*

# CHAPTER I

## MUTUAL AID AMONG ANIMALS

Struggle for existence.—Mutual Aid—a law of Nature and chief factor of progressive evolution.—Invertebrates.—Ants and Bees.—Birds: Hunting and fishing associations.—Sociability.—Mutual protection among small birds.—Cranes; parrots.

THE conception of struggle for existence as a factor of evolution, introduced into science by Darwin and Wallace, has permitted us to embrace an immensely-wide range of phenomena in one single generalization, which soon became the very basis of our philosophical, biological, and sociological speculations. An immense variety of facts:—adaptations of function and structure of organic beings to their surroundings; physiological and anatomical evolution; intellectual progress, and moral development itself, which we formerly used to explain by so many different causes, were embodied by Darwin in one general conception. We understood them as continued endeavours—as a struggle against adverse circumstances—for such a development of individuals, races, species and societies, as would result in the greatest possible fulness, variety, and intensity of life. It may be that at the outset Darwin himself was not fully aware of the generality of the factor

B

which he first invoked for explaining one series only of facts relative to the accumulation of individual variations in incipient species. But he foresaw that the term which he was introducing into science would lose its philosophical and its only true meaning if it were to be used in its narrow sense only—that of a struggle between separate individuals for the sheer means of existence. And at the very beginning of his memorable work he insisted upon the term being taken in its "large and metaphorical sense including dependence of one being on another, and including (which is more important) not only the life of the individual, but success in leaving progeny."[1]

While he himself was chiefly using the term in its narrow sense for his own special purpose, he warned his followers against committing the error (which he seems once to have committed himself) of overrating its narrow meaning. In *The Descent of Man* he gave some powerful pages to illustrate its proper, wide sense. He pointed out how, in numberless animal societies, the struggle between separate individuals for the means of existence disappears, how *struggle* is replaced by *co-operation*, and how that substitution results in the development of intellectual and moral faculties which secure to the species the best conditions for survival. He intimated that in such cases the fittest are not the physically strongest, nor the cunningest, but those who learn to combine so as mutually to support each other, strong and weak alike, for the welfare of the community. "Those communities," he wrote, "which included the greatest number of the most sympathetic members would flourish best, and rear the greatest number of offspring" (2nd edit., p. 163). The term, which

[1] *Origin of Species*, chap. iii.

originated from the narrow Malthusian conception of competition between each and all, thus lost its narrowness in the mind of one who knew Nature.

Unhappily, these remarks, which might have become the basis of most fruitful researches, were overshadowed by the masses of facts gathered for the purpose of illustrating the consequences of a real competition for life. Besides, Darwin never attempted to submit to a closer investigation the relative importance of the two aspects under which the struggle for existence appears in the animal world, and he never wrote the work he proposed to write upon the natural checks to over-multiplication, although that work would have been the crucial test for appreciating the real purport of individual struggle. Nay, on the very pages just mentioned, amidst data disproving the narrow Malthusian conception of struggle, the old Malthusian leaven reappeared—namely, in Darwin's remarks as to the alleged inconveniences of maintaining the " weak in mind and body " in our civilized societies (ch. v.). As if thousands of weak-bodied and infirm poets, scientists, inventors, and reformers, together with other thousands of so-called " fools " and " weak-minded enthusiasts," were not the most precious weapons used by humanity in its struggle for existence by intellectual and moral arms, which Darwin himself emphasized in those same chapters of *Descent of Man*.

It happened with Darwin's theory as it always happens with theories having any bearing upon human relations. Instead of widening it according to his own hints, his followers narrowed it still more. And while Herbert Spencer, starting on independent but closely-allied lines, attempted to widen the inquiry into that great question, " Who are the fittest? " especially in

the appendix to the third edition of the *Data of Ethics*,
the numberless followers of Darwin reduced the notion
of struggle for existence to its narrowest limits. They
came to conceive the animal world as a world of perpetual struggle among half-starved individuals, thirst-
ing for one another's blood. They made modern
literature resound with the war-cry of *woe to the van-
quished*, as if it were the last word of modern biology.
They raised the " pitiless " struggle for personal
advantages to the height of a biological principle
which man must submit to as well, under the menace
of otherwise succumbing in a world based upon mutual
extermination. Leaving aside the economists who
know of natural science but a few words borrowed from
second-hand vulgarizers, we must recognize that even
the most authorized exponents of Darwin's views did
their best to maintain those false ideas. In fact, if we
take Huxley, who certainly is considered as one of the
ablest exponents of the theory of evolution, were we
not taught by him, in a paper on the ' Struggle for
Existence and its Bearing upon Man,' that,

" from the point of view of the moralist, the animal world is
on about the same level as a gladiators' show. The creatures
are fairly well treated, and set to fight; whereby the strongest,
the swiftest, and the cunningest live to fight another day.
The spectator has no need to turn his thumb down, as no
quarter is given."

Or, further down in the same article, did he not tell
us that, as among animals, so among primitive men,

"the weakest and stupidest went to the wall, while the
toughest and shrewdest, those who were best fitted to cope
with their circumstances, but not the best in another way,
survived. Life was a continuous free fight, and beyond the
limited and temporary relations of the family, the Hobbesian
war of each against all was the normal state of existence." [1]

---

[1] *Nineteenth Century*, Feb. 1888, p. 165.

In how far this view of nature is supported by fact, will be seen from the evidence which will be here submitted to the reader as regards the animal world, and as regards primitive man. But it may be remarked at once that Huxley's view of nature had as little claim to be taken as a scientific deduction as the opposite view of Rousseau, who saw in nature but love, peace, and harmony destroyed by the accession of man. In fact, the first walk in the forest, the first observation upon any animal society, or even the perusal of any serious work dealing with animal life (D'Orbigny's, Audubon's, Le Vaillant's, no matter which), cannot but set the naturalist thinking about the part taken by social life in the life of animals, and prevent him from seeing in Nature nothing but a field of slaughter, just as this would prevent him from seeing in Nature nothing but harmony and peace. Rousseau had committed the error of excluding the beak-and-claw fight from his thoughts; and Huxley committed the opposite error; but neither Rousseau's optimism nor Huxley's pessimism can be accepted as an impartial interpretation of nature.

As soon as we study animals—not in laboratories and museums only, but in the forest and the prairie, in the steppe and the mountains—we at once perceive that though there is an immense amount of warfare and extermination going on amidst various species, and especially amidst various classes of animals, there is, at the same time, as much, or perhaps even more, of mutual support, mutual aid, and mutual defence amidst animals belonging to the same species or, at least, to the same society. Sociability is as much a law of nature as mutual struggle. Of course it would be extremely difficult to estimate, however roughly,

the relative numerical importance of both these series
of facts.    But if we resort to an indirect test, and ask
Nature: "Who are the fittest: those who are continually at war with each other, or those who support
one another?" we at once see that those animals
which acquire habits of mutual aid are undoubtedly
the fittest.    They have more chances to survive, and
they attain, in their respective classes, the highest
development of intelligence and bodily organization.
If the numberless facts which can be brought forward
to support this view are taken into account, we may
safely say that mutual aid is as much a law of animal
life as mutual struggle, but that, as a factor of evolution,
it most probably has a far greater importance, inas-
much as it favours the development of such habits and
characters as insure the maintenance and further
development of the species, together with the greatest
amount of welfare and enjoyment of life for the
individual, with the least waste of energy.

Of the scientific followers of Darwin, the first, as far
as I know, who understood the full purport of Mutual
Aid *as a law of Nature and the chief factor of evolution*,
was a well-known Russian zoologist, the late Dean of
the St. Petersburg University, Professor Kessler.    He
developed his ideas in an address which he delivered
in January 1880, a few months before his death, at
a Congress of Russian naturalists; but, like so many
good things published in the Russian tongue only,
that remarkable address remains almost entirely unknown.[1]

[1] Leaving aside the pre-Darwinian writers, like Toussenel, Fée,
and many others, several works containing many striking instances
of mutual aid—chiefly, however, illustrating animal intelligence—
were issued previously to that date.    I may mention those of
Houzeau, *Les facultés mentales des animaux*, 2 vols., Brussels, 1872;

"As a zoologist of old standing," he felt bound to protest against the abuse of a term—the struggle for existence—borrowed from zoology, or, at least, against overrating its importance. Zoology, he said, and those sciences which deal with man, continually insist upon what they call the pitiless law of struggle for existence. But they forget the existence of another law which may be described as the law of mutual aid, which law, at least for the animals, is far more essential than the former. He pointed out how the need of leaving progeny necessarily brings animals together, and, "the more the individuals keep together, the more they mutually support each other, and the more are the chances of the species for surviving, as well as for making further progress in its intellectual development." "All classes of animals," he continued, "and especially the higher ones, practise mutual aid," and he illustrated his idea by examples borrowed from the life of the burying beetles and the social life of birds and some mammalia. The examples were few, as

L. Büchner's *Aus dem Geistesleben der Thiere*, 2nd ed. in 1877; and Maximilian Perty's *Ueber das Seelenleben der Thiere*, Leipzig, 1876. Espinas published his most remarkable work, *Les Sociétés animales*, in 1877, and in that work he pointed out the importance of animal societies, and their bearing upon the preservation of species, and entered upon a most valuable discussion of the origin of societies. In fact, Espinas's book contains all that has been written since upon mutual aid, and many good things besides. If I nevertheless make a special mention of Kessler's address, it is because he raised mutual aid to the height of a law much more important in evolution than the law of mutual struggle. The same ideas were developed next year (in April 1881) by J. Lanessan in a lecture published in 1882 under this title: *La lutte pour l'existence et l'association pour la lutte.* G. Romanes's capital work, *Animal Intelligence*, was issued in 1882, and followed next year by the *Mental Evolution in Animals.* About the same time (1883), Büchner published another work, *Liebe und Liebes-Leben in der Thierwelt*, a second edition of which was issued in 1885. The idea, as seen, was in the air.

might have been expected in a short opening address, but the chief points were clearly stated ; and, after mentioning that in the evolution of mankind mutual aid played a still more prominent part, Professor Kessler concluded as follows :—

"I obviously do not deny the struggle for existence, but I maintain that the progressive development of the animal kingdom, and especially of mankind, is favoured much more by mutual support than by mutual struggle. . . . All organic beings have two essential needs : that of nutrition, and that of propagating the species. The former brings them to a struggle and to mutual extermination, while the needs of maintaining the species bring them to approach one another and to support one another. But I am inclined to think that in the evolution of the organic world—in the progressive modification of organic beings—mutual support among individuals plays a much more important part than their mutual struggle." [1]

The correctness of the above views struck most of the Russian zoologists present, and Syevertsoff, whose work is well known to ornithologists and geographers, supported them and illustrated them by a few more examples. He mentioned some of the species of falcons which have "an almost ideal organization for robbery," and nevertheless are in decay, while other species of falcons, which practise mutual help, do thrive. "Take, on the other side, a sociable bird, the duck," he said ; "it is poorly organized on the whole, but it practises mutual support, and it almost invades the earth, as may be judged from its numberless varieties and species."

The readiness of the Russian zoologists to accept Kessler's views seems quite natural, because nearly all of them have had opportunities of studying the

[1] *Memoirs (Trudy) of the St. Petersburg Society of Naturalists,* vol. xi. 1880.

animal world in the wide uninhabited regions of Northern Asia and East Russia ; and it is impossible to study like regions without being brought to the same ideas. I recollect myself the impression produced upon me by the animal world of Siberia when I explored the Vitim regions in the company of so accomplished a zoologist as my friend Polyakoff was. We both were under the fresh impression of the *Origin of Species*, but we vainly looked for the keen competition between animals of the same species which the reading of Darwin's work had prepared us to expect, even after taking into account the remarks of the third chapter (p. 54). We saw plenty of adaptations for struggling, very often in common, against the adverse circumstances of climate, or against various enemies, and Polyakoff wrote many a good page upon the mutual dependency of carnivores, ruminants, and rodents in their geographical distribution ; we witnessed numbers of facts of mutual support, especially during the migrations of birds and ruminants; but even in the Amur and Usuri regions, where animal life swarms in abundance, facts of real competition and struggle between higher animals of the same species came very seldom under my notice, though I eagerly searched for them. The same impression appears in the works of most Russian zoologists, and it probably explains why Kessler's ideas were so welcomed by the Russian Darwinists, whilst like ideas are not in vogue amidst the followers of Darwin in Western Europe.

The first thing which strikes us as soon as we begin studying the struggle for existence under both its aspects—direct and metaphorical—is the abundance of facts of mutual aid, not only for rearing progeny, as recognized by most evolutionists, but also for the safety of the individual, and for providing it with the necessary food. With many large divisions of the animal kingdom mutual aid is the rule. Mutual aid is met with even amidst the lowest animals, and we must be prepared to learn some day, from the students of microscopical pond-life, facts of unconscious mutual support, even from the life of micro-organisms. Of course, our knowledge of the life of the invertebrates, save the termites, the ants, and the bees, is extremely limited; and yet, even as regards the lower animals, we may glean a few facts of well-ascertained co-operation. The numberless associations of locusts, vanessæ, cicindelæ, cicadæ, and so on, are practically quite unexplored; but the very fact of their existence indicates that they must be composed on about the same principles as the temporary associations of ants or bees for purposes of migration.[1] As to the beetles, we have quite well-observed facts of mutual help amidst the burying beetles (*Necrophorus*). They must have some decaying organic matter to lay their eggs in, and thus to provide their larvæ with food; but that matter must not decay very rapidly. So they are wont to bury in the ground the corpses of all kinds of small animals which they occasionally find in their rambles. As a rule, they live an isolated life, but when one of them has discovered the corpse of a mouse or of a bird, which it hardly could manage to bury itself, it calls four, six, or ten other beetles to perform the operation with united efforts; if necessary, they transport the corpse to a suitable soft ground; and they bury it in a very considerate way, without

[1] See Appendix I.

quarrelling as to which of them will enjoy the privilege of laying its eggs in the buried corpse. And when Gleditsch attached a dead bird to a cross made out of two sticks, or suspended a toad to a stick planted in the soil, the little beetles would in the same friendly way combine their intelligences to overcome the artifice of Man. The same combination of efforts has been noticed among the dung-beetles.

Even among animals standing at a somewhat lower stage of organization we may find like examples. Some land-crabs of the West Indies and North America combine in large swarms in order to travel to the sea and to deposit therein their spawn; and each such migration implies concert, co-operation, and mutual support. As to the big Molucca crab (*Limulus*), I was struck (in 1882, at the Brighton Aquarium) with the extent of mutual assistance which these clumsy animals are capable of bestowing upon a comrade in case of need. One of them had fallen upon its back in a corner of the tank, and its heavy saucepan-like carapace prevented it from returning to its natural position, the more so as there was in the corner an iron bar which rendered the task still more difficult. Its comrades came to the rescue, and for one hour's time I watched how they endeavoured to help their fellow-prisoner. They came two at once, pushed their friend from beneath, and after strenuous efforts succeeded in lifting it upright; but then the iron bar would prevent them from achieving the work of rescue, and the crab would again heavily fall upon its back. After many attempts, one of the helpers would go in the depth of the tank and bring two other crabs, which would begin with fresh forces the same pushing and lifting of their helpless comrade. We stayed in the

Aquarium for more than two hours, and, when leaving, we again came to cast a glance upon the tank : the work of rescue still continued! Since I saw that, I cannot refuse credit to the observation quoted by Dr. Erasmus Darwin—namely, that "the common crab during the moulting season stations as sentinel an unmoulted or hard-shelled individual to prevent marine enemies from injuring moulted individuals in their unprotected state." [1]

Facts illustrating mutual aid amidst the termites, the ants, and the bees are so well known to the general reader, especially through the works of Romanes, L. Büchner, and Sir John Lubbock, that I may limit my remarks to a very few hints. [2] If we take an ants' nest, we not only see that every description of work— rearing of progeny, foraging, building, rearing of aphides, and so on—is performed according to the principles of voluntary mutual aid ; we must also recognize, with Forel, that the chief, the fundamental feature of the life of many species of ants is the fact and the obligation for every ant of sharing its food, already swallowed and partly digested, with every member of the community which may apply for it. Two ants belonging to two different species or to two hostile nests, when they occasionally meet together, will avoid each other. But two ants belonging to the same nest or to the same colony of nests will approach

---

[1] George J. Romanes's *Animal Intelligence*, 1st ed. p. 233.

[2] Pierre Huber's *Recherches sur les fourmis*, Génève, 1810 ; reprinted as *Les fourmis indigènes*, Génève, 1861 ; Forel's *Recherches sur les fourmis de la Suisse*, Zurich, 1874, and J. T. Moggridge's *Harvesting Ants and Trapdoor Spiders*, London, 1873 and 1874, ought to be in the hands of every boy and girl. See also : Blanchard's *Métamorphoses des Insectes*, Paris, 1868 ; J. H. Fabre's *Souvenirs entomologiques*, Paris, 1886 ; Ebrard's *Etudes des mœurs des fourmis*, Génève, 1864 ; Sir John Lubbock's *Ants, Bees, and Wasps*, and so on.

each other, exchange a few movements with the antennæ, and "if one of them is hungry or thirsty, and especially if the other has its crop full . . . it immediately asks for food." The individual thus requested never refuses; it sets apart its mandibles, takes a proper position, and regurgitates a drop of transparent fluid which is licked up by the hungry ant. Regurgitating food for other ants is so prominent a feature in the life of ants (at liberty), and it so constantly recurs both for feeding hungry comrades and for feeding larvæ, that Forel considers the digestive tube of the ants as consisting of two different parts, one of which, the posterior, is for the special use of the individual, and the other, the anterior part, is chiefly for the use of the community. If an ant which has its crop full has been selfish enough to refuse feeding a comrade, it will be treated as an enemy, or even worse. If the refusal has been made while its kinsfolk were fighting with some other species, they will fall back upon the greedy individual with greater vehemence than even upon the enemies themselves. And if an ant has not refused to feed another ant belonging to an enemy species, it will be treated by the kinsfolk of the latter as a friend. All this is confirmed by most accurate observation and decisive experiments.[1]

In that immense division of the animal kingdom which embodies more than one thousand species, and is so numerous that the Brazilians pretend that Brazil belongs to the ants, not to men, competition amidst the members of the same nest, or the colony of nests,

---

[1] Forel's *Recherches*, pp. 244, 275, 278. Huber's description of the process is admirable. It also contains a hint as to the possible origin of the instinct (popular edition, pp. 158, 160). See Appendix II.

does not exist. However terrible the wars between different species, and whatever the atrocities committed at war-time, mutual aid within the community, self-devotion grown into a habit, and very often self-sacrifice for the common welfare, are the rule. The ants and termites have renounced the "Hobbesian war," and they are the better for it. Their wonderful nests, their buildings, superior in relative size to those of man; their paved roads and overground vaulted galleries; their spacious halls and granaries; their corn-fields, harvesting and "malting" of grain;[1] their rational methods of nursing their eggs and larvæ, and of building special nests for rearing the aphides whom Linnæus so picturesquely described as "the cows of the ants"; and, finally, their courage, pluck, and superior intelligence—all these are the natural outcome of the mutual aid which they practise at every stage of their busy and laborious lives. That mode of life also necessarily resulted in the development of another essential feature of the life of ants: the immense development of individual initiative which, in its turn, evidently led to the development of that high and varied intelligence which cannot but strike the human observer.[2]

If we knew no other facts from animal life than what

[1] The agriculture of the ants is so wonderful that for a long time it has been doubted. The fact is now so well proved by Mr. Moggridge, Dr. Lincecum, Mr. MacCook, Col. Sykes, and Dr. Jerdon, that no doubt is possible. See an excellent summary of evidence in Mr. Romanes's work. See also *Die Pilzgaerten einiger Süd-Amerikanischen Ameisen*, by Alf. Moeller, in Schimper's *Botan. Mitth. aus den Tropen*, vi. 1893.

[2] This second principle was not recognized at once. Former observers often spoke of kings, queens, managers, and so on; but since Huber and Forel have published their minute observations, no doubt is possible as to the free scope left for every individual's initiative in whatever the ants do, including their wars.

we know about the ants and the termites, we already
might safely conclude that mutual aid (which leads to
mutual confidence, the first condition for courage) and
individual initiative (the first condition for intellectual
progress) are two factors infinitely more important than
mutual struggle in the evolution of the animal kingdom.
In fact, the ant thrives without having any of the
"protective" features which cannot be dispensed with
by animals living an isolated life. Its colour renders it
conspicuous to its enemies, and the lofty nests of many
species are conspicuous in the meadows and forests. It
is not protected by a hard carapace, and its stinging
apparatus, however dangerous when hundreds of stings
are plunged into the flesh of an animal, is not of a
great value for individual defence; while the eggs and
larvæ of the ants are a dainty for a great number of
the inhabitants of the forests. And yet the ants, in
their thousands, are not much destroyed by the birds,
not even by the ant-eaters, and they are dreaded by
most stronger insects. When Forel emptied a bagful
of ants in a meadow, he saw that "the crickets ran
away, abandoning their holes to be sacked by the
ants; the grasshoppers and the crickets fled in all
directions; the spiders and the beetles abandoned
their prey in order not to become prey themselves;"
even the nests of the wasps were taken by the ants,
after a battle during which many ants perished for the
safety of the commonwealth. Even the swiftest insects
cannot escape, and Forel often saw butterflies, gnats,
flies, and so on, surprised and killed by the ants.
Their force is in mutual support and mutual confidence.
And if the ant—apart from the still higher developed
termites—stands at the very top of the whole class of
insects for its intellectual capacities; if its courage is

only equalled by the most courageous vertebrates ; and if its brain—to use Darwin's words—" is one of the most marvellous atoms of matter in the world, perhaps more so than the brain of man," is it not due to the fact that mutual aid has entirely taken the place of mutual struggle in the communities of ants?

The same is true as regards the bees. These small insects, which so easily might become the prey of so many birds, and whose honey has so many admirers in all classes of animals from the beetle to the bear, also have none of the protective features derived from mimicry or otherwise, without which an isolatedly-living insect hardly could escape wholesale destruction ; and yet, owing to the mutual aid they practise, they obtain the wide extension which we know and the intelligence we admire. By working in common they multiply their individual forces ; by resorting to a temporary division of labour combined with the capacity of each bee to perform every kind of work when required, they attain such a degree of well-being and safety as no isolated animal can ever expect to achieve however strong or well-armed it may be. In their combinations they are often more successful than man, when he neglects to take advantage of a well-planned mutual assistance. Thus, when a new swarm of bees is going to leave the hive in search of a new abode, a number of bees will make a preliminary exploration of the neighbourhood, and if they discover a convenient dwelling-place—say, an old basket, or anything of the kind—they will take possession of it, clean it, and guard it, sometimes for a whole week, till the swarm comes to settle therein. But how many human settlers will perish in new countries simply for not having understood the necessity of combining their

efforts! By combining their individual intelligences
they succeed in coping with adverse circumstances,
even quite unforeseen and unusual, like those bees of
the Paris Exhibition which fastened with their resinous
propolis the shutter to a glass-plate fitted in the wall
of their hive. Besides, they display none of the
sanguinary proclivities and love of useless fighting
with which many writers so readily endow animals.
The sentries which guard the entrance to the hive
pitilessly put to death the robbing bees which attempt
entering the hive; but those stranger bees which
come to the hive by mistake are left unmolested,
especially if they come laden with pollen, or are young
individuals which can easily go astray. There is no
more warfare than is strictly required.

The sociability of the bees is the more instructive as
predatory instincts and laziness continue to exist among
the bees as well, and reappear each time that their
growth is favoured by some circumstances. It is well
known that there always are a number of bees which
prefer a life of robbery to the laborious life of a
worker; and that both periods of scarcity and periods
of an unusually rich supply of food lead to an increase
of the robbing class. When our crops are in and
there remains but little to gather in our meadows and
fields, robbing bees become of more frequent occurrence; while, on the other side, about the sugar plantations of the West Indies and the sugar refineries of
Europe, robbery, laziness, and very often drunkenness
become quite usual with the bees. We thus see that
anti-social instincts continue to exist amidst the bees as
well; but natural selection continually must eliminate
them, because in the long run the practice of solidarity
proves much more advantageous to the species than

c

the development of individuals endowed with predatory inclinations. The cunningest and the shrewdest are eliminated in favour of those who understand the advantages of sociable life and mutual support.

Certainly, neither the ants, nor the bees, nor even the termites, have risen to the conception of a higher solidarity embodying the whole of the species. In that respect they evidently have not attained a degree of development which we do not find even among our political, scientific, and religious leaders. Their social instincts hardly extend beyond the limits of the hive or the nest. However, colonies of no less than two hundred nests, belonging to two different species (*Formica exsecta* and *F. pressilabris*) have been described by Forel on Mount Tendre and Mount Salève; and Forel maintains that each member of these colonies recognizes every other member of the colony, and that they all take part in common defence; while in Pennsylvania Mr. MacCook saw a whole nation of from 1,600 to 1,700 nests of the mound-making ant, all living in perfect intelligence; and Mr. Bates has described the hillocks of the termites covering large surfaces in the "campos"—some of the nests being the refuge of two or three different species, and most of them being connected by vaulted galleries or arcades.[1] Some steps towards the amalgamation of larger divisions of the species for purposes of mutual protection are thus met with even among the invertebrate animals.

Going now over to higher animals, we find far more instances of undoubtedly conscious mutual help for all possible purposes, though we must recognize at once

[1] H. W. Bates, *The Naturalist on the River Amazons*, ii. 59 *seq.*

that our knowledge even of the life of higher animals still remains very imperfect. A large number of facts have been accumulated by first-rate observers, but there are whole divisions of the animal kingdom of which we know almost nothing. Trustworthy information as regards fishes is extremely scarce, partly owing to the difficulties of observation, and partly because no proper attention has yet been paid to the subject. As to the mammalia, Kessler already remarked how little we know about their manners of life. Many of them are nocturnal in their habits ; others conceal themselves underground ; and those ruminants whose social life and migrations offer the greatest interest do not let man approach their herds. It is chiefly upon birds that we have the widest range of information, and yet the social life of very many species remains but imperfectly known. Still, we need not complain about the lack of well-ascertained facts, as will be seen from the following.

I need not dwell upon the associations of male and female for rearing their offspring, for providing it with food during their first steps in life, or for hunting in common ; though it may be mentioned by the way that such associations are the rule even with the least sociable carnivores and rapacious birds ; and that they derive a special interest from being the field upon which tenderer feelings develop even amidst otherwise most cruel animals. It may also be added that the rarity of associations larger than that of the family among the carnivores and the birds of prey, though mostly being the result of their very modes of feeding, can also be explained to some extent as a consequence of the change produced in the animal world by the rapid increase of mankind. At any rate it is worthy

of note that there are species living a quite isolated
life in densely-inhabited regions, while the same species,
or their nearest congeners, are gregarious in uninhabited countries. Wolves, foxes, and several birds
of prey may be quoted as instances in point.

However, associations which do not extend beyond
the family bonds are of relatively small importance in
our case, the more so as we know numbers of associations for more general purposes, such as hunting,
mutual protection, and even simple enjoyment of life.
Audubon already mentioned that eagles occasionally
associate for hunting, and his description of the two
bald eagles, male and female, hunting on the Mississippi, is well known for its graphic powers. But one
of the most conclusive observations of the kind belongs
to Syevertsoff. Whilst studying the fauna of the
Russian Steppes, he once saw an eagle belonging to an
altogether gregarious species (the white-tailed eagle,
*Haliaetos albicilla*) rising high in the air; for half-an-
hour it was describing its wide circles in silence when
at once its piercing voice was heard. Its cry was soon
answered by another eagle which approached it, and
was followed by a third, a fourth, and so on, till nine or
ten eagles came together and soon disappeared. In
the afternoon, Syevertsoff went to the place whereto
he saw the eagles flying; concealed by one of the
undulations of the Steppe, he approached them, and
discovered that they had gathered around the corpse of
a horse. The old ones, which, as a rule, begin the
meal first—such are their rules of propriety—already
were sitting upon the haystacks of the neighbourhood
and kept watch, while the younger ones were continuing the meal, surrounded by bands of crows. From
this and like observations, Syevertsoff concluded that

the white-tailed eagles combine for hunting; when they all have risen to a great height they are enabled, if they are ten, to survey an area of at least twenty-five miles square; and as soon as any one has discovered something, he warns the others.[1] Of course, it might be argued that a simple instinctive cry of the first eagle, or even its movements, would have had the same effect of bringing several eagles to the prey; but in this case there is strong evidence in favour of mutual warning, because the ten eagles came together before descending towards the prey, and Syevertsoff had later on several opportunities of ascertaining that the white-tailed eagles always assemble for devouring a corpse, and that some of them (the younger ones first) always keep watch while the others are eating. In fact, the white-tailed eagle—one of the bravest and best hunters —is a gregarious bird altogether, and Brehm says that when kept in captivity it very soon contracts an attachment to its keepers.

Sociability is a common feature with very many other birds of prey. The Brazilian kite, one of the most "impudent" robbers, is nevertheless a most sociable bird. Its hunting associations have been described by Darwin and other naturalists, and it is a fact that when it has seized upon a prey which is too big, it calls together five or six friends to carry it away. After a busy day, when these kites retire for their night-rest to a tree or to the bushes, they always gather in bands, sometimes coming together from distances of ten or more miles, and they often are joined by several other vultures, especially the percnopters, "their true friends," D'Orbigny says. In

---

[1] N. Syevertsoff, *Periodical Phenomena in the Life of Mammalia, Birds, and Reptiles of Voronèje*, Moscow, 1855 (in Russian).

another continent, in the Transcaspian deserts, they
have, according to Zarudnyi, the same habit of
nesting together. The sociable vulture, one of the
strongest vultures, has received its very name from
its love of society. They live in numerous bands,
and decidedly enjoy society ; numbers of them join
in their high flights for sport. " They live in very
good friendship," Le Vaillant says, "and in the
same cave I sometimes found as many as three
nests close together." [1]  The Urubú vultures of
Brazil are as, or perhaps even more, sociable than
rooks. [2]  The little Egyptian vultures live in close
friendship. They play in bands in the air, they come
together to spend the night, and in the morning they
all go together to search for their food, and never does
the slightest quarrel arise among them ; such is the
testimony of Brehm, who had plenty of opportunities
of observing their life. The red-throated falcon is
also met with in numerous bands in the forests of
Brazil, and the kestrel (*Tinnunculus cenchris*), when
it has left Europe, and has reached in the winter
the prairies and forests of Asia, gathers in numerous
societies. In the Steppes of South Russia it is (or
rather was) so sociable that Nordmann saw them in
numerous bands, with other falcons (*Falco tinnunculus,
F. œsulon,* and *F. subbuteo*), coming together every
fine afternoon about four o'clock, and enjoying their
sports till late in the night. They set off flying, all at
once, in a quite straight line, towards some determined
point, and, having reached it, immediately returned
over the same line, to repeat the same flight. [3]

---

[1] A. Brehm, *Life of Animals,* iii. 477; all quotations after the
French edition.        [2] Bates, p. 151.
[3] *Catalogue raisonné des oiseaux de la faune pontique,* in Démidoff's

To take flights in flocks for the mere pleasure of
the flight, is quite common among all sorts of birds.
"In the Humber district especially," Ch. Dixon
writes, "vast flights of dunlins often appear upon the
mud-flats towards the end of August, and remain for
the winter. . . . The movements of these birds are
most interesting, as a vast flock wheels and spreads
out or closes up with as much precision as drilled
troops. Scattered among them are many odd stints
and sanderlings and ringed-plovers." [1]

It would be quite impossible to enumerate here the
various hunting associations of birds; but the fishing
associations of the pelicans are certainly worthy of
notice for the remarkable order and intelligence displayed by these clumsy birds. They always go fishing
in numerous bands, and after having chosen an
appropriate bay, they form a wide half-circle in face
of the shore, and narrow it by paddling towards the
shore, catching all fish that happen to be enclosed in
the circle. On narrow rivers and canals they even
divide into two parties, each of which draws up on a
half-circle, and both paddle to meet each other, just as
if two parties of men dragging two long nets should
advance to capture all fish taken between the nets
when both parties come to meet. As the night comes
they fly to their resting-places—always the same for
each flock—and no one has ever seen them fighting
for the possession of either the bay or the resting-
place. In South America they gather in flocks of

---

*Voyage;* abstracts in Brehm, iii. 360. During their migrations birds
of prey often associate. One flock, which H. Seebohm saw crossing
the Pyrenees, represented a curious assemblage of "eight kites, one
crane, and a peregrine falcon" (*The Birds of Siberia,* 1901, p. 417).

[1] *Birds in the Northern Shires,* p. 207.

from forty to fifty thousand individuals, part of which enjoy sleep while the others keep watch, and others again go fishing.[1]   And finally, I should be doing an injustice to the much-calumniated house-sparrows if I did not mention how faithfully each of them shares any food it discovers with all members of the society to which it belongs.   The fact was known to the Greeks, and it has been transmitted to posterity how a Greek orator once exclaimed (I quote from memory):— " While I am speaking to you a sparrow has come to tell to other sparrows that a slave has dropped on the floor a sack of corn, and they all go there to feed upon the grain."   The more, one is pleased to find this observation of old confirmed in a recent little book by Mr. Gurney, who does not doubt that the house-sparrows always inform each other as to where there is some food to steal ; he says, " When a stack has been thrashed ever so far from the yard, the sparrows in the yard have always had their crops full of the grain."[2]   True, the sparrows are extremely particular in keeping their domains free from the invasions of strangers ; thus the sparrows of the Jardin du Luxembourg bitterly fight all other sparrows which may attempt to enjoy their turn of the garden and its visitors ; but within their own communities they fully practise mutual support, though occasionally there will be of course some quarrelling even amongst the best friends.

Hunting and feeding in common is so much the habit in the feathered world that more quotations hardly would be needful : it must be considered as an

---

[1] Max. Perty, *Ueber das Seelenleben der Thiere* (Leipzig, 1876), pp. 87, 103.

[2] G. H. Gurney, *The House-Sparrow* (London, 1885), p. 5.

established fact. As to the force derived from such associations, it is self-evident. The strongest birds of prey are powerless in face of the associations of our smallest bird pets. Even eagles—even the powerful and terrible booted eagle, and the martial eagle, which is strong enough to carry away a hare or a young antelope in its claws—are compelled to abandon their prey to bands of those beggars the kites, which give the eagle a regular chase as soon as they see it in possession of a good prey. The kites will also give chase to the swift fishing-hawk, and rob it of the fish it has captured; but no one ever saw the kites fighting together for the possession of the prey so stolen. On the Kerguelen Island, Dr. Coües saw the *Buphagus*—the sea-hen of the sealers—pursue gulls to make them disgorge their food, while, on the other side, the gulls and the terns combined to drive away the sea-hen as soon as it came near to their abodes, especially at nesting-time.[1] The little, but extremely swift lapwings (*Vanellus cristatus*) boldly attack the birds of prey. "To see them attacking a buzzard, a kite, a crow, or an eagle, is one of the most amusing spectacles. One feels that they are sure of victory, and one sees the anger of the bird of prey. In such circumstances they perfectly support one another, and their courage grows with their numbers.[2] The lapwing has well merited the name of a "good mother" which the Greeks gave to it, for it never fails to protect other aquatic birds from the attacks of their enemies. But even the little white wagtails (*Motacilla alba*), whom we well know in our gardens and whose

[1] Dr. Elliot Coües, *Birds of the Kerguelen Island*, in Smithsonian Miscellaneous Collections, vol. xiii. No. 2, p. 11.
[2] Brehm, iv. 567.

whole length hardly attains eight inches, compel the sparrow-hawk to abandon its hunt. "I often admired their courage and agility," the old Brehm wrote, "and I am persuaded that the falcon alone is capable of capturing any of them. . . . When a band of wagtails has compelled a bird of prey to retreat, they make the air resound with their triumphant cries, and after that they separate." They thus come together for the special purpose of giving chase to their enemy, just as we see it when the whole bird-population of a forest has been raised by the news that a nocturnal bird has made its appearance during the day, and all together —birds of prey and small inoffensive singers—set to chase the stranger and make it return to its concealment.

What an immense difference between the force of a kite, a buzzard or a hawk, and such small birds as the meadow-wagtail ; and yet these little birds, by their common action and courage, prove superior to the powerfully-winged and armed robbers ! In Europe, the wagtails not only chase the birds of prey which might be dangerous to them, but they chase also the fishing-hawk "rather for fun than for doing it any harm ; " while in India, according to Dr. Jerdon's testimony, the jackdaws chase the gowinda-kite " for simple matter of amusement." Prince Wied saw the Brazilian eagle *urubitinga* surrounded by numberless flocks of toucans and cassiques (a bird nearly akin to our rook), which mocked it. " The eagle," he adds, "usually supports these insults very quietly, but from time to time it will catch one of these mockers."' In all such cases the little birds, though very much inferior in force to the bird of prey, prove superior to it by their common action.[1]

¹ As to the house-sparrows, a New Zealand observer, Mr. T. W.

However, the most striking effects of common life for the security of the individual, for its enjoyment of life, and for the development of its intellectual capacities, are seen in two great families of birds, the cranes and the parrots. The cranes are extremely sociable and live in most excellent relations, not only with their congeners, but also with most aquatic birds. Their prudence is really astonishing, so also their intelligence; they grasp the new conditions in a moment, and act accordingly. Their sentries always keep watch around a flock which is feeding or resting, and the hunters know well how difficult it is to approach them. If man has succeeded in surprising them, they will never return to the same place without having sent out one single scout first, and a party of scouts afterwards; and when the reconnoitring party returns and reports that there is no danger, a second group of scouts is sent out to verify the first report, before the whole band moves. With kindred species the cranes contract real friendship; and in captivity there is no bird, save the also sociable and highly-intelligent parrot, which enters into such real friend-ship with man. "It sees in man, not a master, but a friend, and endeavours to manifest it," Brehm concludes from a wide personal experience. The

Kirk, described as follows the attack of these "impudent" birds upon an "unfortunate" hawk:—"He heard one day a most unusual noise, as though all the small birds of the country had joined in one grand quarrel. Looking up, he saw a large hawk (C. gouldi—a carrion feeder) being buffeted by a flock of sparrows. They kept dashing at him in scores, and from all points at once. The unfortun-ate hawk was quite powerless. At last, approaching some scrub, the hawk dashed into it and remained there, while the sparrows congregated in groups round the bush, keeping up a constant chatter-ing and noise" (Paper read before the New Zealand Institute; *Nature*, Oct. 10, 1891).

crane is in continual activity from early in the morning till late in the night; but it gives a few hours
only in the morning to the task of searching its
food, chiefly vegetable. All the remainder of the
day is given to society life. "It picks up small
pieces of wood or small stones, throws them in the
air and tries to catch them; it bends its neck, opens
its wings, dances, jumps, runs about, and tries to
manifest by all means its good disposition of mind,
and always it remains graceful and beautiful." [1] As it
lives in society it has almost no enemies, and though
Brehm occasionally saw one of them captured by a
crocodile, he wrote that except the crocodile he knew
no enemies of the crane. It eschews all of them by
its proverbial prudence; and it attains, as a rule, a
very old age. No wonder that for the maintenance
of the species the crane need not rear a numerous
offspring; it usually hatches but two eggs. As to its
superior intelligence, it is sufficient to say that all
observers are unanimous in recognizing that its intellectual capacities remind one very much of those of
man.

The other extremely sociable bird, the parrot, stands,
as known, at the very top of the whole feathered
world for the development of its intelligence. Brehm
has so admirably summed up the manners of life of
the parrot, that I cannot do better than translate the
following sentence :—

"Except in the pairing season, they live in very numerous
societies or bands. They choose a place in the forest to stay
there, and thence they start every morning for their hunting
expeditions. The members of each band remain faithfully
attached to each other, and they share in common good or

[1] Brehm, iv. 671 *seq.*

bad luck. All together they repair in the morning to a field, or to a garden, or to a tree, to feed upon fruits. They post sentries to keep watch over the safety of the whole band, and are attentive to their warnings. In case of danger, all take to flight, mutually supporting each other, and all simultaneously return to their resting-place. In a word, they always live closely united."

They enjoy society of other birds as well. In India, the jays and crows come together from many miles round, to spend the night in company with the parrots in the bamboo thickets. When the parrots start hunting, they display the most wonderful intelligence, prudence, and capacity of coping with circumstances. Take, for instance, a band of white cacadoos in Australia. Before starting to plunder a corn-field, they first send out a reconnoitring party which occupies the highest trees in the vicinity of the field, while other scouts perch upon the intermediate trees between the field and the forest and transmit the signals. If the report runs "All right," a score of cacadoos will separate from the bulk of the band, take a flight in the air, and then fly towards the trees nearest to the field. They also will scrutinize the neighbourhood for a long while, and only then will they give the signal for general advance, after which the whole band starts at once and plunders the field in no time. The Australian settlers have the greatest difficulties in beguiling the prudence of the parrots; but if man, with all his art and weapons, has succeeded in killing some of them, the cacadoos become so prudent and watchful that they henceforward baffle all stratagems.[1]

There can be no doubt that it is the practice of life in society which enables the parrots to attain that very high level of almost human intelligence and almost

[1] R. Lendenfeld, in *Der zoologische Garten*, 1889.

human feelings which we know in them. Their high
intelligence has induced the best naturalists to describe
some species, namely the grey parrot, as the "bird-
man." As to their mutual attachment it is known that
when a parrot has been killed by a hunter, the others
fly over the corpse of their comrade with shrieks of
complaints and "themselves fall the victims of their
friendship," as Audubon said; and when two captive
parrots, though belonging to two different species,
have contracted mutual friendship, the accidental
death of one of the two friends has sometimes been
followed by the death from grief and sorrow of the
other friend. It is no less evident that in their
societies they find infinitely more protection than they
possibly might find in any ideal development of beak
and claw. Very few birds of prey or mammals dare
attack any but the smaller species of parrots, and
Brehm is absolutely right in saying of the parrots, as
he also says of the cranes and the sociable monkeys,
that they hardly have any enemies besides men; and
he adds: "It is most probable that the larger parrots
succumb chiefly to old age rather than die from the
claws of any enemies." Only man, owing to his still
more superior intelligence and weapons, also derived
from association, succeeds in partially destroying
them. Their very longevity would thus appear as a
result of their social life. Could we not say the same
as regards their wonderful memory, which also must
be favoured in its development by society-life and by
longevity accompanied by a full enjoyment of bodily
and mental faculties till a very old age?

As seen from the above, the war of each against all
is not *the* law of nature. Mutual aid is as much a law
of nature as mutual struggle, and that law will become

still more apparent when we have analyzed some other associations of birds and those of the mammalia. A few hints as to the importance of the law of mutual aid for the evolution of the animal kingdom have already been given in the preceding pages; but their purport will still better appear when, after having given a few more illustrations, we shall be enabled presently to draw therefrom our conclusions.

# CHAPTER II

## MUTUAL AID AMONG ANIMALS (*continued*)

Migrations of birds.—Breeding associations.—Autumn societies.—Mammals: small number of unsociable species.—Hunting associations of wolves, lions, etc.—Societies of rodents; of ruminants; of monkeys.—Mutual Aid in the struggle for life.—Darwin's arguments to prove the struggle for life within the species.—Natural checks to over-multiplication.—Supposed extermination of intermediate links.—Elimination of competition in Nature.

As soon as spring comes back to the temperate zone, myriads and myriads of birds which are scattered over the warmer regions of the South come together in numberless bands, and, full of vigour and joy, hasten northwards to rear their offspring. Each of our hedges, each grove, each ocean cliff, and each of the lakes and ponds with which Northern America, Northern Europe, and Northern Asia are dotted tell us at that time of the year the tale of what mutual aid means for the birds; what force, energy, and protection it confers to every living being, however feeble and defenceless it otherwise might be. Take, for instance, one of the numberless lakes of the Russian and Siberian Steppes. Its shores are peopled with myriads of aquatic birds, belonging to at least a score of different species, all living in perfect peace—all protecting one another.

"For several hundred yards from the shore the air is filled with gulls and terns, as with snow-flakes on a winter day. Thousands of plovers and sand-coursers run over the beach, searching their food, whistling, and simply enjoying life. Further on, on almost each wave, a duck is rocking, while higher up you notice the flocks of the Casarki ducks. Exuberant life swarms everywhere."[1]

And here are the robbers—the strongest, the most cunning ones, those "ideally organized for robbery." And you hear their hungry, angry, dismal cries as for hours in succession they watch the opportunity of snatching from this mass of living beings one single unprotected individual.   But as soon as they approach, their presence is signalled by dozens of voluntary sentries, and hundreds of gulls and terns set to chase the robber.   Maddened by hunger, the robber soon abandons his usual precautions : he suddenly dashes into the living mass ; but, attacked from all sides, he again is compelled to retreat.   From sheer despair he falls upon the wild ducks ; but the intelligent, social birds rapidly gather in a flock and fly away if the robber is an erne ; they plunge into the lake if it is a falcon ;  or they raise a cloud of water-dust and bewilder the assailant if it is a kite.[2]   And while life continues to swarm on the lake, the robber flies away with cries of anger, and looks out for carrion, or for a young bird or a field-mouse not yet used to obey in time the warnings of its comrades.   In the face of an exuberant life, the ideally-armed robber must be satisfied with the off-fall of that life.

Further north, in the Arctic archipelagoes,

"you may sail along the coast for many miles and see all the ledges, all the cliffs and corners of the mountain-sides, up to

---

[1] Syevertsoff's *Periodical Phenomena*, p. 251.
[2] Seyfferlitz, quoted by Brehm, iv. 760.

a height of from two to five hundred feet, literally covered
with sea-birds, whose white breasts show against the dark
rocks as if the rocks were closely sprinkled with chalk specks.
The air, near and far, is, so to say, full with fowls." [1]

Each of such "bird-mountains" is a living illustration
of mutual aid, as well as of the infinite variety of
characters, individual and specific, resulting from
social life.  The oyster-catcher is renowned for its
readiness to attack the birds of prey.  The barge is
known for its watchfulness, and it easily becomes the
leader of more placid birds.  The turnstone, when
surrounded by comrades belonging to more energetic
species, is a rather timorous bird; but it undertakes
to keep watch for the security of the commonwealth
when surrounded by smaller birds.  Here you have
the dominative swans; there, the extremely sociable
kittiwake-gulls, among whom quarrels are rare and
short;  the prepossessing polar guillemots, which
continually caress each other;  the egoist she-goose,
who has repudiated the orphans of a killed comrade;
and, by her side, another female who adopts any one's
orphans, and now paddles surrounded by fifty or sixty
youngsters, whom she conducts and cares for as if
they all were her own breed.  Side by side with the
penguins, which steal one another's eggs, you have
the dotterels, whose family relations are so "charming
and touching" that even passionate hunters recoil
from shooting a female surrounded by her young
ones;  or the eider-ducks, among which (like the
velvet-ducks, or the *coroyas* of the Savannahs) several
females hatch together in the same nest;  or the lums,
which sit in turn upon a common covey.  Nature is

---

[1] *The Arctic Voyages of A. E. Nordenskjöld,* London, 1879, p. 135.
See also the powerful description of the St. Kilda Islands by Mr.
Dixon (quoted by Seebohm), and nearly all books of Arctic travel.

variety itself, offering all possible varieties of characters, from the basest to the highest: and that is why she cannot be depicted by any sweeping assertion. Still less can she be judged from the moralist's point of view, because the views of the moralist are themselves a result—mostly unconscious—of the observation of Nature.[1]

Coming together at nesting-time is so common with most birds that more examples are scarcely needed. Our trees are crowned with groups of crows' nests; our hedges are full of nests of smaller birds; our farmhouses give shelter to colonies of swallows; our old towers are the refuge of hundreds of nocturnal birds; and pages might be filled with the most charming descriptions of the peace and harmony which prevail in almost all these nesting associations. As to the protection derived by the weakest birds from their unions, it is evident. That excellent observer, Dr. Coües, saw, for instance, the little cliff-swallows nesting in the immediate neighbourhood of the prairie falcon (*Falco polyargus*). The falcon had its nest on the top of one of the minarets of clay which are so common in the cañons of Colorado, while a colony of swallows nested just beneath. The little peaceful birds had no fear of their rapacious neighbour; they never let it approach to their colony. They immediately surrounded it and chased it, so that it had to make off at once.[2]

---

[1] See Appendix III.

[2] Elliot Coües, in *Bulletin U.S. Geol. Survey of Territories*, iv. No. 7, pp. 556, 579, etc. Among the gulls (*Larus argentatus*), Polyakoff saw on a marsh in Northern Russia, that the nesting-grounds of a very great number of these birds were always patrolled by one male, which warned the colony of the approach of danger. All birds rose in such case and attacked the enemy with great vigour.

Life in societies does not cease when the nesting period is over; it begins then in a new form. The young broods gather in societies of youngsters, generally including several species. Social life is practised at that time chiefly for its own sake—partly for security, but chiefly for the pleasures derived from it. So we see in our forests the societies formed by the young nuthatchers (*Sitta cæsia*), together with titmouses, chaffinches, wrens, tree-creepers, or some wood-peckers.[1] In Spain the swallow is met with in company with kestrels, fly-catchers, and even pigeons. In the Far West of America the young horned larks live in large societies, together with another lark (Sprague's), the skylark, the Savannah sparrow, and several species of buntings and longspurs.[2] In fact, it would be much easier to describe the species which live isolated than to simply name those species which join the autumnal societies of young birds—not for hunting or nesting purposes, but simply to enjoy life in society and to spend their time in plays and sports, after having given a few hours every day to find their daily food.

And, finally, we have that immense display of mutual aid among birds—their migrations—which I dare not even enter upon in this place. Sufficient to say that birds which have lived for months in small

---

The females, which had five or six nests together on each knoll of the marsh, kept a certain order in leaving their nests in search of food. The fledglings, which otherwise are extremely unprotected and easily become the prey of the rapacious birds, were never left alone ("Family Habits among the Aquatic Birds," in *Proceedings of the Zool. Section of St. Petersburg Soc. of Nat.*, Dec. 17, 1874).

[1] Brehm Father, quoted by A. Brehm, iv. 34 *seq.* See also White's *Natural History of Selborne*, Letter XI.

[2] Dr. Coües, *Birds of Dakota and Montana*, in *Bulletin U.S. Survey of Territories*, iv. No. 7.

bands scattered over a wide territory gather in
thousands; they come together at a given place, for
several days in succession, before they start, and they
evidently discuss the particulars of the journey. Some
species will indulge every afternoon in flights preparatory to the long passage. All wait for their tardy
congeners, and finally they start in a certain well-
chosen direction—a fruit of accumulated collective
experience—the strongest flying at the head of the
band, and relieving one another in that difficult task.
They cross the seas in large bands consisting of both
big and small birds, and when they return next spring
they repair to the same spot, and, in most cases, each
of them takes possession of the very same nest which
it had built or repaired the previous year.[1]

This subject is so vast, and yet so imperfectly
studied; it offers so many striking illustrations of
mutual-aid habits, subsidiary to the main fact of
migration—each of which would, however, require a
special study—that I must refrain from entering here
into more details. I can only cursorily refer to the
numerous and animated gatherings of birds which
take place, always on the same spot, before they
begin their long journeys north or south, as also
those which one sees in the north, after the birds have
arrived at their breeding-places on the Yenisei or in
the northern counties of England. For many days in

---

[1] It has often been intimated that larger birds may occasionally
*transport* some of the smaller birds when they cross together the
Mediterranean, but the fact still remains doubtful. On the other
side, it is certain that some smaller birds join the bigger ones for
migration. The fact has been noticed several times, and it was
recently confirmed by L. Buxbaum at Raunheim. He saw several
parties of cranes which had larks flying in the midst and on both
sides of their migratory columns (*Der zoologische Garten*, 1886, p.
133).

succession—sometimes one month—they will come together every morning for one hour, before flying in search of food—perhaps discussing the spot where they are going to build their nests.[1]  And if, during the migration, their columns are overtaken by a storm, birds of the most different species will be brought together by common misfortune.  The birds which are not exactly migratory, but slowly move northwards and southwards with the seasons, also perform these peregrinations in flocks.  So far from migrating isolately, in order to secure for each separate individual the advantages of better food or shelter which are to be found in another district—they always wait for each other, and gather in flocks, before they move north or south, in accordance with the season.[2]

Going now over to mammals, the first thing which strikes us is the overwhelming numerical predominance of social species over those few carnivores which do not associate.  The plateaus, the Alpine tracts, and the Steppes of the Old and New World are stocked with herds of deer, antelopes, gazelles, fallow deer, buffaloes, wild goats and sheep, all of which are sociable animals.  When the Europeans came to settle in America, they found it so densely peopled with buffaloes, that pioneers had to stop their advance when a column of migrating buffaloes came to cross the route they followed ; the march past of the dense

---

[1] H. Seebohm and Ch. Dixon both mention this habit.

[2] The fact is well known to every field-naturalist, and with reference to England several examples may be found in Charles Dixon's *Among the Birds in Northern Shires*.  The chaffinches arrive during winter in vast flocks ; and about the same time, *i.e.* in November, come flocks of bramblings; redwings also frequent the same places "in similar large companies," and so on (pp. 165, 166).

column lasting sometimes for two and three days.
And when the Russians took possession of Siberia
they found it so densely peopled with deer, antelopes,
squirrels, and other sociable animals, that the very
conquest of Siberia was nothing but a hunting expedition which lasted for two hundred years; while the
grass plains of Eastern Africa are still covered with
herds composed of zebra, the hartebeest, and other
antelopes.

Not long ago the small streams of Northern
America and Northern Siberia were peopled with
colonies of beavers, and up to the seventeenth century
like colonies swarmed in Northern Russia. The flat
lands of the four great continents are still covered with
countless colonies of mice, ground-squirrels, marmots,
and other rodents. In the lower latitudes of Asia and
Africa the forests are still the abode of numerous
families of elephants, rhinoceroses, and numberless
societies of monkeys. In the far north the reindeer
aggregate in numberless herds; while still further
north we find the herds of the musk-oxen and numberless bands of polar foxes. The coasts of the ocean are
enlivened by flocks of seals and morses; its waters, by
shoals of sociable cetaceans; and even in the depths
of the great plateau of Central Asia we find herds of
wild horses, wild donkeys, wild camels, and wild
sheep. All these mammals live in societies and
nations sometimes numbering hundreds of thousands
of individuals, although now, after three centuries of
gunpowder civilization, we find but the *débris* of the
immense aggregations of old. How trifling, in comparison with them, are the numbers of the carnivores!
And how false, therefore, is the view of those who
speak of the animal world as if nothing were to be

seen in it but lions and hyenas plunging their bleeding teeth into the flesh of their victims! One might as well imagine that the whole of human life is nothing but a succession of war massacres.

Association and mutual aid are the rule with mammals. We find social habits even among the carnivores, and we can only name the cat tribe (lions, tigers, leopards, etc.) as a division the members of which decidedly prefer isolation to society, and are but seldom met with even in small groups. And yet, even among lions "this is a very common practice to hunt in company."[1] The two tribes of the civets (*Viverridæ*) and the weasels (*Mustelidæ*) might also be characterized by their isolated life, but it is a fact that during the last century the common weasel was more sociable than it is now; it was seen then in larger groups in Scotland and in the Unterwalden canton of Switzerland. As to the great tribe of the dogs, it is eminently sociable, and association for hunting purposes may be considered as eminently characteristic of its numerous species. It is well known, in fact, that wolves gather in packs for hunting, and Tschudi left an excellent description of how they draw up in a half-circle, surround a cow which is grazing on a mountain slope, and then, suddenly appearing with a loud barking, make it roll in the abyss.[2] Audubon, in the thirties, also saw the Labrador wolves hunting in packs, and one pack following a man to his cabin, and killing the dogs. During severe winters the packs of wolves grow so numerous as to become a danger for human settlements, as was the case in France some five-and-forty

---

[1] S. W. Baker, *Wild Beasts*, etc., vol. i. p. 316.
[2] Tschudi, *Thierleben der Alpenwelt*, p. 404.

years ago. In the Russian Steppes they never attack
the horses otherwise than in packs ; and yet they have
to sustain bitter fights, during which the horses
(according to Kohl's testimony) sometimes assume
offensive warfare, and in such cases, if the wolves do
not retreat promptly, they run the risk of being
surrounded by the horses and killed by their hoofs.
The prairie-wolves (*Canis latrans*) are known to
associate in bands of from twenty to thirty individuals
when they chase a buffalo occasionally separated from
its herd.[1] Jackals, which are most courageous and
may be considered as one of the most intelligent
representatives of the dog tribe, always hunt in packs ;
thus united, they have no fear of the bigger carnivores.[2]
As to the wild dogs of Asia (the *Kholzuns*, or *Dholes*),
Williamson saw their large packs attacking all larger
animals save elephants and rhinoceroses, and over-
powering bears and tigers. Hyenas always live in
societies and hunt in packs, and the hunting organizations of the painted lycaons are highly praised by
Cumming. Nay, even foxes, which, as a rule, live
isolated in our civilized countries, have been seen
combining for hunting purposes.[3] As to the polar fox,
it is—or rather was in Steller's time—one of the most
sociable animals ; and when one reads Steller's description of the war that was waged by Behring's unfortunate crew against these intelligent small animals,
one does not know what to wonder at most: the
extraordinary intelligence of the foxes and the mutual
aid they displayed in digging out food concealed under
cairns, or stored upon a pillar (one fox would climb on

[1] Houzeau's *Études*, ii. 463.
[2] For their hunting associations see Sir E. Tennant's *Natural
History of Ceylon*, quoted in Romanes's *Animal Intelligence*, p. 432.
[3] See Emil Hüter's letter in L. Büchner's *Liebe*.

its top and throw the food to its comrades beneath), or the cruelty of man, driven to despair by the numerous packs of foxes. Even some bears live in societies where they are not disturbed by man. Thus Steller saw the black bear of Kamtchatka in numerous packs, and the polar bears are occasionally found in small groups. Even the unintelligent insectivores do not always disdain association.[1]

However, it is especially with the rodents, the ungulata, and the ruminants that we find a highly-developed practice of mutual aid. The squirrels are individualist to a great extent. Each of them builds its own comfortable nest, and accumulates its own provision. Their inclinations are towards family life, and Brehm found that a family of squirrels is never so happy as when the two broods of the same year can join together with their parents in a remote corner of a forest. And yet they maintain social relations. The inhabitants of the separate nests remain in a close intercourse, and when the pine-cones become rare in the forest they inhabit, they emigrate in bands. As to the black squirrels of the Far West, they are eminently sociable. Apart from the few hours given every day to foraging, they spend their lives in playing in numerous parties. And when they multiply too rapidly in a region, they assemble in bands, almost as numerous as those of locusts, and move southwards, devastating the forests, the fields, and the gardens; while foxes, polecats, falcons, and nocturnal birds of prey follow their thick columns and live upon the individuals remaining behind. The ground-squirrel— a closely-akin genus—is still more sociable. It is given to hoarding, and stores up in its subterranean

[1] See Appendix IV.

halls large amounts of edible roots and nuts, usually
plundered by man in the autumn. According to some
observers, it must know something of the joys of a
miser. And yet it remains sociable. It always lives
in large villages, and Audubon, who opened some
dwellings of the hackee in the winter, found several
individuals in the same apartment; they must have
stored it with common efforts.

The large tribe of the marmots, which includes the
three large genuses of *Arctomys, Cynomys,* and *Sper-
mophilus,* is still more sociable and still more intelligent.
They also prefer having each one its own dwelling;
but they live in big villages. That terrible enemy of
the crops of South Russia—the *souslik*—of which
some ten millions are exterminated every year by man
alone, lives in numberless colonies; and while the
Russian provincial assemblies gravely discuss the
means of getting rid of this enemy of society, it enjoys
life in its thousands in the most joyful way. Their
play is so charming that no observer could refrain
from paying them a tribute of praise, and from mentioning the melodious concerts arising from the sharp
whistlings of the males and the melancholic whistlings
of the females, before—suddenly returning to his
citizen's duties—he begins inventing the most diabolic
means for the extermination of the little robbers. All
kinds of rapacious birds and beasts of prey having
proved powerless, the last word of science in this
warfare is the inoculation of cholera! The villages of
the prairie-dogs in America are one of the loveliest
sights. As far as the eye can embrace the prairie, it
sees heaps of earth, and on each of them a prairie-dog
stands, engaged in a lively conversation with its
neighbours by means of short barkings. As soon as

the approach of man is signalled, all plunge in a moment into their dwellings ; all have disappeared as by enchantment. But if the danger is over, the little creatures soon reappear. Whole families come out of their galleries and indulge in play. The young ones scratch one another, they worry one another, and display their gracefulness while standing upright, and in the meantime the old ones keep watch. They go visiting one another, and the beaten footpaths which connect all their heaps testify to the frequency of the visitations. In short, the best naturalists have written some of their best pages in describing the associations of the prairie-dogs of America, the marmots of the Old World, and the polar marmots of the Alpine regions. And yet, I must make, as regards the marmots, the same remark as I have made when speaking of the bees. They have maintained their fighting instincts, and these instincts reappear in captivity. But in their big associations, in the face of free Nature, the unsociable instincts have no opportunity to develop, and the general result is peace and harmony.

Even such harsh animals as the rats, which continually fight in our cellars, are sufficiently intelligent not to quarrel when they plunder our larders, but to aid one another in their plundering expeditions and migrations, and even to feed their invalids. As to the beaver-rats or musk-rats of Canada, they are extremely sociable. Audubon could not but admire " their peaceful communities, which require only being left in peace to enjoy happiness." Like all sociable animals, they are lively and playful, they easily combine with other species, and they have attained a very high degree of intellectual development. In their villages, always disposed on the shores of lakes and rivers, they take into account the changing level of water; their dome-shaped houses, which are built of beaten clay inter-woven with reeds, have separate corners for organic refuse, and their halls are well carpeted at winter-time; they are warm, and, nevertheless, well ventilated. As to the beavers, which are endowed, as known, with a most sympathetic character, their astounding dams and villages, in which generations live and die without knowing of any enemies but the otter and man, so wonderfully illustrate what mutual aid can achieve for the security of the species, the develop-ment of social habits, and the evolution of intelligence, that they are familiar to all interested in animal life. Let me only remark that with the beavers, the musk-rats, and some other rodents, we already find the feature which will also be distinctive of human com-munities—that is, work in common.

I pass in silence the two large families which include the jerboa, the chinchilla, the *biscacha*, and the *tushkan*, or underground hare of South Russia, though all these small rodents might be taken as excellent illustrations of the pleasures derived by animals from social life.[1] Precisely, the pleasures; because it is extremely diffi-cult to say what brings animals together—the needs of mutual protection, or simply the pleasure of feeling surrounded by their congeners. At any rate, our common hares, which do not gather in societies for life

[1] With regard to the viscacha it is very interesting to note that these highly-sociable little animals not only live peaceably together in each village, but that whole villages visit each other at nights. Sociability is thus extended to the whole species—not only to a given society, or to a nation, as we saw it with the ants. When the farmer destroys a viscacha-burrow, and buries the inhabitants under a heap of earth, other viscachas—we are told by Hudson—"come from a distance to dig out those that are buried alive" (*l.c.*, p. 311). This is a widely-known fact in La Plata, verified by the author.

in common, and which are not even endowed with
intense parental feelings, cannot live without coming
together for play.    Dietrich de Winckell, who is
considered to be among the best acquainted with the
habits of hares, describes them as passionate players,
becoming so intoxicated by their play that a hare has
been known to take an approaching fox for a playmate.[1]
As to the rabbit, it lives in societies, and its family
life is entirely built upon the image of the old patriarchal family ; the young ones being kept in absolute
obedience to the father and even the grandfather.[2]
And here we have the example of two very closely-
allied species which cannot bear each other—not
because they live upon nearly the same food, as like
cases are too often explained, but most probably
because the passionate, eminently-individualist hare
cannot make friends with that placid, quiet, and sub-
missive creature, the rabbit.    Their tempers are too
widely different not to be an obstacle to friendship.

Life in societies is again the rule with the large
family of horses, which includes the wild horses and
donkeys of Asia, the zebras, the mustangs, the *cimar-
rones* of the Pampas, and the half-wild horses of
Mongolia and Siberia.    They all live in numerous
associations made up of many studs, each of which
consists of a number of mares under the leadership of
a male.    These numberless inhabitants of the Old
and the New World, badly organized on the whole for
resisting both their numerous enemies and the adverse
conditions of climate, would soon have disappeared

[1] *Handbuch für Jäger und Jagdberechtigte*, quoted by Brehm, ii.
223.
[2] Buffon's *Histoire Naturelle*.

from the surface of the earth were it not for their sociable spirit. When a beast of prey approaches them, several studs unite at once; they repulse the beast and sometimes chase it: and neither the wolf nor the bear, not even the lion, can capture a horse or even a zebra as long as they are not detached from the herd. When a drought is burning the grass in the prairies, they gather in herds of sometimes 10,000 individuals strong, and migrate. And when a snow-storm rages in the Steppes, each stud keeps close together, and repairs to a protected ravine. But if confidence disappears, or the group has been seized by panic, and disperses, the horses perish and the survivors are found after the storm half dying from fatigue. Union is their chief arm in the struggle for life, and man is their chief enemy. Before his increasing numbers the ancestors of our domestic horse (the *Equus Przewalskii*, so named by Polyakoff) have preferred to retire to the wildest and least accessible plateaus on the outskirts of Thibet, where they continue to live, surrounded by carnivores, under a climate as bad as that of the Arctic regions, but in a region inaccessible to man.[1]

Many striking illustrations of social life could be taken from the life of the reindeer, and especially of

[1] In connection with the horses it is worthy of notice that the quagga zebra, which never comes together with the dauw zebra, nevertheless lives on excellent terms, not only with ostriches, which are very good sentries, but also with gazelles, several species of antelopes, and gnus. We thus have a case of mutual dislike between the quagga and the dauw which cannot be explained by competition for food. The fact that the quagga lives together with ruminants feeding on the same grass as itself excludes that hypothesis, and we must look for some incompatibility of character, as in the case of the hare and the rabbit. Cf., among others, Clive Phillips-Wolley's *Big Game Shooting* (Badminton Library), which contains excellent illustrations of various species living together in East Africa.

that large division of ruminants which might include the roebucks, the fallow deer, the antelopes, the gazelles, the ibex, and, in fact, the whole of the three numerous families of the Antelopides, the Caprides, and the Ovides. Their watchfulness over the safety of their herds against attacks of carnivores; the anxiety displayed by all individuals in a herd of chamois as long as all of them have not cleared a difficult passage over rocky cliffs; the adoption of orphans; the despair of the gazelle whose mate, or even comrade of the same sex, has been killed; the plays of the youngsters, and many other features, could be mentioned. But perhaps the most striking illustration of mutual support is given by the occasional migrations of fallow deer, such as I saw once on the Amur. When I crossed the high plateau and its border ridge, the Great Khingan, on my way from Transbaikalia to Merghen, and further travelled over the high prairies on my way to the Amur, I could ascertain how thinly-peopled with fallow deer these mostly uninhabited regions are.[1] Two years later I was travelling up the Amur, and by the end of October reached the lower end of that picturesque gorge which the Amur pierces in the Dousse-alin (Little Khingan) before it enters the lowlands where it joins the Sungari. I found the Cossacks in the villages of that gorge in the greatest excitement, because thousands and thousands of fallow deer were crossing the Amur where it is narrowest, in order to reach the lowlands. For several days in succession,

[1] Our Tungus hunter, who was going to marry, and therefore was prompted by the desire of getting as many furs as he possibly could, was beating the hill-sides all day long on horseback in search of deer. His efforts were not rewarded by even so much as one fallow deer killed every day; and he was an excellent hunter.

upon a length of some forty miles up the river, the
Cossacks were butchering the deer as they crossed the
Amur, in which already floated a good deal of ice.
Thousands were killed every day, and the exodus
nevertheless continued.   Like migrations were never
seen either before or since, and this one must have
been called for by an early and heavy snow-fall in the
Great Khingan, which compelled the deer to make a
desperate attempt at reaching the lowlands in the east
of the Dousse mountains.   Indeed, a few days later
the Dousse-alin was also buried under snow two or
three feet deep.   Now, when one imagines the
immense territory (almost as big as Great Britain)
from which the scattered groups of deer must have
gathered for a migration which was undertaken under
the pressure of exceptional circumstances, and realizes
the difficulties which had to be overcome before all the
deer came to the common idea of crossing the Amur
further south, where it is narrowest, one cannot but
deeply admire the amount of sociability displayed by
these intelligent animals.   The fact is not the less
striking if we remember that the buffaloes of North
America displayed the same powers of combination.
One saw them grazing in great numbers in the plains,
but these numbers were made up by an infinity of small
groups which never mixed together.   And yet, when
necessity arose, all groups, however scattered over an
immense territory, came together and made up those
immense columns, numbering hundreds of thousands
of individuals, which I mentioned on a preceding
page.

    I also ought to say a few words at least about the
" compound families " of the elephants, their mutual
attachment, their deliberate ways in posting sentries,

E

and the feelings of sympathy developed by such a life of close mutual support.[1] I might mention the sociable feelings of those disreputable creatures the wild boars, and find a word of praise for their powers of association in the case of an attack by a beast of prey.[2] The hippopotamus and the rhinoceros, too, would occupy a place in a work devoted to animal sociability. Several striking pages might be given to the sociability and mutual attachment of the seals and the walruses ; and finally, one might mention the most excellent feelings existing among the sociable cetaceans. But I have to say yet a few words about the societies of monkeys, which acquire an additional interest from their being the link which will bring us to the societies of primitive men.

It is hardly needful to say that those mammals, which stand at the very top of the animal world and most approach man by their structure and intelligence, are eminently sociable. Evidently we must be prepared to meet with all varieties of character and habits in so great a division of the animal kingdom which includes hundreds of species. But, all things considered, it must be said that sociability, action in common, mutual protection, and a high development of those feelings which are the necessary outcome of social life, are characteristic of most monkeys and apes. From the smallest species to the biggest ones, sociability is a rule to which we know but a few

[1] According to Samuel W. Baker, elephants combine in larger groups than the "compound family." "I have frequently observed," he wrote, "in the portion of Ceylon known as the Park Country, the tracks of elephants in great numbers which have evidently been considerable herds that have joined together in a general retreat from a ground which they considered insecure " (*Wild Beasts and their Ways*, vol. i. p. 102).

[2] Pigs, attacked by wolves, do the same (Hudson, *l. c.*).

exceptions. The nocturnal apes prefer isolated life ; the capuchins (*Cebus capucinus*), the monos, and the howling monkeys live but in small families ; and the orang-outans have never been seen by A. R. Wallace otherwise than either solitary or in very small groups of three or four individuals, while the gorillas seem never to join in bands. But all the remainder of the monkey tribe—the chimpanzees, the sajous, the sakis, the mandrills, the baboons, and so on—are sociable in the highest degree. They live in great bands, and even join with other species than their own. Most of them become quite unhappy when solitary. The cries of distress of each one of the band immediately bring together the whole of the band, and they boldly repulse the attacks of most carnivores and birds of prey. Even eagles do not dare attack them. They plunder our fields always in bands—the old ones taking care for the safety of the commonwealth. The little tee-tees, whose childish sweet faces so much struck Humboldt, embrace and protect one another when it rains, rolling their tails over the necks of their shivering comrades. Several species display the greatest solicitude for their wounded, and do not abandon a wounded comrade during a retreat till they have ascertained that it is dead and that they are helpless to restore it to life. Thus James Forbes narrated in his *Oriental Memoirs* a fact of such resistance in reclaiming from his hunting party the dead body of a female monkey that one fully understands why " the witnesses of this extraordinary scene resolved never again to fire at one of the monkey race." [1] In some species several individuals will combine to overturn a stone in order to search for ants' eggs under it. The

---

[1] Romanes's *Animal Intelligence*, p. 472.

hamadryas not only post sentries, but have been seen making a chain for the transmission of the spoil to a safe place ; and their courage is well known. Brehm's description of the regular fight which his caravan had to sustain before the hamadryas would let it resume its journey in the valley of the Mensa, in Abyssinia, has become classical.[1] The playfulness of the tailed apes and the mutual attachment which reigns in the families of chimpanzees also are familiar to the general reader. And if we find among the highest apes two species, the orang-outan and the gorilla, which are not sociable, we must remember that both—limited as they are to very small areas, the one in the heart of Africa, and the other in the two islands of Borneo and Sumatra— have all the appearance of being the last remnants of formerly much more numerous species. The gorilla at least seems to have been sociable in olden times, if the apes mentioned in the *Periplus* really were gorillas.

We thus see, even from the above brief review, that life in societies is no exception in the animal world ; it is the rule, the law of Nature, and it reaches its fullest development with the higher vertebrates. Those species which live solitary, or in small families only, are relatively few, and their numbers are limited. Nay, it appears very probable that, apart from a few exceptions, those birds and mammals which are not gregarious now, were living in societies before man multiplied on the earth and waged a permanent war against them, or destroyed the sources from which

---

[1] Brehm, i. 82 ; Darwin's *Descent of Man*, ch. iii. The Kozloff expedition of 1899–1901 have also had to sustain in Northern Thibet a similar fight.

they formerly derived food. - " On ne s'associe pas
pour mourir," was the sound remark of Espinas ; and
Houzeau, who knew the animal world of some parts of
America when it was not yet affected by man, wrote
to the same effect.

Association is found in the animal world at all
degrees of evolution ; and, according to the grand
idea of Herbert Spencer, so brilliantly developed in
Perrier's *Colonies Animales*, colonies are at the very
origin of evolution in the animal kingdom.  But, in
proportion as we ascend the scale of evolution, we see
association growing more and more conscious.  It
loses its purely physical character, it ceases to be
simply instinctive, it becomes reasoned.  With the
higher vertebrates it is periodical, or is resorted to for
the satisfaction of a given want—propagation of the
species, migration, hunting, or mutual defence.  It
even becomes occasional, when birds associate against
a robber, or mammals combine, under the pressure of
exceptional circumstances, to emigrate.  In this last
case, it becomes a voluntary deviation from habitual
moods of life.  The combination sometimes appears in
two or more degrees—the family first, then the group,
and finally the association of groups, habitually
scattered, but uniting in case of need, as we saw it
with the bisons and other ruminants.  It also takes
higher forms, guaranteeing more independence to
the individual without depriving it of the benefits of
social life.  With most rodents the individual has its
own dwelling, which it can retire to when it prefers
being left alone ; but the dwellings are laid out in
villages and cities, so as to guarantee to all inhabitants
the benefits and joys of social life.  And finally, in
several species, such as rats, marmots, hares, etc.,

sociable life is maintained notwithstanding the quarrel-
some or otherwise egotistic inclinations of the isolated
individual. Thus it is not imposed, as is the case with
ants and bees, by the very physiological structure of
the individuals; it is cultivated for the benefits of
mutual aid, or for the sake of its pleasures. And this,
of course, appears with all possible gradations and
with the greatest variety of individual and specific
characters—the very variety of aspects taken by social
life being a consequence, and for us a further proof, of
its generality.[1]

Sociability—that is, the need of the animal of
associating with its like—the love of society for society's
sake, combined with the "joy of life," only now begins
to receive due attention from the zoologists.[2] We
know at the present time that all animals, beginning
with the ants, going on to the birds, and ending with
the highest mammals, are fond of plays, wrestling,
running after each other, trying to capture each other,
teasing each other, and so on. And while many plays
are, so to speak, a school for the proper behaviour of
the young in mature life, there are others, which, apart
from their utilitarian purposes, are, together with
dancing and singing, mere manifestations of an excess
of forces—"the joy of life," and a desire to communicate in some way or another with other individuals of

---

[1] The more strange was it to read in the previously-mentioned
article by Huxley the following paraphrase of a well-known sentence
of Rousseau: "The first men who substituted mutual peace for that
of mutual war—whatever the motive which impelled them to take
that step—*created society*" (*Nineteenth Century*, Feb. 1888, p. 165).
Society has *not* been created by man; it is anterior to man.

[2] Such monographs as the chapter on " Music and Dancing in
Nature" which we have in Hudson's *Naturalist on the La Plata*, and
Carl Gross' *Play of Animals*, have already thrown a considerable
light upon an instinct which is absolutely universal in Nature.

the same or of other species—in short, a manifestation of *sociability proper*, which is a distinctive feature of all the animal world.[1] Whether the feeling be fear, experienced at the appearance of a bird of prey, or "a fit of gladness" which bursts out when the animals are in good health and especially when young, or merely the desire of giving play to an excess of impressions and of vital power—the necessity of communicating impressions, of playing, of chattering, or of simply feeling the proximity of other kindred living beings pervades Nature, and is, as much as any other physiological function, a distinctive feature of life and impressionability. This need takes a higher development and attains a more beautiful expression in mammals, especially amidst their young, and still more among the birds ; but it pervades all Nature, and has been fully observed by the best naturalists, including Pierre Huber, even amongst the ants, and it is evidently the same instinct which brings together the big columns of butterflies which have been referred to already.

The habit of coming together for dancing and of decorating the places where the birds habitually perform their dances is, of course, well known from the pages that Darwin gave to this subject in *The Descent of Man* (ch. xiii.). Visitors of the London Zoological Gardens also know the bower of the satin bower-bird. But this habit of dancing seems to be much more widely spread than was formerly believed, and Mr. W.

[1] Not only numerous species of birds possess the habit of assembling together—in many cases always at the same spot—to indulge in antics and dancing performances, but W. H. Hudson's experience is that nearly all mammals and birds (" probably there are really *no* exceptions ") indulge frequently in more or less regular or set performances with or without sound, or composed of sound exclusively (p. 264).

Hudson gives in his master-work on La Plata the most interesting description, which must be read in the original, of complicated dances, performed by quite a number of birds : rails, jacanas, lapwings, and so on.

The habit of singing in concert, which exists in several species of birds, belongs to the same category of social instincts. It is most strikingly developed with the chakar (*Chauna chavarria*), to which the English have given the most unimaginative misnomer of "crested screamer." These birds sometimes assemble in immense flocks, and in such cases they frequently sing all in concert. W. H. Hudson found them once in countless numbers, ranged all round a pampas lake in well-defined flocks, of about 500 birds in each flock.

"Presently," he writes, "one flock near me began singing, and continued their powerful chant for three or four minutes ; when they ceased the next flock took up the strains, and after it the next, and so on, until once more the notes of the flocks on the opposite shore came floating strong and clear across the water—then passed away, growing fainter and fainter, until once more the sound approached me travelling round to my side again."

On another occasion the same writer saw a whole plain covered with an endless flock of chakars, not in close order, but scattered in pairs and small groups. About nine o'clock in the evening, "suddenly the entire multitude of birds covering the marsh for miles around burst forth in a tremendous evening song. . . . It was a concert well worth riding a hundred miles to hear." [1]   It may be added that like all sociable animals, the chakar easily becomes tame and grows very attached to man. " They are mild-tempered birds, and

----

[1] For the choruses of monkeys, see Brehm.

very rarely quarrel "—we are told—although they are well provided with formidable weapons. Life in societies renders these weapons useless.

That life in societies is the most powerful weapon in the struggle for life, taken in its widest sense, has been illustrated by several examples on the foregoing pages, and could be illustrated by any amount of evidence, if further evidence were required. Life in societies enables the feeblest insects, the feeblest birds, and the feeblest mammals to resist, or to protect themselves from, the most terrible birds and beasts of prey; it permits longevity; it enables the species to rear its progeny with the least waste of energy and to maintain its numbers albeit a very slow birth-rate; it enables the gregarious animals to migrate in search of new abodes. Therefore, while fully admitting that force, swiftness, protective colours, cunningness, and endurance to hunger and cold, which are mentioned by Darwin and Wallace, are so many qualities making the individual, or the species, the fittest under certain circumstances, we maintain that under *any* circumstances sociability is the greatest advantage in the struggle for life. Those species which willingly or unwillingly abandon it are doomed to decay; while those animals which know best how to combine, have the greatest chances of survival and of further evolution, although they may be inferior to others in *each* of the faculties enumerated by Darwin and Wallace, save the intellectual faculty. The highest vertebrates, and especially mankind, are the best proof of this assertion. As to the intellectual faculty, while every Darwinist will agree with Darwin that it is the most powerful arm in the struggle for life, and the most powerful factor of further evolution, he also will admit that intelligence is an eminently social faculty. Language, imitation, and accumulated experience are so many elements of growing intelligence of which the unsociable animal is deprived. Therefore we find, at the top of each class of animals, the ants, the parrots, and the monkeys, all combining the greatest sociability with the highest development of intelligence. The fittest are thus the most sociable animals, and sociability appears as the chief factor of evolution, both directly, by securing the well-being of the species while diminishing the waste of energy, and indirectly, by favouring the growth of intelligence.

Moreover, it is evident that life in societies would be utterly impossible without a corresponding development of social feelings, and, especially, of a certain collective sense of justice growing to become a habit. If every individual were constantly abusing its personal advantages without the others interfering in favour of the wronged, no society-life would be possible. And feelings of justice develop, more or less, with all gregarious animals. Whatever the distance from which the swallows or the cranes come, each one returns to the nest it has built or repaired last year. If a lazy sparrow intends appropriating the nest which a comrade is building, or even steals from it a few sprays of straw, the group interferes against the lazy comrade; and it is evident that without such interference being the rule, no nesting associations of birds could exist. Separate groups of penguins have separate resting-places and separate fishing abodes, and do not fight for them. The droves of cattle in Australia have particular spots to which each group repairs to rest, and from which it never deviates; and so on.[1] We have any numbers of direct observations of the peace that prevails in the nesting associations of birds, the villages of the rodents, and the herds of grass-eaters ; while, on the other side, we know of few sociable animals which so continually quarrel as the rats in our cellars do, or as the morses, which fight for the possession of a sunny place on the shore. Sociability thus puts a limit to physical struggle, and leaves room for the development of better moral feelings. The high development of parental love in all classes of animals, even with lions and tigers, is generally known. As to the young birds and mammals whom we continually see associating, sympathy —not love—attains a further development in their associations. Leaving aside the really touching facts of mutual attachment and compassion which have been recorded as regards domesticated animals and with animals kept in captivity, we have a number of well-certified facts of compassion between wild animals at liberty. Max Perty and L. Büchner have given a number of such facts.[2] J. C. Wood's narrative of a weasel which came to pick up and to carry away an injured comrade enjoys a well-merited popularity.[3] So also the observation of Captain Stansbury on his journey to Utah which is quoted by Darwin ; he saw a blind pelican which was fed, and well fed, by other pelicans upon fishes which had to be brought from

[1] Haygarth, *Bush Life in Australia*, p. 58.

[2] To quote but a few instances, a wounded badger was carried away by another badger suddenly appearing on the scene ; rats have been seen feeding a blind couple (*Seelenleben der Thiere*, p. 64 seq.). Brehm himself saw two crows feeding in a hollow tree a third crow which was wounded ; its wound was several weeks old (*Hausfreund*, 1874, 715 ; Büchner's *Liebe*, 203). Mr. Blyth saw Indian crows feeding two or three blind comrades ; and so on.

[3] *Man and Beast*, p. 344.

a distance of thirty miles.[1] And when a herd of vicunas was hotly pursued by hunters, H. A. Weddell saw more than once during his journey to Bolivia and Peru, the strong males covering the retreat of the herd and lagging behind in order to protect the retreat. As to facts of compassion with wounded comrades, they are continually mentioned by all field zoologists. Such facts are quite natural. Compassion is a necessary outcome of social life. But compassion also means a considerable advance in general intelligence and sensibility. It is the first step towards the development of higher moral sentiments. It is, in its turn, a powerful factor of further evolution.

If the views developed on the preceding pages are correct, the question necessarily arises, in how far are they consistent with the theory of struggle for life as it has been developed by Darwin, Wallace, and their followers? and I will now briefly answer this important question. First of all, no naturalist will doubt that the idea of a struggle for life carried on through organic nature is the greatest generalization of our century. Life *is* struggle; and in that struggle the fittest survive. But the answers to the questions, " By which arms is this struggle chiefly carried on ?" and "Who are the fittest in the struggle?" will widely differ according to the importance given to the two different aspects of the struggle : the direct one, for food and safety among separate individuals, and the struggle which Darwin described as " metaphorical "—the struggle, very often collective, against adverse circumstances. No one will deny that

[1] L. H. Morgan, *The American Beaver*, 1868, p. 272; *Descent of Man*, ch. iv.

there is, within each species, a certain amount of real competition for food—at least, at certain periods. But the question is, whether competition is carried on to the extent admitted by Darwin, or even by Wallace; and whether this competition has played, in the evolution of the animal kingdom, the part assigned to it.

The idea which permeates Darwin's work is certainly one of real competition going on within each animal group for food, safety, and possibility of leaving an offspring. He often speaks of regions being stocked with animal life to their full capacity, and from that overstocking he infers the necessity of competition. But when we look in his work for real proofs of that competition, we must confess that we do not find them sufficiently convincing. If we refer to the paragraph entitled "Struggle for Life most severe between Individuals and Varieties of the same Species," we find in it none of that wealth of proofs and illustrations which we are accustomed to find in whatever Darwin wrote. The struggle between individuals of the same species is not illustrated under that heading by even one single instance: it is taken as granted; and the competition between closely-allied animal species is illustrated by but five examples, out of which one, at least (relating to the two species of thrushes), now proves to be doubtful.[1] But when we look for more details in order

---

[1] One species of swallow is said to have caused the decrease of another swallow species in North America; the recent increase of the missel-thrush in Scotland has caused the decrease of the song-thrush; the brown rat has taken the place of the black rat in Europe; in Russia the small cockroach has everywhere driven before it its greater congener; and in Australia the imported hive-bee is rapidly exterminating the small stingless bee. Two other cases, but relative to domesticated animals, are mentioned in the preceding paragraph. While recalling these same facts, A. R. Wallace remarks in a footnote relative to the Scottish thrushes: "Prof. A. Newton, however,

to ascertain how far the decrease of one species was really occasioned by the increase of the other species, Darwin, with his usual fairness, tells us :

"We can dimly see why the competition should be most severe between allied forms which fill nearly the same place in nature ; but probably in no case could we precisely say why one species has been victorious over another in the great battle of life."

As to Wallace, who quotes the same facts under a slightly-modified heading ("Struggle for Life between closely-allied Animals and Plants *often* most severe "), he makes the following remark (italics are mine), which gives quite another aspect to the facts above quoted. He says :

"In *some* cases, no doubt, there is actual war between the two, the stronger killing the weaker ; *but this is by no means necessary*, and there may be cases in which the weaker species, physically, may prevail by its power of more rapid multiplication, its better withstanding vicissitudes of climate, or its greater cunning in escaping the attacks of common enemies."

In such cases what is described as competition may be no competition at all. One species succumbs, not because it is exterminated or starved out by the other species, but because it does not well accommodate itself to new conditions, which the other does. The term "struggle for life" is again used in its metaphorical sense, and may have no other. As to the real compe-

---

informs me that these species do not interfere in the way here stated" (*Darwinism*, p. 34). As to the brown rat, it is known that, owing to its amphibian habits, it usually stays in the lower parts of human dwellings (low cellars, sewers, etc.), as also on the banks of canals and rivers ; it also undertakes distant migrations in numberless bands. The black rat, on the contrary, prefers staying in our dwellings themselves, under the floor, as well as in our stables and barns. It thus is much more exposed to be exterminated by man ; and we cannot maintain, with any approach to certainty, that the black rat is being either exterminated or starved out by the brown rat and not by man.

tition between individuals of the same species, which is illustrated in another place by the cattle of South America during a period of drought, its value is impaired by its being taken from among domesticated animals. Bisons emigrate in like circumstances in order to avoid competition. However severe the struggle between plants—and this is amply proved—we cannot but repeat Wallace's remark to the effect that "plants live where they can," while animals have, to a great extent, the power of choice of their abode. So that we again are asking ourselves, To what extent does competition really exist within each animal species? Upon what is the assumption based?

The same remark must be made concerning the indirect argument in favour of a severe competition and struggle for life within each species, which may be derived from the "extermination of transitional varieties," so often mentioned by Darwin. It is known that for a long time Darwin was worried by the difficulty which he saw in the absence of a long chain of intermediate forms between closely-allied species, and that he found the solution of this difficulty in the supposed extermination of the intermediate forms.[1] However, an attentive reading of the different chapters in which Darwin and Wallace speak of this subject soon brings one to the conclusion that the word "extermination" does not mean real extermination;

[1] "But it may be urged that when several closely-allied species inhabit the same territory, we surely ought to find at the present time many transitional forms. . . . By my theory these allied species are descended from a common parent; and during the process of modification, each has become adapted to the conditions of life of its own region, and has supplanted and exterminated its original parent-form and all the transitional varieties between its past and present states" (*Origin of Species*, 6th ed. p. 134); also p. 137, 296 (all paragraph "On Extinction").

the same remark which Darwin made concerning his expression : "struggle for existence," evidently applies to the word "extermination" as well. It can by no means be understood in its direct sense, but must be taken "in its metaphoric sense."

If we start from the supposition that a given area is stocked with animals to its fullest capacity, and that a keen competition for the sheer means of existence is consequently going on between all the inhabitants— each animal being compelled to fight against all its congeners in order to get its daily food—then the appearance of a new and successful variety would certainly mean in many cases (though not always) the appearance of individuals which are enabled to seize more than their fair share of the means of existence ; and the result would be that those individuals would starve both the parental form which does not possess the new variation and the intermediate forms which do not possess it in the same degree. It may be that at the outset, Darwin understood the appearance of new varieties under this aspect ; at least, the frequent use of the word "extermination" conveys such an impression. But both he and Wallace knew Nature too well not to perceive that this is by no means the only possible and necessary course of affairs.

If the physical and the biological conditions of a given area, the extension of the area occupied by a given species, and the habits of all the members of the latter remained unchanged—then the sudden appearance of a new variety might mean the starving out and the extermination of all the individuals which were not endowed in a sufficient degree with the new feature by which the new variety is characterized. But such a combination of conditions is precisely what we do not see in Nature. Each species is continually tending to enlarge its abode ; migration to new abodes is the rule with the slow snail, as with the swift bird ; physical changes are continually going on in every given area ; and new varieties among animals consist in an immense number of cases—perhaps in the majority—not in the growth of new weapons for snatching the food from the mouth of its congeners—food is only one out of a hundred of various conditions of existence—but, as Wallace himself shows in a charming paragraph on the " divergence of characters " (*Darwinism*, p. 107), in forming new habits, moving to new abodes, and taking to new sorts of food. In all such cases there will be no extermination, even no competition—the new adaptation being *a relief from competition, if it ever existed ;* and yet there will be, after a time, an absence of intermediate links, in consequence of a mere survival of those which are best fitted for the new conditions—as surely as under the hypothesis of extermination of the parental form. It hardly need be added that if we admit, with Spencer, all the Lamarckians, and Darwin himself, the modifying influence of the surroundings upon the species, there remains still less necessity for the extermination of the intermediate forms.

The importance of migration and of the consequent isolation of groups of animals, for the origin of new varieties and ultimately of new species, which was indicated by Moritz Wagner, was fully recognized by Darwin himself. Subsequent researches have only accentuated the importance of this factor, and they have shown how the largeness of the area occupied by a given species—which Darwin considered with full reason so important for the appearance of new varieties —can be combined with the isolation of parts of the species, in consequence of local geological changes, or of local barriers. It would be impossible to enter here into the discussion of this wide question, but a few remarks will do to illustrate the combined action of these agencies. It is known that portions of a given species will often take to a new sort of food. The squirrels, for instance, when there is a scarcity of cones in the larch forests, remove to the fir-tree forests, and this change of food has certain well-known physiological effects on the squirrels. If this change of habits does not last—if next year the cones are again plentiful in the dark larch woods—no new variety of squirrels will evidently arise from this cause. But if part of the wide area occupied by the squirrels begins to have its physical characters altered—in consequence of, let us say, a milder climate or desiccation, which both bring about an increase of the pine forests in proportion to the larch woods—and if some other conditions concur to induce the squirrels to dwell on the outskirts of the desiccating region—we shall have then a new variety, *i. e.* an incipient new species of squirrels, without there having been anything that would deserve the name of extermination among the squirrels. A larger proportion of squirrels of the new, better-adapted variety would survive every year, and the intermediate links would die *in the course of time*, without having been starved out by Malthusian competitors. This is exactly what we see going on during the great physical changes which are accomplished over large areas in Central Asia, owing to the desiccation which is going on there since the glacial period.

To take another example, it has been proved by geologists that the present wild horse (*Equus Przewalski*) has slowly been evolved during the later parts of the Tertiary and the Quaternary period, but that during this succession of ages its ancestors were *not* confined to some given, limited area of the globe. They wandered over both the Old and New World, returning, in all probability, after a time to the pastures which they had, in the course of their migrations, formerly left.[1] Consequently, if we do not find now, in Asia, all the intermediate links between the present wild, horse and its Asiatic Post-Tertiary ancestors, this does not mean at all that the intermediate links have been exterminated. No such extermination has ever taken place. No exceptional mortality may even have occurred among the ancestral species : the individuals which belonged to intermediate varieties and species have died in the usual course of events—often amidst plentiful food, and their remains were buried all over the globe.

In short, if we carefully consider this matter, and carefully re-read what Darwin himself wrote upon this subject, we see that if the word "extermination" be used at all in connection with transitional varieties, it must be used in its metaphoric sense. As to "competition," this expression, too, is continually used by Darwin (see, for instance, the paragraph " On Extinction ") as an image, or as a way-of-speaking, rather than with the intention of conveying the idea of a real competition between two portions of the same species for the means of existence. At any rate, the absence of intermediate forms is no argument in favour of it.

In reality, the chief argument in favour of a keen

[1] According to Madame Marie Pavloff, who has made a special study of this subject, they migrated from Asia to Africa, stayed there some time, and returned next to Asia. Whether this double migration be confirmed or not, the fact of a former extension of the ancestor of our horse over Asia, Africa, and America is settled beyond doubt.

competition for the means of existence continually
going on within every animal species is—to use Professor Geddes' expression—the "arithmetical argument"
borrowed from Malthus.

But this argument does not prove it at all. We
might as well take a number of villages in South-East
Russia, the inhabitants of which enjoy plenty of food,
but have no sanitary accommodation of any kind ; and
seeing that for the last eighty years the birth-rate was
sixty in the thousand, while the population is now
what it was eighty years ago, we might conclude that
there has been a terrible competition between the
inhabitants. But the truth is that from year to year
the population remained stationary, for the simple
reason that one-third of the new-born died before reaching their sixth month of life ; one-half died within the
next four years, and out of each hundred born, only
seventeen or so reached the age of twenty. The
new-comers went away before having grown to be
competitors. It is evident that if such is the case
with men, it is still more the case with animals. In
the feathered world the destruction of the eggs goes
on on such a tremendous scale that eggs are the chief
food of several species in the early summer ; not to
say a word of the storms, the inundations which
destroy nests by the million in America, and the
sudden changes of weather which are fatal to the
young mammals. Each storm, each inundation, each
visit of a rat to a bird's nest, each sudden change of
temperature, take away those competitors which appear
so terrible in theory.

As to the facts of an extremely rapid increase of
horses and cattle in America, of pigs and rabbits in
New Zealand, and even of wild animals imported from

Europe (where their numbers are kept down by man, not by competition), they rather seem opposed to the theory of over-population. If horses and cattle could so rapidly multiply in America, it simply proved that, however numberless the buffaloes and other ruminants were at ·that time in the New World, its grass-eating population was far below what the prairies could maintain. If millions of intruders have found plenty of food without starving out the former population of the prairies, we must rather conclude that the Europeans found a *want* of grass-eaters in America, not an excess. And we have good reasons to believe that want of animal population is the natural state of things all over the world, with but a few temporary exceptions to the rule. The actual numbers of animals in a given region are determined, not by the highest feeding capacity of the region, but by what it is every year under the most unfavourable conditions. So that, for that reason alone, competition hardly can be a normal condition ; but other causes intervene as well to cut down the animal population below even that low standard. If we take the horses and cattle which are grazing all the winter through in the Steppes of Transbaikalia, we find them very lean and exhausted at the end of the winter. But they grow exhausted not because there is not enough food for all of them—the grass buried under a thin sheet of snow is everywhere in abundance—but because of the difficulty of getting it from beneath the snow, and this difficulty is the same for all horses alike. Besides, days of glazed frost are common in early spring, and if several such days come in succession the horses grow still more exhausted. But then comes a snow-storm, which compels the already weakened animals to remain without any food for several days, and very great numbers of them die. The losses during the spring are so severe that if the season has been more inclement than usual they are even not repaired by the new breeds—the more so as *all* horses are exhausted, and the young foals are born in a weaker condition. The numbers of horses and cattle thus always remain beneath what they otherwise might be ; all the year round there is food for five or ten times as many animals, and yet their population increases extremely slowly. But as soon as the Buriate owner makes ever so small a provision of hay in the steppe, and throws it open during days of glazed frost, or heavier snow-fall, he immediately sees the increase of his herd. Almost all free grass-eating animals and many rodents in Asia and America being in very much the same conditions, we can safely say that their numbers are *not* kept down by competition ; that at no time of the year they can struggle for food, and that if they never reach anything approaching to over-population, the cause is in the climate, not in competition.

The importance of natural checks to over-multiplication, and especially their bearing upon the competition hypothesis, seems never to have been taken into due account. The checks, or rather some of them, are mentioned, but their action is seldom studied in detail. However, if we compare the action of the natural checks with that of competition, we must recognize at once that the latter sustains no comparison whatever with the other checks. Thus, Mr. Bates mentions the really astounding numbers of winged ants which are destroyed during their exodus. The dead or half-dead bodies of the formica de fuego (*Myrmica sævissima*) which had been blown into the

river during a gale "were heaped in a line an inch or
two in height and breadth, the line continuing without
interruption for miles at the edge of the water."[1]
Myriads of ants are thus destroyed amidst a nature
which might support a hundred times as many ants as
are actually living. Dr. Altum, a German forester,
who wrote a very interesting book about animals
injurious to our forests, also gives many facts showing
the immense importance of natural checks. He says
that a succession of gales or cold and damp weather
during the exodus of the pine-moth (*Bombyx pini*)
destroy it to incredible amounts, and during the spring
of 1871 all these moths disappeared at once, probably
killed by a succession of cold nights.[2] Many like
examples relative to various insects could be quoted
from various parts of Europe. Dr. Altum also mentions the bird-enemies of the pine-moth, and the
immense amount of its eggs destroyed by foxes; but
he adds that the parasitic fungi which periodically
infest it are a far more terrible enemy than any bird,
because they destroy the moth over very large areas
at once. As to various species of mice (*Mus sylvati-
cus*, *Arvicola arvalis*, and *A. agrestis*), the same author
gives a long list of their enemies, but he remarks:
"However, the most terrible enemies of mice are not
other animals, but such sudden changes of weather as
occur almost every year." Alternations of frost and
warm weather destroy them in numberless quantities;
"one single sudden change can reduce thousands of
mice to the number of a few individuals." On the
other side, a warm winter, or a winter which gradually

---

[1] *The Naturalist on the River Amazons*, ii. 85, 95.
[2] Dr. B. Altum, *Waldbeschädigungen durch Thiere und Gegenmittel*
(Berlin, 1889), pp. 207 *seq.*

steps in, make them multiply in menacing proportions,
notwithstanding every enemy; such was the case in
1876 and 1877.[1] Competition, in the case of mice,
thus appears a quite trifling factor when compared with
weather. Other facts to the same effect are also given
as regards squirrels.

As to birds, it is well known how they suffer from
sudden changes of weather. Late snow-storms are as
destructive of bird-life on the English moors, as they
are in Siberia; and Ch. Dixon saw the red grouse
so pressed during some exceptionally severe winters,
that they quitted the moors in numbers, "and we
have then known them actually to be taken in the
streets of Sheffield. Persistent wet," he adds, "is
almost as fatal to them."

On the other side, the contagious diseases which
continually visit most animal species destroy them in
such numbers that the losses often cannot be repaired
for many years, even with the most rapidly-multiply-
ing animals. Thus, some sixty years ago, the *sousliks*
suddenly disappeared in the neighbourhood of Sarepta,
in South-Eastern Russia, in consequence of some
epidemics; and for years no *sousliks* were seen in that
neighbourhood. It took many years before they
became as numerous as they formerly were.[2]

Like facts, all tending to reduce the importance
given to competition, could be produced in numbers.[3]
Of course, it might be replied, in Darwin's words, that
nevertheless each organic being "at some period of
its life, during some season of the year, during each
generation or at intervals, has to struggle for life and

---

[1] Dr. B. Altum, *ut supra*, pp. 13 and 187.

[2] A. Becker in the *Bulletin de la Société des Naturalistes de Moscou*,
1889, p. 625.

[3] See Appendix V.

to suffer great destruction," and that the fittest survive
during such periods of hard struggle for life. But if
the evolution of the animal world were based exclusively, or even chiefly, upon the survival of the fittest
during periods of calamities ; if natural selection were
limited in its action to periods of exceptional drought,
or sudden changes of temperature, or inundations,
retrogression would be the rule in the animal world.
Those who survive a famine, or a severe epidemic of
cholera, or small-pox, or diphtheria, such as we see
them in uncivilized countries, are neither the strongest,
nor the healthiest, nor the most intelligent.  No
progress could be based on those survivals—the less
so as all survivors usually come out of the ordeal with
an impaired health, like the Transbaikalian horses just
mentioned, or the Arctic crews, or the garrison of a
fortress which has been compelled to live for a few
months on half rations, and comes out of its experience
with a broken health, and subsequently shows a quite
abnormal mortality.  All that natural selection can do
in times of calamities is to spare the individuals
endowed with the greatest endurance for privations of
all kinds.  So it does among the Siberian horses and
cattle.  They *are* enduring ; they can feed upon the
Polar birch in case of need ; they resist cold and
hunger.  But no Siberian horse is capable of carrying
half the weight which a European horse carries with
ease ; no Siberian cow gives half the amount of milk
given by a Jersey cow, and no natives of uncivilized
countries can bear a comparison with Europeans.
They may better endure hunger and cold, but their
physical force is very far below that of a well-fed
European, and their intellectual progress is despair-
ingly slow.  " Evil cannot be productive of good," as

Tchernyshevsky wrote in a remarkable essay upon Darwinism.[1]

Happily enough, competition is not the rule either in the animal world or in mankind. It is limited among animals to exceptional periods, and natural selection finds better fields for its activity. Better conditions are created by the *elimination of competition* by means of mutual aid and mutual support.[2] In the great struggle for life—for the greatest possible fulness and intensity of life with the least waste of energy—natural selection continually seeks out the ways precisely for avoiding competition as much as possible. The ants combine in nests and nations; they pile up their stores, they rear their cattle—and thus avoid competition; and natural selection picks out of the ants' family the species which know best how to avoid competition, with its unavoidably deleterious consequences. Most of our birds slowly move southwards as the winter comes, or gather in numberless societies and undertake long journeys—and thus avoid competition. Many rodents fall asleep when the time comes that competition should set in ; while other rodents store food for the winter, and gather in large villages for obtaining the necessary protection when at work. The reindeer, when the lichens are dry in the interior of the continent, migrate towards the sea. Buffaloes cross an immense continent in order to find plenty of food. And the beavers, when they grow numerous on a

[1] *Russkaya Mysl*, Sept. 1888 : "The Theory of Beneficency of Struggle for Life, being a Preface to various Treatises on Botanics, Zoology, and Human Life," by an Old Transformist.

[2] "One of the most frequent modes in which Natural Selection acts is, by adapting some individuals of a species to a somewhat different mode of life, whereby they are able to seize unappropriated places in Nature" (*Origin of Species*, p. 145)—in other words, to avoid competition.

river, divide into two parties, and go, the old ones
down the river, and the young ones up the river—
and avoid competition. And when animals can neither
fall asleep, nor migrate, nor lay in stores, nor themselves grow their food like the ants, they do what the
titmouse does, and what Wallace (*Darwinism*, ch. v.)
has so charmingly described : they resort to new
kinds of food—and thus, again, avoid competition.[1]

"Don't compete !—competition is always injurious
to the species, and you have plenty of resources to
avoid it !" That is the *tendency* of nature, not always
realized in full, but always present. That is the
watchword which comes to us from the bush, the
forest, the river, the ocean. "Therefore combine—
practise mutual aid ! That is the surest means for
giving to each and to all the greatest safety, the best
guarantee of existence and progress, bodily, intellectual, and moral." That is what Nature teaches us ;
and that is what all those animals which have attained
the highest position in their respective classes have
done. That is also what man—the most primitive
man—has been doing ; and that is why man has
reached the position upon which we stand now, as we
shall see in the subsequent chapters devoted to mutual
aid in human societies.

[1] See Appendix VI.

# CHAPTER III

## MUTUAL AID AMONG SAVAGES

Supposed war of each against all.—Tribal origin of human society.—Late appearance of the separate family.—Bushmen and Hottentots.—Australians, Papuas.—Eskimos, Aleoutes.—Features of savage life difficult to understand for the European.—The Dayak's conception of justice.—Common law.

THE immense part played by mutual aid and mutual support in the evolution of the animal world has been briefly analyzed in the preceding chapters. We have now to cast a glance upon the part played by the same agencies in the evolution of mankind. We saw how few are the animal species which live an isolated life, and how numberless are those which live in societies, either for mutual defence, or for hunting and storing up food, or for rearing their offspring, or simply for enjoying life in common. We also saw that, though a good deal of warfare goes on between different classes of animals, or different species, or even different tribes of the same species, peace and mutual support are the rule within the tribe or the species; and that those species which best know how to combine, and to avoid competition, have the best chances of survival and of a further progressive development. They prosper, while the unsociable species decay.

It is evident that it would be quite contrary to all that we know of nature if men were an exception to

so general a rule : if a creature so defenceless as man
was at his beginnings should have found his protection
and his way to progress, not in mutual support, like
other animals, but in a reckless competition for personal advantages, with no regard to the interests
of the species.   To a mind accustomed to the idea
of unity in nature, such a proposition appears utterly
indefensible.   And yet, improbable and unphilosophical
as it is, it has never found a lack of supporters.   There
always were writers who took a pessimistic view of
mankind.   They knew it, more or less superficially,
through their own limited experience ; they knew of
history what the annalists, always watchful of wars,
cruelty, and oppression, told of it, and little more
besides ; and they concluded that mankind is nothing
but a loose aggregation of beings, always ready to
fight with each other, and only prevented from so
doing by the intervention of some authority.

Hobbes took that position ; and while some of his
eighteenth-century followers endeavoured to prove
that at no epoch of its existence—not even in its
most primitive condition—mankind lived in a state
of perpetual warfare ; that men have been sociable
even in "the state of nature," and that want of
knowledge, rather than the natural bad inclinations
of man, brought humanity to all the horrors of its
early historical life,—his idea was, on the contrary,
that the so-called "state of nature" was nothing but
a permanent fight between individuals, accidentally
huddled together by the mere caprice of their bestial
existence.   True, that science has made some progress
since Hobbes's time, and that we have safer ground
to stand upon than the speculations of Hobbes or
Rousseau.   But the Hobbesian philosophy has plenty

of admirers still; and we have had of late quite a
school of writers who, taking possession of Darwin's
terminology rather than of his leading ideas, made
of it an argument in favour of Hobbes's views upon
primitive man, and even succeeded in giving them a
scientific appearance. Huxley, as is known, took the
lead of that school, and in a paper written in 1888
he represented primitive men as a sort of tigers or
lions, deprived of all ethical conceptions, fighting out
the struggle for existence to its bitter end, and living
a life of "continual free fight"; to quote his own
words—"beyond the limited and temporary relations
of the family, the Hobbesian war of each against all
was the normal state of existence."[1]

It has been remarked more than once that the chief
error of Hobbes, and the eighteenth-century philosophers as well, was to imagine that mankind began
its life in the shape of small straggling families, something like the "limited and temporary" families of the
bigger carnivores, while in reality it is now positively
known that such was *not* the case. Of course, we
have no direct evidence as to the modes of life of the
first man-like beings. We are not yet settled even
as to the time of their first appearance, geologists
being inclined at present to see their traces in the
pliocene, or even the miocene, deposits of the Tertiary
period. But we have the indirect method which permits us to throw some light even upon that remote
antiquity. A most careful investigation into the social
institutions of the lowest races has been carried on
during the last forty years, and it has revealed among
the present institutions of primitive folk some traces
of still older institutions which have long disap-

[1] *Nineteenth Century*, February 1888, p. 165.

peared, but nevertheless left unmistakable traces of their previous existence. A whole science devoted to the embryology of human institutions has thus developed in the hands of Bachofen, MacLennan, Morgan, Edward B. Tylor, Maine, Post, Kovalevsky, Lubbock, and many others. And that science has established beyond any doubt that mankind did *not* begin its life in the shape of small isolated families.

Far from being a primitive form of organization, the family is a very late product of human evolution. As far as we can go back in the palæo-ethnology of mankind, we find men living in societies—in tribes similar to those of the highest mammals; and an extremely slow and long evolution was required to bring these societies to the gentile, or clan organization, which, in its turn, had to undergo another, also very long evolution, before the first germs of family, polygamous or monogamous, could appear. Societies, bands, or tribes—not families—were thus the primitive form of organization of mankind and its earliest ancestors. That is what ethnology has come to after its painstaking researches. And in so doing it simply came to what might have been foreseen by the zoologist. None of the higher mammals, save a few carnivores and a few undoubtedly-decaying species of apes (orang-outans and gorillas), live in small families, isolatedly straggling in the woods. All others live in societies. And Darwin so well understood that isolately-living apes never could have developed into man-like beings, that he was inclined to consider man as descended from some comparatively weak *but social species*, like the chimpanzee, rather than from some stronger but unsociable species, like the gorilla.[1]

[1] *The Descent of Man*, end of ch. ii. pp. 63 and 64 of the 2nd edition.

Zoology and palæo-ethnology are thus agreed in considering that the band, not the family, was the earliest form of social life. The first human societies simply were a further development of those societies which constitute the very essence of life of the higher animals.[1]

If we now go over to positive evidence, we see that the earliest traces of man, dating from the glacial or the early post-glacial period, afford unmistakable proofs of man having lived even then in societies. Isolated finds of stone implements, even from the old stone age, are very rare; on the contrary, wherever one flint implement is discovered others are sure to be found, in most cases in very large quantities. At a time when men were dwelling in caves, or under occasionally protruding rocks, in company with mammals now extinct, and hardly succeeded in making the roughest sorts of flint hatchets, they already knew the advantages of life in societies. In the valleys of the tributaries of the Dordogne, the surface of the rocks is in some places entirely covered with caves which were inhabited by palæolithic men.[2] Sometimes the cave-dwellings are superposed in storeys,

[1] Anthropologists who fully endorse the above views as regards man nevertheless intimate, sometimes, that the apes live in polygamous families, under the leadership of "a strong and jealous male." I do not know how far that assertion is based upon conclusive observation. But the passage from Brehm's *Life of Animals*, which is sometimes referred to, can hardly be taken as very conclusive. It occurs in his general description of monkeys; but his more detailed descriptions of separate species either contradict it or do not confirm it. Even as regards the cercopithèques, Brehm is affirmative in saying that they "nearly always live in bands, and very seldom in families" (French edition, p. 59). As to other species, the very numbers of their bands, always containing many males, render the "polygamous family" more than doubtful Further observation is evidently wanted.

[2] Lubbock, *Prehistoric Times*, fifth edition, 1890.

and they certainly recall much more the nesting colonies of swallows than the dens of carnivores. As to the flint implements discovered in those caves, to use Lubbock's words, "one may say without exaggeration that they are numberless." The same is true of other palæolithic stations. It also appears from Lartet's investigations that the inhabitants of the Aurignac region in the south of France partook of tribal meals at the burial of their dead. So that men lived in societies, and had germs of a tribal worship, even at that extremely remote epoch.

The same is still better proved as regards the later part of the stone age. Traces of neolithic man have been found in numberless quantities, so that we can reconstitute his manner of life to a great extent. When the ice-cap (which must have spread from the Polar regions as far south as middle France, middle Germany, and middle Russia, and covered Canada as well as a good deal of what is now the United States) began to melt away, the surfaces freed from ice were covered, first, with swamps and marshes, and later on with numberless lakes.[1] Lakes filled all depressions of the valleys before their waters dug out those permanent channels which, during a subsequent epoch, became our rivers. And wherever we explore, in Europe, Asia, or America, the shores of the literally numberless lakes of that period, whose proper name would be the Lacustrine period, we find traces of neolithic man.

---

[1] That extension of the ice-cap is admitted by most of the geologists who have specially studied the glacial age. The Russian Geological Survey already has taken this view as regards Russia, and most German specialists maintain it as regards Germany. The glaciation of most of the central plateau of France will not fail to be recognized by the French geologists, when they pay more attention to the glacial deposits altogether.

They are so numerous that we can only wonder at the
relative density of population at that time. The
"stations" of neolithic man closely follow each other
on the terraces which now mark the shores of the old
lakes. And at each of those stations stone implements
appear in such numbers, that no doubt is possible as
to the length of time during which they were inhabited
by rather numerous tribes. Whole workshops of flint
implements, testifying of the numbers of workers who
used to come together, have been discovered by the
archæologists.

Traces of a more advanced period, already characterized by the use of some pottery, are found in the
shell-heaps of Denmark. They appear, as is well
known, in the shape of heaps from five to ten feet
thick, from 100 to 200 feet wide, and 1,000 feet or
more in length, and they are so common along some
parts of the sea-coast that for a long time they were
considered as natural growths. And yet they "contain
*nothing* but what has been in some way or other
subservient to the use of man," and they are so densely
stuffed with products of human industry that, during a
two days' stay at Milgaard, Lubbock dug out no less
than 191 pieces of stone-implements and four fragments
of pottery.[1] The very size and extension of the shell-
heaps prove that for generations and generations the
coasts of Denmark were inhabited by hundreds of
small tribes which certainly lived as peacefully together
as the Fuegian tribes, which also accumulate like shell-
heaps, are living in our own times.

As to the lake-dwellings of Switzerland, which
represent a still further advance in civilization, they
yield still better evidence of life and work in societies.

[1] *Prehistoric Times*, pp. 232 and 242.

It is known that even during the stone age the shores of the Swiss lakes were dotted with a succession of villages, each of which consisted of several huts, and was built upon a platform supported by numberless pillars in the lake. No less than twenty-four, mostly stone age villages, were discovered along the shores of Lake Leman, thirty-two in the Lake of Constance, forty-six in the Lake of Neuchâtel, and so on ; and each of them testifies to the immense amount of labour which was spent in common by the tribe, not by the family. It has even been asserted that the life of the lake-dwellers must have been remarkably free of warfare. And so it probably was, especially if we refer to the life of those primitive folk who live until the present time in similar villages built upon pillars on the sea coasts.

It is thus seen, even from the above rapid hints, that our knowledge of primitive man is not so scanty after all, and that, so far as it goes, it is rather opposed than favourable to the Hobbesian speculations. Moreover, it may be supplemented, to a great extent, by the direct observation of such primitive tribes as now stand on the same level of civilization as the inhabitants of Europe stood in prehistoric times.

That these primitive tribes which we find now are *not* degenerated specimens of mankind who formerly knew a higher civilization, as it has occasionally been maintained, has sufficiently been proved by Edward B. Tylor and Lubbock. However, to the arguments already opposed to the degeneration theory, the following may be added. Save a few tribes clustering in the less-accessible highlands, the " savages " represent a girdle which encircles the more or less civilized nations, and they occupy the extremities of our con-

tinents, most of which have retained still, or recently
were bearing, an early post-glacial character.　Such
are the Eskimos and their congeners in Greenland,
Arctic America, and Northern Siberia ; and, in the
Southern hemisphere, the Australians, the Papuas, the
Fuegians, and, partly, the Bushmen ; while within the
civilized area, like primitive folk are only found in the
Himalayas, the highlands of Australasia, and the
plateaus of Brazil.　Now it must be borne in mind
that the glacial age did not come to an end at once
over the whole surface of the earth.　It still continues
in Greenland.　Therefore, at a time when the littoral
regions of the Indian Ocean, the Mediterranean, or
the Gulf of Mexico already enjoyed a warmer climate,
and became the seats of higher civilizations, immense
territories in middle Europe, Siberia, and Northern
America, as well as in Patagonia, Southern Africa,
and Southern Australasia, remained in early post-
glacial conditions which rendered them inaccessible to
the civilized nations of the torrid and sub-torrid zones.
They were at that time what the terrible *urmans* of
North-West Siberia are now, and their population,
inaccessible to and untouched by civilization, retained
the characters of early post-glacial man.　Later on,
when desiccation rendered these territories more suitable for agriculture, they were peopled with more
civilized immigrants ; and while part of their previous
inhabitants were assimilated by the new settlers, another
part migrated further, and settled where we find them.
The territories they inhabit now are still, or recently
were, sub-glacial, as to their physical features ; their
arts and implements are those of the neolithic age ;
and, notwithstanding their racial differences, and the
distances which separate them, their modes of life and

social institutions bear a striking likeness. So we cannot but consider them as fragments of the early post-glacial population of the now civilized area.

The first thing which strikes us as soon as we begin studying primitive folk is the complexity of the organization of marriage relations under which they are living. With most of them the family, in the sense we attribute to it, is hardly found in its germs. But they are by no means loose aggregations of men and women coming in a disorderly manner together in conformity with their momentary caprices. All of them are under a certain organization, which has been described by Morgan in its general aspects as the "gentile," or clan organization.[1]

To tell the matter as briefly as possible, there is

[1] Bachofen, *Das Mutterrecht,* Stuttgart, 1861 ; Lewis H. Morgan, *Ancient Society, or Researches in the Lines of Human Progress from Savagery through Barbarism to Civilization,* New York, 1877 ; J. F. MacLennan, *Studies in Ancient History,* 1st series, new edition, 1886 ; 2nd series, 1896 ; L. Fison and A. W. Howitt, *Kamilaroi and Kurnai,* Melbourne. These four writers—as has been very truly remarked by Giraud Teulon,—starting from different facts and different general ideas, and following different methods, have come to the same conclusion. To Bachofen we owe the notion of the maternal family and the maternal succession ; to Morgan—the system of kinship, Malayan and Turanian, and a highly-gifted sketch of the main phases of human evolution ; to MacLennan—the law of exogeny ; and to Fison and Howitt—the cuadro, or scheme, of the conjugal societies in Australia. All four end in establishing the same fact of the tribal origin of the family. When Bachofen first drew attention to the maternal family, in his epoch-making work, and Morgan described the clan-organization,—both concurring to the almost general extension of these forms and maintaining that the marriage laws lie at the very basis of the consecutive steps of human evolution, they were accused of exaggeration. However, the most careful researches prosecuted since, by a phalanx of students of ancient law, have proved that all races of mankind bear traces of having passed through similar stages of development of marriage laws, such as we now see in force among certain savages. See the works of Post, Dargun, Kovalevsky, Lubbock, and their numerous followers : Lippert, Mucke, etc.

little doubt that mankind has passed at its beginnings through a stage which may be described as that of "communal marriage"; that is, the whole tribe had husbands and wives in common with but little regard to consanguinity. But it is also certain that some restrictions to that free intercourse were imposed at a very early period. Inter-marriage was soon prohibited between the sons of one mother and her sisters, grand-daughters, and aunts. Later on it was prohibited between the sons and daughters of the same mother, and further limitations did not fail to follow. The idea of a *gens*, or clan, which embodied all presumed descendants from one stock (or rather all those who gathered in one group) was evolved, and marriage within the clan was entirely prohibited. It still remained "communal," but the wife or the husband had to be taken from another clan. And when a gens became too numerous, and subdivided into several gentes, each of them was divided into classes (usually four), and marriage was permitted only between certain well-defined classes. That is the stage which we find now among the Kamilaroi-speaking Australians. As to the family, its first germs appeared amidst the clan organization. A woman who was captured in war from some other clan, and who formerly would have belonged to the whole gens, could be kept at a later period by the capturer, under certain obligations towards the tribe. She may be taken by him to a separate hut, after she had paid a certain tribute to the clan, and thus constitute within the gens a separate family, the appearance of which evidently was opening a quite new phase of civilization.[1]

Now, if we take into consideration that this com-

[1] See Appendix VII.

plicated organization developed among men who stood
at the lowest known degree of development, and that
it maintained itself in societies knowing no kind of
authority besides the authority of public opinion, we
at once see how deeply inrooted social instincts must
have been in human nature, even at its lowest stages.
A savage who is capable of living under such an
organization, and of freely submitting to rules which
continually clash with his personal desires, certainly is
not a beast devoid of ethical principles and knowing
no rein to its passions.   But the fact becomes still
more striking if we consider the immense antiquity
of the clan organization.   It is now known that the
primitive Semites, the Greeks of Homer, the prehistoric Romans, the Germans of Tacitus, the early
Celts and the early Slavonians, all have had their own
period of clan organization, closely analogous to that
of the Australians, the Red Indians, the Eskimos, and
other inhabitants of the "savage girdle." [1] So we
must admit that either the evolution of marriage laws
went on on the same lines among all human races,
or the rudiments of the clan rules were developed
among some common ancestors of the Semites, the
Aryans, the Polynesians, etc., before their differentia-
tion into separate races took place, and that these rules
were maintained, until now, among races long ago
separated from the common stock.  .Both alternatives
imply, however, an equally striking tenacity of the

---

[1] For the Semites and the Aryans, see especially Prof. Maxim
Kovalevsky's *Primitive Law* (in Russian), Moscow, 1886 and 1887.
Also his lectures delivered at Stockholm (*Tableau des origines et
de l'évolution de la famille et de la propriété*, Stockholm, 1890),
which represents an admirable review of the whole question.  Cf.
also A. Post *Die Geschlechtsgenossenschaft der Urzeit*, Oldenburg,
1875

institution—such a tenacity that no assaults of the
individual could break it down through the scores of
thousands of years that it was in existence. The very
persistence of the clan organization shows how utterly
false it is to represent primitive mankind as a disorderly
agglomeration of individuals, who only obey their
individual passions, and take advantage of their personal
force and cunningness against all other representatives
of the species. Unbridled individualism is a modern
growth, but it is not characteristic of primitive mankind.[1]

Going now over to the existing savages, we may
begin with the Bushmen, who stand at a very low
level of development—so low indeed that they have
no dwellings and sleep in holes dug in the soil,
occasionally protected by some screens. It is known
that when Europeans settled in their territory and
destroyed deer, the Bushmen began stealing the
settlers' cattle, whereupon a war of extermination, too
horrible to be related here, was waged against them.

[1] It would be impossible to enter here into a discussion of the
origin of the marriage restrictions. Let me only remark that a
division into groups, similar to Morgan's *Hawaian*, exists among
birds; the young broods live together separately from their parents.
A like division might probably be traced among some mammals as
well. As to the prohibition of relations between brothers and sisters,
it is more likely to have arisen, not from speculations about the bad
effects of consanguinity, which speculations really do not seem
probable, but to avoid the too-easy precocity of like marriages.
Under close cohabitation it must have become of imperious necessity.
I must also remark that in discussing the origin of new customs
altogether, we must keep in mind that the savages, like us, have their
"thinkers" and *savants*—wizards, doctors, prophets, etc.—whose
knowledge and ideas are in advance upon those of the masses.
United as they are in their secret unions (another almost universal
feature) they are certainly capable of exercising a powerful influence,
and of enforcing customs the utility of which may not yet be
recognized by the majority of the tribe.

Five hundred Bushmen were slaughtered in 1774,
three thousand in 1808 and 1809 by the Farmers'
Alliance, and so on. They were poisoned like rats,
killed by hunters lying in ambush before the carcass of
some animal, killed wherever met with.[1]  So that our
knowledge of the Bushmen, being chiefly borrowed
from those same people who exterminated them, is
necessarily limited. But still we know that when the
Europeans came, the Bushmen lived in small tribes
(or clans), sometimes federated together; that they
used to hunt in common, and divided the spoil without
quarrelling; that they never abandoned their wounded,
and displayed strong affection to their comrades.
Lichtenstein has a most touching story about a Bushman, nearly drowned in a river, who was rescued by
his companions. They took off their furs to cover
him, and shivered themselves; they dried him, rubbed
him before the fire, and smeared his body with warm
grease till they brought him back to life. And when
the Bushmen found, in Johan van der Walt, a man
who treated them well, they expressed their thankful-
ness by a most touching attachment to that man.[2]
Burchell and Moffat both represent them as good-
hearted, disinterested, true to their promises, and
grateful,[3] all qualities which could develop only by
being practised within the tribe. As to their love to
children, it is sufficient to say that when a European
wished to secure a Bushman woman as a slave, he

---

[1] Col. Collins, in Philips' *Researches in South Africa*, London,
1828.   Quoted by Waitz, ii. 334.
[2] Lichtenstein's *Reisen im südlichen Afrika*, ii. pp. 92, 97.   Berlin,
1811.
[3] Waitz, *Anthropologie der Naturvölker*, ii. pp. 335 *seq*.   See also
Fritsch's *Die Eingeboren Afrika's*, Breslau, 1872, pp. 386 *seq*.; and
*Drei Jahre in Süd-Afrika*. Also W. Bleck, *A Brief Account of
Bushmen Folklore*, Capetown, 1875.

stole her child : the mother was sure to come into slavery to share the fate of her child.[1]

The same social manners characterize the Hottentots, who are but a little more developed than the Bushmen. Lubbock describes them as "the filthiest animals," and filthy they really are. A fur suspended to the neck and worn till it falls to pieces is all their dress ; their huts are a few sticks assembled together and covered with mats, with no kind of furniture within. And though they kept oxen and sheep, and seem to have known the use of iron before they made acquaintance with the Europeans, they still occupy one of the lowest degrees of the human scale. And yet those who knew them highly praised their sociability and readiness to aid each other. If anything is given to a Hottentot, he at once divides it among all present—a habit which, as is known, so much struck Darwin among the Fuegians. He cannot eat alone, and, however hungry, he calls those who pass by to share his food. And when Kolben expressed his astonishment thereat, he received the answer : " That is Hottentot manner." But this is not Hottentot manner only : it is an all but universal habit among the "savages." Kolben, who knew the Hottentots well and did not pass by their defects in silence, could not praise their tribal morality highly enough.

" Their word is sacred," he wrote. They know " nothing of the corruptness and faithless arts of Europe." " They live in great tranquillity and are seldom at war with their neighbours." They are " all kindness and goodwill to one another. One of the greatest pleasures of the Hottentots certainly lies in their gifts and good offices to one another." " The integrity of the Hottentots, their strictness and celerity in the

---

[1] Elisée Reclus, *Géographie Universelle*, xiii. 475.

exercise of justice, and their chastity, are things in which they excel all or most nations in the world." [1]

Tachart, Barrow, and Moodie [2] fully confirm Kolben's testimony. Let me only remark that when Kolben wrote that "they are certainly the most friendly, the most liberal and the most benevolent people to one another that ever appeared on the earth" (i. 332), he wrote a sentence which has continually appeared since in the description of savages. When first meeting with primitive races, the Europeans usually make a caricature of their life ; but when an intelligent man has stayed among them for a longer time, he generally describes them as the "kindest" or "the gentlest" race on the earth. These very same words have been applied to the Ostyaks, the Samoyedes, the Eskimos, the Dayaks, the Aleoutes, the Papuas, and so on, by the highest authorities. I also remember having read them applied to the Tunguses, the Tchuktchis, the Sioux, and several others. The very frequency of that high commendation already speaks volumes in itself.

The natives of Australia do not stand on a higher level of development than their South African brothers. Their huts are of the same character ; very often simple screens are the only protection against cold winds. In their food they are most indifferent : they devour horribly putrefied corpses, and cannibalism is resorted to in times of scarcity. When first discovered by Europeans, they had no implements but in stone or bone, and these were of the roughest description.

---

[1] P. Kolben, *The Present State of the Cape of Good Hope*, translated from the German by Mr. Medley, London, 1731, vol. i. pp. 59, 71, 333, 336, etc.

[2] Quoted in Waitz's *Anthropologie*, ii. 335 *seq.*

Some tribes had even no canoes, and did not know barter-trade. And yet, when their manners and customs were carefully studied, they proved to be living under that elaborate clan organization which I have mentioned on a preceding page.[1]

The territory they inhabit is usually allotted between the different gentes or clans; but the hunting and fishing territories of each clan are kept in common, and the produce of fishing and hunting belongs to the whole clan; so also the fishing and hunting implements.[2] The meals are taken in common. Like many other savages, they respect certain regulations as to the seasons when certain gums and grasses may be collected.[3] As to their morality altogether, we cannot do better than transcribe the following answers given to the questions of the Paris Anthropological Society by Lumholtz, a missionary who sojourned in North Queensland:[4]—

"The feeling of friendship is known among them; it is strong. Weak people are usually supported; sick people are very well attended to; they never are abandoned or killed. These tribes are cannibals, but they very seldom eat members of their own tribe (when immolated on religious principles, I suppose); they eat strangers only. The parents love their children, play with them, and pet them. Infanticide meets

---

[1] The natives living in the north of Sydney, and speaking the Kamilaroi language, are best known under this aspect, through the capital work of Lorimer Fison and A. W. Howitt, *Kamilaroi and Kurnai*, Melbourne, 1880. See also A. W. Howitt's "Further Note on the Australian Class Systems," in *Journal of the Anthropological Institute*, 1889, vol. xviii. p. 31, showing the wide extension of the same organization in Australia.

[2] *The Folklore, Manners, etc., of Australian Aborigines*, Adelaide, 1879, p. 11.

[3] Grey's *Journals of Two Expeditions of Discovery in North-West and Western Australia*, London, 1841, vol. ii. pp. 237, 298.

[4] *Bulletin de la Société d'Anthropologie*, 1888, vol. xi. p. 652. I abridge the answers.

with common approval.   Old people are very well treated,
never put to death.   No religion, no idols, only a fear of
death.   Polygamous marriage.   Quarrels arising within the
tribe are settled by means of duels fought with wooden
swords and shields.   No slaves; no culture of any kind; no
pottery; no dress, save an apron sometimes worn by women.
The clan consists of two hundred individuals, divided into
four classes of men and four of women; marriage being only
permitted within the usual classes, and never within the gens."

For the Papuas, closely akin to the above, we have
the testimony of G. L. Bink, who stayed in New
Guinea, chiefly in Geelwink Bay, from 1871 to 1883.
Here is the essence of his answers to the same
questioner : [1]—

"They are sociable and cheerful ; they laugh very much.
Rather timid than courageous.   Friendship is relatively strong
among persons belonging to different tribes, and still stronger
within the tribe.   A friend will often pay the debt of his
friend, the stipulation being that the latter will repay it without interest to the children of the lender.   They take care of
the ill and the old; old people are never abandoned, and in
no case are they killed—unless it be a slave who was ill for
a long time.   War prisoners are sometimes eaten.   The
children are very much petted and loved.   Old and feeble war
prisoners are killed, the others are sold as slaves.   They have
no religion, no gods, no idols, no authority of any description ;
the oldest man in the family is the judge.   In cases of adul-
tery a fine is paid, and part of it goes to the *negoria* (the
community).   The soil is kept in common, but the crop
belongs to those who have grown it   They have pottery, and
know barter-trade—the custom being that the merchant gives
them the goods, whereupon they return to their houses and
bring the native goods required by the merchant; if the
latter cannot be obtained, the European goods are returned.[2]
They are head-hunters, and in so doing they prosecute blood

---

[1] *Bulletin de la Société d'Anthropologie*, 1888, vol. xi. p. 386.
[2] The same is the practice with the Papuas of Kaimani Bay, who
have a high reputation of honesty.   "It never happens that the
Papua be untrue to his promise," Finsch says in *Neuguinea und seine
Bewohner*, Bremen, 1865, p. 829.

revenge. 'Sometimes,' Finsch says, 'the affair is referred to the Rajah of Namotatte, who terminates it by imposing a fine.'"

When well treated, the Papuas are very kind. Miklukho-Maclay landed on the eastern coast of New Guinea, followed by one single man, stayed for two years among tribes reported to be cannibals, and left them with regret ; he returned again to stay one year more among them, and never had he any conflict to complain of. True that his rule was *never*—under no pretext whatever—to say anything which was not truth, nor make any promise which he could not keep. These poor creatures, who even do not know how to obtain fire, and carefully maintain it in their huts, live under their primitive communism, without any chiefs ; and within their villages they have no quarrels worth speaking of. They work in common, just enough to get the food of the day ; they rear their children in common ; and in the evenings they dress themselves as coquettishly as they can, and dance. Like all savages, they are fond of dancing. Each village has its *barla*, or *balai*—the " long house," " longue maison," or " grande maison "—for the unmarried men, for social gatherings, and for the discussion of common affairs—again a trait which is common to most inhabit-ants of the Pacific Islands, the Eskimos, the Red Indians, and so on. Whole groups of villages are on friendly terms, and visit each other *en bloc*.

Unhappily, feuds are not uncommon—not in con-sequence of "overstocking of the area," or "keen competition," and like inventions of a mercantile cen-tury, but chiefly in consequence of superstition. As soon as any one falls ill, his friends and relatives come together, and deliberately discuss who might be the

cause of the illness. All possible enemies are considered, every one confesses of his own petty quarrels,
and finally the real cause is discovered. An enemy
from the next village has called it down, and a raid
upon that village is decided upon. Therefore, feuds
are rather frequent, even between the coast villages,
not to say a word of the cannibal mountaineers who
are considered as real witches and enemies, though,
on a closer acquaintance, they prove to be exactly the
same sort of people as their neighbours on the sea-
coast.[1]

Many striking pages could be written about the
harmony which prevails in the villages of the Polynesian inhabitants of the Pacific Islands. But they
belong to a more advanced stage of civilization. So
we shall now take our illustrations from the far north.
I must mention, however, before leaving the Southern
Hemisphere, that even the Fuegians, whose reputation
has been so bad, appear under a much better light
since they begin to be better known. A few French
missionaries who stay among them "know of no act
of malevolence to complain of." In their clans,
consisting of from 120 to 150 souls, they practise the
same primitive communism as the Papuas ; they share
everything in common, and treat their old people very
well. Peace prevails among these tribes.[2]

With the Eskimos and their nearest congeners, the
Thlinkets, the Koloshes, and the Aleoutes, we find
one of the nearest illustrations of what man may have
been during the glacial age. Their implements hardly

---

[1] *Izvestia* of the Russian Geographical Society, 1880, pp. 161 *seq.*
Few books of travel give a better insight into the petty details of the
daily life of savages than these scraps from Maklay's note-books.

[2] L. F. Martial, in *Mission Scientifique au Cap Horn*, Paris, 1883,
vol. i. pp. 183-201.

differ from those of palæolithic man, and some of their
tribes do not yet know fishing : they simply spear the
fish with a kind of harpoon.[1] They know the use of
iron, but they receive it from the Europeans, or find it
on wrecked ships. Their social organization is of a
very primitive kind, though they already have emerged
from the stage of "communal marriage," even under
the gentile restrictions. They live in families, but the
family bonds are often broken ; husbands and wives
are often exchanged.[2] The families, however, remain
united in clans, and how could it be otherwise ? How
could they sustain the hard struggle for life unless by
closely combining their forces ? So they do, and the
tribal bonds are closest where the struggle for life
is hardest, namely, in North-East Greenland. The
"long house" is their usual dwelling, and several
families lodge in it, separated from each other by small
partitions of ragged furs, with a common passage in
the front. Sometimes the house has the shape of a
cross, and in such case a common fire is kept in the
centre. The German Expedition which spent a winter
close by one of those "long houses" could ascertain
that "no quarrel disturbed the peace, no dispute arose
about the use of this narrow space" throughout the
long winter. "Scolding, or even unkind words, are
considered as a misdemeanour, if not produced under
the legal form of process, namely, the nith-song."[3]
Close cohabitation and close interdependence are

[1] Captain Holm's Expedition to East Greenland.

[2] In Australia whole clans have been seen exchanging all their
wives, in order to conjure a calamity (Post, *Studien zur Entwicklungsgeschichte des Familienrechts*, 1890, p. 342). More brotherhood
is their specific against calamities.

[3] Dr. H. Rink, *The Eskimo Tribes*, p. 26 (*Meddelelser om Grön-
land*, vol. xi. 1887).

sufficient for maintaining century after century that deep respect for the interests of the community which is characteristic of Eskimo life. Even in the larger communities of Eskimos, "public opinion formed the real judgment-seat, the general punishment consisting in the offenders being shamed in the eyes of the people." [1]

Eskimo life is based upon communism. What is obtained by hunting and fishing belongs to the clan. But in several tribes, especially in the West, under the influence of the Danes, private property penetrates into their institutions. However, they have an original means for obviating the inconveniences arising from a personal accumulation of wealth which would soon destroy their tribal unity. When a man has grown rich, he convokes the folk of his clan to a great festival, and, after much eating, distributes among them all his fortune. On the Yukon river, Dall saw an Aleoute family distributing in this way ten guns, ten full fur dresses, 200 strings of beads, numerous blankets, ten wolf furs, 200 beavers, and 500 zibelines. After that they took off their festival dresses, gave them away, and, putting on old ragged furs, addressed a few words to their kinsfolk, saying that though they are now poorer than any one of them, they have won their friendship. [2] Like distributions of wealth appear

---

[1] Dr. Rink, *loc. cit.* p. 24. Europeans, grown in the respect of Roman law, are seldom capable of understanding that force of tribal authority. "In fact," Dr. Rink writes, "it is not the exception, but the rule, that white men who have stayed for ten or twenty years among the Eskimo, return without any real addition to their knowledge of the traditional ideas upon which their social state is based. The white man, whether a missionary or a trader, is firm in his dogmatic opinion that the most vulgar European is better than the most distinguished native."—*The Eskimo Tribes*, p. 31.

[2] Dall, *Alaska and its Resources*, Cambridge, U.S., 1870.

to be a regular habit with the Eskimos, and to take place at a certain season, after an exhibition of all that has been obtained during the year.[1] In my opinion these distributions reveal a very old institution, contemporaneous with the first apparition of personal wealth; they must have been a means for re-establishing equality among the members of the clan, after it had been disturbed by the enrichment of the few. The periodical redistribution of land and the periodical abandonment of all debts which took place in historical times with so many different races (Semites, Aryans, etc.), must have been a survival of that old custom. And the habit of either burying with the dead, or destroying upon his grave, all that belonged to him personally—a habit which we find among all primitive races—must have had the same origin. In fact, while everything that belongs *personally* to the dead is burnt or broken upon his grave, nothing is destroyed of what belonged to him in common with the tribe, such as boats, or the communal implements of fishing. The destruction bears upon personal property alone. At a later epoch this habit becomes a religious ceremony: it receives a mystical interpretation, and is imposed by religion, when public opinion alone proves incapable of enforcing its general observance. And, finally, it is substituted by either burning simple models of the dead man's property (as in China), or by simply carrying his property to the grave and

---

[1] Dall saw it in Alaska, Jacobsen at Ignitok in the vicinity of the Bering Strait. Gilbert Sproat mentions it among the Vancouver Indians; and Dr. Rink, who describes the periodical exhibitions just mentioned, adds: "The principal use of the accumulation of personal wealth is for *periodically* distributing it." He also mentions (*loc. cit.* p. 31) "the destruction of property for the same purpose" (of maintaining equality).

taking it back to his house after the burial ceremony
is over—a habit which still prevails with the Europeans as regards swords, crosses, and other marks of
public distinction.[1]

The high standard of the tribal morality of the
Eskimos has often been mentioned in general literature. Nevertheless the following remarks upon the
manners of the Aleoutes—nearly akin to the Eskimos
—will better illustrate savage morality as a whole.
They were written, after a ten years' stay among the
Aleoutes, by a most remarkable man—the Russian
missionary, Veniaminoff. I sum them up, mostly in
his own words :—

Endurability (he wrote) is their chief feature. It is simply
colossal. Not only do they bathe every morning in the frozen
sea, and stand naked on the beach, inhaling the icy wind, but
their endurability, even when at hard work on insufficient
food, surpasses all that can be imagined. During a protracted
scarcity of food, the Aleoute cares first for his children ; he
gives them all he has, and himself fasts. They are not
inclined to stealing ; that was remarked even by the first
Russian immigrants. Not that they never steal ; every
Aleoute would confess having sometime stolen something,
but it is always a trifle ; the whole is so childish. The
attachment of the parents to their children is touching,
though it is never expressed in words or pettings. The
Aleoute is with difficulty moved to make a promise, but once
he has made it he will keep it whatever may happen. (An
Aleoute made Veniaminoff a gift of dried fish, but it was
forgotten on the beach in the hurry of the departure. He
took it home. The next occasion to send it to the missionary
was in January ; and in November and December there was
a great scarcity of food in the Aleoute encampment. But
the fish was never touched by the starving people, and in
January it was sent to its destination.) Their code of
morality is both varied and severe. It is considered shame-
ful to be afraid of unavoidable death ; to ask pardon from
an enemy ; to die without ever having killed an enemy ; to

---

[1] See Appendix VIII.

be convicted of stealing; to capsize a boat in the harbour; to be afraid of going to sea in stormy weather; to be the first in a party on a long journey to become an invalid in case of scarcity of food; to show greediness when spoil is divided, in which case every one gives his own part to the greedy man to shame him; to divulge a public secret to his wife; being two persons on a hunting expedition, not to offer the best game to the partner; to boast of his own deeds, especially of invented ones; to scold any one in scorn. Also to beg; to pet his wife in other people's presence, and to dance with her; to bargain personally: selling must always be made through a third person, who settles the price. For a woman it is a shame not to know sewing, dancing and all kinds of woman's work; to pet her husband and children, or even to speak to her husband in the presence of a stranger.[1]

Such is Aleoute morality, which might also be further illustrated by their tales and legends. Let me also add that when Veniaminoff wrote (in 1840) one murder only had been committed since the last century in a population of 60,000 people, and that among 1,800 Aleoutes not one single common law offence had been known for forty years. This will not seem strange if we remark that scolding, scorning, and the use of rough words are absolutely unknown in Aleoute life. Even their children never fight, and never abuse each other in words. All they may say is, "Your mother does not know sewing," or "Your father is blind of one eye."[2]

[1] Veniaminoff, *Memoirs relative to the District of Unalashka* (Russian), 3 vols. St. Petersburg, 1840. Extracts, in English, from the above are given in Dall's *Alaska*. A like description of the Australians' morality is given in *Nature*, xlii. p. 639.

[2] It is most remarkable that several writers (Middendorff, Schrenk, O. Finsch) described the Ostyaks and Samoyedes in almost the same words. Even when drunken, their quarrels are insignificant. "For a hundred years one single murder has been committed in the *tundra*;" "their children never fight;" "anything may be left for years in the tundra, even food and gin, and nobody will touch it;" and so on. Gilbert Sproat "*never* witnessed a fight between two sober natives" of the Aht Indians of Vancouver Island. "Quarrelling is also rare among their children." (Rink, *loc. cit.*) And so on.

Many features of savage life remain, however, a puzzle to Europeans. The high development of tribal solidarity and the good feelings with which primitive folk are animated towards each other, could be illustrated by any amount of reliable testimony. And yet it is not the less certain that those same savages practise infanticide; that in some cases they abandon their old people, and that they blindly obey the rules of blood-revenge. We must then explain the co-existence of facts which, to the European mind, seem so contradictory at the first sight. I have just mentioned how the Aleoute father starves for days and weeks, and gives everything eatable to his child; and how the Bushman mother becomes a slave to follow her child; and I might fill pages with illustrations of the really *tender* relations existing among the savages and their children. Travellers continually mention them incidentally. Here you read about the fond love of a mother; there you see a father wildly running through the forest and carrying upon his shoulders his child bitten by a snake; or a missionary tells you the despair of the parents at the loss of a child whom he had saved, a few years before, from being immolated at its birth; you learn that the "savage" mothers usually nurse their children till the age of four, and that, in the New Hebrides, on the loss of a specially beloved child, its mother, or aunt, will kill herself to take care of it in the other world.[1] And so on.

Like facts are met with by the score; so that, when we see that these same loving parents practise

---

[1] Gill, quoted in Gerland and Waitz's *Anthropologie*, v. 641. See also pp. 636–640, where many facts of parental and filial love are quoted.

infanticide, we are bound to recognize that the habit
(whatever its ulterior transformations may be) took
its origin under the sheer pressure of necessity, as an
obligation towards the tribe, and a means for rearing
the already growing children.  The savages, as a rule,
do not "multiply without stint," as some English
writers put it.  On the contrary, they take all kinds
of measures for diminishing the birth-rate.  A whole
series of restrictions, which Europeans certainly would
find extravagant, are imposed to that effect, and they
are strictly obeyed.  But notwithstanding that, primitive folk cannot rear all their children.  However, it
has been remarked that as soon as they succeed in
increasing their regular means of subsistence, they
at once begin to abandon the practice of infanticide.
On the whole, the parents obey that obligation reluct-
antly, and as soon as they can afford it they resort to
all kinds of compromises to save the lives of their
new-born.  As has been so well pointed out by my
friend Elie Reclus,[1] they invent the lucky and unlucky
days of births, and spare the children born on the lucky
days ; they try to postpone the sentence for a few
hours, and then say that if the baby has lived one day
it must live all its natural life.[2]  They hear the cries
of the little ones coming from the forest, and maintain
that, if heard, they forbode a misfortune for the tribe ;
and as they have no baby-farming nor *crèches* for getting rid of the children, every one of them recoils
before the necessity of performing the cruel sentence ;
they prefer to expose the baby in the wood rather
than to take its life by violence.  Ignorance, not
cruelty, maintains infanticide ; and, instead of moraliz-

---

[1] *Primitive Folk*, London, 1891.
[2] Gerland, *loc. cit.* v. 636.

ing the savages with sermons, the missionaries would
do better to follow the example of Veniaminoff, who,
every year till his old age, crossed the sea of Okhotsk
in a miserable boat, or travelled on dogs among his
Tchuktchis, supplying them with bread and fishing
implements.  He thus had really stopped infanticide.

The same is true as regards what superficial
observers describe as parricide.  We just now saw
that the habit of abandoning old people is not so
widely spread as some writers have maintained it to
be.  It has been extremely exaggerated, but it is
occasionally met with among nearly all savages;
and in such cases it has the same origin as the
exposure of children.  When a "savage" feels that
he is a burden to his tribe; when every morning
his share of food is taken from the mouths of the
children—and the little ones are not so stoical as their
fathers: they cry when they are hungry; when every
day he has to be carried across the stony beach, or
the virgin forest, on the shoulders of younger people—
there are no invalid carriages, nor destitutes to wheel
them in savage lands—he begins to repeat what the
old Russian peasants say until now-a-day: "*Tchujoi
vek zayedayu, Pora na pokoi!*" ("I live other people's
life: it is time to retire!")  And he retires.  He
does what the soldier does in a similar case.  When
the salvation of his detachment depends upon its
further advance, and he can move no more, and knows
that he must die if left behind, the soldier implores
his best friend to render him the last service before
leaving the encampment.  And the friend, with shivering hands, discharges his gun into the dying body.
So the savages do.  The old man asks himself to die;
he himself insists upon this last duty towards the

community, and obtains the consent of the tribe; he
digs out his grave; he invites his kinsfolk to the last
parting meal.　His father has done so, it is now his
turn; and he parts with his kinsfolk with marks of
affection.　The savage so much considers death as
part of his *duties* towards his community, that he not
only refuses to be rescued (as Moffat has told), but
when a woman who had to be immolated on her
husband's grave was rescued by missionaries, and was
taken to an island, she escaped in the night, crossed
a broad sea-arm, swimming and rejoined her tribe, to
die on the grave.[1]　It has become with them a matter
of religion.　But the savages, as a rule, are so reluctant
to take any one's life otherwise than in fight, that none
of them will take upon himself to shed human blood, and
they resort to all kinds of stratagems, which have been
so falsely interpreted.　In most cases, they abandon
the old man in the wood, after having given him more
than his share of the common food.　Arctic expeditions
have done the same when they no more could carry
their invalid comrades.　"Live a few days more!
*may be* there will be some unexpected rescue!"

West European men of science, when coming across
these facts, are absolutely unable to stand them; they
cannot reconcile them with a high development of
tribal morality, and they prefer to cast a doubt upon
the exactitude of absolutely reliable observers, instead
of trying to explain the parallel existence of the two
sets of facts: a high tribal morality together with the
abandonment of the parents and infanticide.　But if
these same Europeans were to tell a savage that
people, extremely amiable, fond of their own children,
and so impressionable that they cry when they see a

---

[1] Erskine, quoted in Gerland and Waitz's *Anthropologie*, v. 640.

misfortune simulated on the stage, are living in Europe
within a stone's throw from dens in which children
die from sheer want of food, the savage, too, would
not understand them. I remember how vainly I tried
to make some of my Tungus friends understand our
civilization of individualism : they could not, and they
resorted to the most fantastical suggestions. The
fact is that a savage, brought up in ideas of a tribal
solidarity in everything for bad and for good, is as
incapable of understanding a "moral" European, who
knows nothing of that solidarity, as the average
European is incapable of understanding the savage.
But if our scientist had lived amidst a half-starving
tribe which does not possess among them all one
man's food for so much as a few days to come, he
probably might have understood their motives. So
also the savage, if he had stayed among us, and
received our education, may be, would understand our
European indifference towards our neighbours, and our
Royal Commissions for the prevention of "baby-
farming." "Stone houses make stony hearts," the
Russian peasants say. But he ought to live in a
stone house first.

Similar remarks must be made as regards cannibalism. Taking into account all the facts which were
brought to light during a recent controversy on this
subject at the Paris Anthropological Society, and
many incidental remarks scattered throughout the
"savage" literature, we are bound to recognize that
that practice was brought into existence by sheer
necessity ; but that it was further developed by superstition and religion into the proportions it attained in
Fiji or in Mexico. It is a fact that until this day
many savages are compelled to devour corpses in the

most advanced state of putrefaction, and that in
cases of absolute scarcity some of them have had to
disinter and to feed upon human corpses, even during
an epidemic.   These are ascertained facts.   But if
we now transport ourselves to the conditions which
man had to face during the glacial period, in a damp
and cold climate, with but little vegetable food at his
disposal ; if we take into account the terrible ravages
which scurvy still makes among underfed natives, and
remember that meat and fresh blood are the only
restoratives which they know, we must admit that
man, who formerly was a granivorous animal, became
a flesh-eater during the glacial period.   He found
plenty of deer at that time, but deer often migrate
in the Arctic regions, and sometimes they entirely
abandon a territory for a number of years.   In such
cases his last resources disappeared.   During like
hard trials, cannibalism has been resorted to even by
Europeans, and it was resorted to by the savages.
Until the present time, they occasionally devour the
corpses of their own dead : they must have devoured
then the corpses of those who had to die.   Old people
died, convinced that by their death they were render-
ing a last service to the tribe.   This is why cannibalism is represented by some savages as of divine origin,
as something that has been ordered by a messenger
from the sky.   But later on it lost its character of
necessity, and survived as a superstition.   Enemies
had to be eaten in order to inherit their courage ; and,
at a still later epoch, the enemy's eye or heart was
eaten for the same purpose ; while among other tribes,
already having a numerous priesthood and a developed
mythology, evil gods, thirsty for human blood, were
invented, and human sacrifices required by the priests

to appease the gods.    In this religious phase of its existence, cannibalism attained its most revolting characters.    Mexico is a well-known example ; and in Fiji, where the king could eat any one of his subjects, we also find a mighty cast of priests, a complicated theology,[1] and a full development of autocracy. Originated by necessity, cannibalism became, at a later period, a religious institution, and in this form it survived long after it had disappeared from among tribes which certainly practised it in former times, but did not attain the theocratical stage of evolution. The same remark must be made as regards infanticide and the abandonment of parents.    In some cases they also have been maintained as a survival of olden times, as a religiously-kept tradition of the past.

I will terminate my remarks by mentioning another custom which also is a source of most erroneous conclusions.    I mean the practice of blood-revenge.    All savages are under the impression that blood shed must be revenged by blood.    If any one has been killed, the murderer must die ; if any one has been wounded, the aggressor's blood must be shed.    There is no exception to the rule, not even for animals ; so the hunter's blood is shed on his return to the village when he has shed the blood of an animal.    That is the savages' conception of justice—a conception which yet prevails in Western Europe as regards murder. Now, when both the offender and the offended belong to the same tribe, the tribe and the offended person settle the affair.[2]    But when the offender belongs to

---

[1] W. T. Pritchard, *Polynesian Reminiscences*, London, 1866, p. 363.

[2] It is remarkable, however, that in case of a sentence of death, nobody will take upon himself to be the executioner.    Every one

another tribe, and that tribe, for one reason or another, refuses a compensation, then the offended tribe decides to take the revenge itself. Primitive folk so much consider every one's acts as a tribal affair, dependent upon tribal approval, that they easily think the clan responsible for every one's acts. Therefore, the due revenge may be taken upon any member of the offender's clan or relatives.[1] It may often happen, however, that the retaliation goes further than the offence. In trying to inflict a wound, they may kill the offender, or wound him more than they intended to do, and this becomes a cause for a new feud, so that the primitive legislators were careful in requiring the retaliation to be limited to an eye for an eye, a tooth for a tooth, and blood for blood.[2]

It is remarkable, however, that with most primitive folk like feuds are infinitely rarer than might be expected ; though with some of them they may attain abnormal proportions, especially with mountaineers

---

throws his stone, or gives his blow with the hatchet, carefully avoiding to give a mortal blow. At a later epoch, the priest will stab the victim with a sacred knife. Still later, it will be the king, until civilization invents the hired hangman. See Bastian's deep remarks upon this subject in *Der Mensch in der Geschichte*, iii. *Die Blutrache*, pp. 1-36. A remainder of this tribal habit, I am told by Professor E. Nys, has survived in military executions till our own times. In the middle portion of the nineteenth century it was the habit to load the rifles of the twelve soldiers called out for shooting the condemned victim, with eleven ball-cartridges and one blank cartridge. As the soldiers never knew who of them had the latter, each one could console his disturbed conscience by thinking that he was not one of the murderers.

[1] In Africa, and elsewhere too, it is a widely-spread habit, that if a theft has been committed, the next clan has to restore the equivalent of the stolen thing, and then look itself for the thief. A. H. Post, *Afrikanische Jurisprudenz*, Leipzig, 1887, vol. i. p. 77.

[2] See Prof. M. Kovalevsky's *Modern Customs and Ancient Law* (Russian), Moscow, 1886, vol. ii., which contains many important considerations upon this subject.

who have been driven to the highlands by foreign invaders, such as the mountaineers of Caucasia, and especially those of Borneo—the Dayaks. With the Dayaks—we were told lately—the feuds had gone so far that a young man could neither marry nor be proclaimed of age before he had secured the head of an enemy. This horrid practice was fully described in a modern English work.[1] It appears, however, that this affirmation was a gross exaggeration. Moreover, Dayak "head-hunting" takes quite another aspect when we learn that the supposed "head-hunter" is not actuated at all by personal passion. He acts under what he considers as a moral obligation towards his tribe, just as the European judge who, in obedience to the same, evidently wrong, principle of "blood for blood," hands over the condemned murderer to the hangman. Both the Dayak and the judge would even feel remorse if sympathy moved them to spare the murderer. That is why the Dayaks, apart from the murders they commit when actuated by their conception of justice, are depicted, by all those who know them, as a most sympathetic people. Thus Carl Bock, the same author who has given such a terrible picture of head-hunting, writes:

"As regards morality, I am bound to assign to the Dayaks a high place in the scale of civilization. . . . Robberies and theft are entirely unknown among them. They also are very truthful. . . . If I did not always get the 'whole truth,' I

---

[1] See Carl Bock, *The Head-Hunters of Borneo*, London, 1881. I am told, however, by Sir Hugh Law, who was for a long time Governor of Borneo, that the "head-hunting" described in this book is grossly exaggerated. Altogether, my informant speaks of the Dayaks in exactly the same sympathetic terms as Ida Pfeiffer. Let me add that Mary Kingsley speaks in her book on West Africa in the same sympathetic terms of the Fans, who had been represented formerly as the most "terrible cannibals."

always got, at least, nothing but the truth from them. I
wish I could say the same of the Malays" (pp. 209 and 210).

Bock's testimony is fully corroborated by that of Ida
Pfeiffer. " I fully recognized," she wrote, "that I should
be pleased longer to travel among them. I usually
found them honest, good, and reserved . . . much
more so than any other nation I know."[1]  Stoltze used
almost the same language when speaking of them.
The Dayaks usually have but one wife, and treat her
well. They are very sociable, and every morning the
whole clan goes out for fishing, hunting, or gardening,
in large parties. Their villages consist of big huts,
each of which is inhabited by a dozen families, and
sometimes by several hundred persons, peacefully
living together. They show great respect for their
wives, and are fond of their children; and when one
of them falls ill, the women nurse him in turn. As a
rule they are very moderate in eating and drinking.
Such is the Dayak in his real daily life.

It would be a tedious repetition if more illustrations
from savage life were given. Wherever we go we
find the same sociable manners, the same spirit of
solidarity. And when we endeavour to penetrate into
the darkness of past ages, we find the same tribal life,
the same associations of men, however primitive, for
mutual support. Therefore, Darwin was quite right
when he saw in man's social qualities the chief factor
for his further evolution, and Darwin's vulgarizers
are entirely wrong when they maintain the contrary.

The small strength and speed of man (he wrote), his want
of natural weapons, etc., are more than counterbalanced, firstly,

---

[1] Ida Pfeiffer, *Meine zweite Weltreise*, Wien, 1856, vol. i. pp. 116
*seq.* See also Müller and Temminch's *Dutch Possessions in Archi-
pelagic India*, quoted by Elisée Reclus, in *Géographie Universelle*, xiii.

by his intellectual faculties (which, he remarked on another page, have been chiefly or even exclusively gained for the benefit of the community); and secondly, *by his social qualities*, which led him to give and receive aid from his fellow men.[1]

In the last century the "savage" and his "life in the state of nature" were idealized. But now men of science have gone to the opposite extreme, especially since some of them, anxious to prove the animal origin of man, but not conversant with the social aspects of animal life, began to charge the savage with all imaginable "bestial" features. It is evident, however, that this exaggeration is even more unscientific than Rousseau's idealization. The savage is not an ideal of virtue, nor is he an ideal of "savagery." But the primitive man has one quality, elaborated and maintained by the very necessities of his hard struggle for life—he identifies his own existence with that of his tribe ; and without that quality mankind never would have attained the level it has attained now.

Primitive folk, as has been already said, so much identify their lives with that of the tribe, that each of their acts, however insignificant, is considered as a tribal affair. Their whole behaviour is regulated by an infinite series of unwritten rules of propriety which are the fruit of their common experience as to what is good or bad—that is, beneficial or harmful for their own tribe. Of course, the reasonings upon which their rules of propriety are based sometimes are absurd in the extreme. Many of them originate in superstition ; and altogether, in whatever the savage does, he sees but the immediate consequences of his acts ; he cannot foresee their indirect and ulterior con-

---

[1] *Descent of Man*, second ed., pp. 63, 64.

sequences—thus simply exaggerating a defect with which Bentham reproached civilized legislators. But, absurd or not, the savage obeys the prescriptions of the common law, however inconvenient they may be. He obeys them even more blindly than the civilized man obeys the prescriptions of the written law. His common law is his religion; it is his very habit of living. The idea of the clan is always present to his mind; and self-restriction and self-sacrifice in the interest of the clan are of daily occurrence. If the savage has infringed one of the smaller tribal rules, he is prosecuted by the mockeries of the women. If the infringement is grave, he is tortured day and night by the fear of having called a calamity upon his tribe. If he has wounded by accident any one of his own clan, and thus has committed the greatest of all crimes, he grows quite miserable : he runs away in the woods, and is ready to commit suicide, unless the tribe absolves him by inflicting upon him a physical pain and sheds some of his own blood.[1] Within the tribe everything is shared in common ; every morsel of food is divided among all present ; and if the savage is alone in the woods, he does not begin eating before he has loudly shouted thrice an invitation to any one who may hear his voice to share his meal.[2]

In short, within the tribe the rule of "each for all" is supreme, so long as the separate family has not yet broken up the tribal unity. But that rule is not ex-tended to the neighbouring clans, or tribes, even when they are federated for mutual protection. Each tribe, or clan, is a separate unity. Just as among mammals

---

[1] See Bastian's *Mensch in der Geschichte*, iii. p. 7. Also Grey, *loc. cit.* ii. p. 238.

[2] Miklukho-Maclay, *loc. cit.* Same habit with the Hottentots.

and birds, the territory is roughly allotted among separate tribes, and, except in times of war, the boundaries are respected. On entering the territory of his neighbours one must show that he has no bad intentions. The louder one heralds his coming, the more confidence he wins ; and if he enters a house, he must deposit his hatchet at the entrance. But no tribe is bound to share its food with the others : it may do so or it may not. Therefore the life of the savage is divided into two sets of actions, and appears under two different ethical aspects : the relations within the tribe, and the relations with the outsiders ; and (like our international law) the "inter-tribal" law widely differs from the common law. Therefore, when it comes to a war the most revolting cruelties may be considered as so many claims upon the admiration of the tribe. This double conception of morality passes through the whole evolution of mankind, and maintains itself until now. We Europeans have realized some progress—not immense, at any rate—in eradicating that double conception of ethics ; but it also must be said that while we have in some measure extended our ideas of solidarity—in theory, at least—over the nation, and partly over other nations as well, we have lessened the bonds of solidarity within our own nations, and even within our own families.

The appearance of a separate family amidst the clan necessarily disturbs the established unity. A separate family means separate property and accumulation of wealth. We saw how the Eskimos obviate its inconveniences ; and it is one of the most interesting studies to follow in the course of ages the different institutions (village communities, guilds, and so on) by means of which the masses endeavoured to maintain the tribal unity, notwithstanding the agencies which were at work to break it down. On the other hand, the first rudiments of knowledge which appeared at an extremely remote epoch, when they confounded themselves with witchcraft, also became a power ·in the hands of the individual which could be used against the tribe. They were carefully kept in secrecy, and transmitted to the initiated only, in the secret societies of witches, shamans, and priests, which we find among all savages. By the same time, wars and invasions created military authority, as also castes of warriors, whose associations or clubs acquired great powers. However, at no period of man's life were wars the *normal* state of existence. While warriors exterminated each other, and the priests celebrated their massacres, the masses continued to live their daily life, they prosecuted their daily toil. And it is one of the most interesting studies to follow that life of the masses ; to study the means by which they maintained their own social organization, which was based upon their own conceptions of equity, mutual aid, and mutual support—of common law, in a word, even when they were submitted to the most ferocious theocracy or autocracy in the State.

# CHAPTER IV

## MUTUAL AID AMONG THE BARBARIANS

The great migrations.—New organization rendered necessary.—
The village community.—Communal work.—Judicial procedure.—
Inter-tribal law.—Illustrations from the life of our contemporaries.—
Buryates.—Kabyles.—Caucasian mountaineers.—African stems.

IT is not possible to study primitive mankind without being deeply impressed by the sociability it has displayed since its very first steps in life. Traces of human societies are found in the relics of both the oldest and the later stone age ; and, when we come to observe the savages whose manners of life are still those of neolithic man, we find them closely bound together by an extremely ancient clan organization which enables them to combine their individually weak forces, to enjoy life in common, and to progress. Man is no exception in nature. He also is subject to the great principle of Mutual Aid which grants the best chances of survival to those who best support each other in the struggle for life. These were the conclusions arrived at in the previous chapters.

However, as soon as we come to a higher stage of civilization, and refer to history which already has something to say about that stage, we are bewildered by the struggles and conflicts which it reveals. The old bonds seem entirely to be broken. Stems are seen to fight against stems, tribes against tribes, individuals

against individuals; and out of this chaotic contest
of hostile forces, mankind issues divided into castes,
enslaved to despots, separated into States always ready
to wage war against each other. And, with this history
of mankind in his hands, the pessimist philosopher
triumphantly concludes that warfare and oppression
are the very essence of human nature ; that the warlike and predatory instincts of man can only be
restrained within certain limits by a strong authority
which enforces peace and thus gives an opportunity to
the few and nobler ones to prepare a better life for
humanity in times to come.

And yet, as soon as the every-day life of man during
the historical period is submitted to a closer analysis —
and so it has been, of late, by many patient students
of very early institutions—it appears at once under
quite a different aspect. Leaving aside the preconceived ideas of most historians and their pronounced
predilection for the dramatic aspects of history, we see
that the very documents they habitually peruse are
such as to exaggerate the part of human life given to
struggles and to underrate its peaceful moods. The
bright and sunny days are lost sight of in the gales and
storms. Even in our own time, the cumbersome records
which we prepare for the future historian, in our Press,
our law courts, our Government offices, and even in
our fiction and poetry, suffer from the same one-sided-
ness. They hand down to posterity the most minute
descriptions of every war, every battle and skirmish,
every contest and act of violence, every kind of individual suffering ; but they hardly bear any trace of the
countless acts of mutual support and devotion which
every one of us knows from his own experience ; they
hardly take notice of what makes the very essence of

our daily life—our social instincts and manners. No wonder, then, if the records of the past were so imperfect. The annalists of old never failed to chronicle the petty wars and calamities which harassed their contemporaries; but they paid no attention whatever to the life of the masses, although the masses chiefly used to toil peacefully while the few indulged in fighting. The epic poems, the inscriptions on monuments, the treaties of peace—nearly all historical documents bear the same character; they deal with breaches of peace, not with peace itself. So that the best-intentioned historian unconsciously draws a distorted picture of the times he endeavours to depict; and, to restore the real proportion between conflict and union, we are now bound to enter into a minute analysis of thousands of small facts and faint indications accidentally preserved in the relics of the past; to interpret them with the aid of comparative ethnology; and, after having heard so much about what used to divide men, to reconstruct stone by stone the institutions which used to unite them.

Ere long history will have to be re-written on new lines, so as to take into account these two currents of human life and to appreciate the part played by each of them in evolution. But in the meantime we may avail ourselves of the immense preparatory work recently done towards restoring the leading features of the second current, so much neglected. From the better-known periods of history we may take some illustrations of the life of the masses, in order to indicate the part played by mutual support during those periods; and, in so doing, we may dispense (for the sake of brevity) from going as far back as the Egyptian, or even the Greek and Roman antiquity. For, in fact, the evolution of mankind has not had the character of one unbroken series. Several times civilization came to an end in one given region, with one given race, and began anew elsewhere, among other races. But at each fresh start it began again with the same clan institutions which we have seen among the savages. So that if we take the last start of our own civilization, when it began afresh in the first centuries of our era, among those whom the Romans called the "barbarians," we shall have the whole scale of evolution, beginning with the gentes and ending in the institutions of our own time. To these illustrations the following pages will be devoted.

Men of science have not yet settled upon the causes which some two thousand years ago drove whole nations from Asia into Europe and resulted in the great migrations of barbarians which put an end to the West Roman Empire. One cause, however, is naturally suggested to the geographer as he contemplates the ruins of populous cities in the deserts of Central Asia, or follows the old beds of rivers now disappeared and the wide outlines of lakes now reduced to the size of mere ponds. It is desiccation : a quite recent desiccation, continued still at a speed which we formerly were not prepared to admit.[1] Against it

[1] Numberless traces of post-pliocene lakes, now disappeared, are found over Central, West, and North Asia. Shells of the same species as those now found in the Caspian Sea are scattered over the surface of the soil as far East as half-way to Lake Aral, and are found in recent deposits as far north as Kazan. Traces of Caspian Gulfs, formerly taken for old beds of the Amu, intersect the Turcoman territory. Deduction must surely be made for temporary, periodical oscillations. But with all that, desiccation is evident, and it progresses at a formerly unexpected speed. Even in the relatively wet parts of South-West Siberia, the succession of reliable surveys, recently

man was powerless. When the inhabitants of North-West Mongolia and East Turkestan saw that water was abandoning them, they had no course open to them but to move down the broad valleys leading to the lowlands, and to thrust westwards the inhabitants of the plains.[1] Stems after stems were thus thrown into Europe, compelling other stems to move and to remove for centuries in succession, westwards and eastwards, in search of new and more or less permanent abodes. Races were mixing with races during those migrations, aborigines with immigrants, Aryans with Ural-Altayans; and it would have been no wonder if the social institutions which had kept them together in their mother-countries had been totally wrecked during the stratification of races which took place in Europe and Asia. But they were *not* wrecked ; they simply underwent the modification which was required by the new conditions of life.

The Teutons, the Celts, the Scandinavians, the Slavonians, and others, when they first came in contact with the Romans, were in a transitional state of social organization. The clan unions, based upon a real or supposed common origin, had kept them together for many thousands of years in succession. But these unions could answer their purpose so long only as there were no separate families within the gens or

---

published by Yadrintseff, shows that villages have grown up on what was, eighty years ago, the bottom of one of the lakes of the Tchany group ; while the other lakes of the same group, which covered hundreds of square miles some fifty years ago, are now mere ponds. In short, the desiccation of North-West Asia goes on at a rate which must be measured by centuries, instead of by the geological units of time of which we formerly used to speak.

[1] Whole civilizations had thus disappeared, as is proved now by the remarkable discoveries in Mongolia on the Orkhon, and in the depressions of Lukchun (by Dmitri Clements), and of Lob-nor (by Sven Hedin).

clan itself. However, for causes already mentioned, the separate patriarchal family had slowly but steadily developed within the clans, and in the long run it evidently meant the individual accumulation of wealth and power, and the hereditary transmission of both. The frequent migrations of the barbarians and the ensuing wars only hastened the division of the gentes into separate families, while the dispersing of stems and their mingling with strangers offered singular facilities for the ultimate disintegration of those unions which were based upon kinship. The barbarians thus stood in a position of either seeing their clans dissolved into loose aggregations of families, of which the wealthiest, especially if combining sacerdotal functions or military repute with wealth, would have succeeded in imposing their authority upon the others; or of finding out some new form of organization based upon some new principle.

Many stems had no force to resist disintegration: they broke up and were lost for history. But the more vigorous ones did not disintegrate. They came out of the ordeal with a new organization—the *village community*—which kept them together for the next fifteen centuries or more. The conception of a common *territory*, appropriated or protected by common efforts, was elaborated, and it took the place of the vanishing conceptions of common descent. The common gods gradually lost their character of ancestors and were endowed with a local territorial character. They became the gods or saints of a given locality; " the land " was identified with its inhabitants. Territorial unions grew up instead of the consanguine unions of old, and this new organization evidently offered many advantages under the given circumstances. It recognized the independence of the family and even emphasized it, the village community disclaiming all rights of interference in what was going on within the family enclosure; it gave much more freedom to personal initiative; it was not hostile in principle to union between men of different descent, and it maintained at the same time the necessary cohesion of action and thought, while it was strong enough to oppose the dominative tendencies of the minorities of wizards, priests, and professional or distinguished warriors. Consequently it became the primary cell of future organization, and with many nations the village community has retained this character until now.

It is now known, and scarcely contested, that the village community was not a specific feature of the Slavonians, nor even of the ancient Teutons. It prevailed in England during both the Saxon and Norman times, and partially survived till the last century;[1] it was at the bottom of the social organization of old Scotland, old Ireland, and old Wales. In France, the communal possession and the communal allotment of arable land by the village folkmote persisted from the first centuries of our era till the times of Turgot, who found the folkmotes "too noisy" and therefore abolished them. It survived Roman rule in

---

[1] If I follow the opinions of (to name modern specialists only) Nasse, Kovalevsky, and Vinogradov, and not those of F. Seebohm (Mr. Denman Ross can only be named for the sake of completeness), it is not only because of the deep knowledge and concordance of views of these three writers, but also on account of their perfect knowledge of the village community altogether—a knowledge the want of which is much felt in the otherwise remarkable work of Mr. Seebohm. The same remark applies, in a still higher degree, to the most elegant writings of Fustel de Coulanges, whose opinions and passionate interpretations of old texts are confined to himself.

Italy, and revived after the fall of the Roman Empire.
It was the rule with the Scandinavians, the Slavonians,
the Finns (in the *pittäyä*, as also, probably, the *kihlakunta*), the Coures, and the Lives. The village
community in India—past and present, Aryan and
non-Aryan—is well known through the epoch-making
works of Sir Henry Maine; and Elphinstone has
described it among the Afghans. We also find it
in the Mongolian *oulous*, the Kabyle *thaddart*, the
Javanese *dessa*, the Malayan *kota* or *tofa*, and under a
variety of names in Abyssinia, the Soudan, in the
interior of Africa, with natives of both Americas, with
all the small and large tribes of the Pacific archipelagoes.
In short, we do not know one single human race or
one single nation which has not had its period of
village communities. This fact alone disposes of the
theory according to which the village community in
Europe would have been a servile growth. It is
anterior to serfdom, and even servile submission was
powerless to break it. It was a universal phase of
evolution, a natural outcome of the clan organization,
with all those stems, at least, which have played, or
play still, some part in history.[1]

---

[1] The literature of the village community is so vast that but a few
works can be named. Those of Sir Henry Maine, F. Seebohm, and
Walter's *Das alte Wallis* (Bonn, 1859), are well-known popular
sources of information about Scotland, Ireland, and Wales. For
France, P. Viollet, *Précis de l'histoire du droit français: Droit privé*,
1886, and several of his monographs in *Bibl. de l'École des Chartes;*
Babeau, *Le Village sous l'ancien régime* (the *mir* in the eighteenth
century), third edition, 1887; Bonnemère, Doniol, etc. For Italy
and Scandinavia, the chief works are named in Laveleye's *Primitive
Property*, German version by K. Bücher. For the Finns, Rein's
*Föreläsningar*, i. 16; Koskinen, *Finnische Geschichte*, 1874, and
various monographs. For the Lives and Coures, Prof. Lutchitzky in
*Severnyi Vestnik*, 1891. For the Teutons, besides the well-known
works of Maurer, Sohm (*Altdeutsche Reichs- und Gerichts-Verfassung*),

It was a natural growth, and an absolute uniformity in its structure was therefore not possible. As a rule, it was a union between families considered as of common descent and owning a certain territory in common. But with some stems, and under certain circumstances, the families used to grow very numerous before they threw off new buds in the shape of new families ; five, six, or seven generations continued to live under the same roof, or within the same enclosure, owning their joint household and cattle in common, and taking their meals at the common hearth. They kept in such case to what ethnology knows as the "joint family," or the "undivided household," which we still see all over China, in India, in the South Slavonian *zadruga*, and occasionally find in Africa, in America, in Denmark, in North Russia, and West France.[1] With other stems, or in other circumstances,

---

also Dahn (*Urzeit, Völkerwanderung, Langobardische Studien*), Janssen, Wilh. Arnold, etc. For India, besides H. Maine and the works he names, Sir John Phear's *Aryan Village*. For Russia and South Slavonians, see Kavelin, Posnikoff, Sokolovsky, Kovalevsky, Efimenko, Ivanisheff, Klaus, etc. (copious bibliographical index up to 1880 in the *Sbornik svedeniy ob obschinye* of the Russ. Geog. Soc.). For general conclusions, besides Laveleye's *Propriété*, Morgan's *Ancient Society*, Lippert's *Kulturgeschichte*, Post, Dargun, etc., also the lectures of M. Kovalevsky (*Tableau des origines et de l'évolution de la famille et de la propriété*, Stockholm, 1890). Many special monographs ought to be mentioned ; their titles may be found in the excellent lists given by P. Viollet in *Droit privé* and *Droit public.* For other races, see subsequent notes.

[1] Several authorities are inclined to consider the joint household as an intermediate stage between the clan and the village community ; and there is no doubt that in very many cases village communities have grown up out of undivided families. Nevertheless, I consider the joint household as a fact of a different order. We find it within the gentes ; on the other hand, we cannot affirm that joint families have existed at any period without belonging either to a gens or to a village community, or to a *Gau*. I conceive the early village communities as slowly originating directly from the gentes, and consisting, according to racial and local circumstances, either of several joint

not yet well specified, the families did not attain the
same proportions ; the grandsons, and occasionally the
sons, left the household as soon as they were married,
and each of them started a new cell of his own. But,
joint or not, clustered together or scattered in the
woods, the families remained united into village
communities ; several villages were grouped into
tribes ; and the tribes joined into confederations.
Such was the social organization which developed
among the so-called "barbarians," when they began to
settle more or less permanently in Europe.

A very long evolution was required before the
gentes, or clans, recognized the separate existence of
a patriarchal family in a separate hut ; but even after
that had been recognized, the clan, as a rule, knew no
personal inheritance of property. The few things
which might have belonged personally to the individual
were either destroyed on his grave or buried with him.
The village community, on the contrary, fully recognized
the private accumulation of wealth within the family
and its hereditary transmission. But wealth was conceived exclusively in the shape of *movable* property,
including cattle, implements, arms, and the dwelling-
house which—"like all things that can be destroyed
by fire"—belonged to the same category.[1] As to
private property in land, the village community did
not, and could not, recognize anything of the kind,

families, or of both joint and simple families, or (especially in the
case of new settlements) of simple families only. If this view be
correct, we should not have the right of establishing the series : gens,
compound family, village community—the second member of the
series having not the same ethnological value as the two others.
See Appendix IX.

[1] Stobbe, *Beiträge zur Geschichte des deutschen Rechtes*, p. 62.

and, as a rule, it does not recognize it now. The land was the common property of the tribe, or of the whole stem, and the village community itself owned its part of the tribal territory so long only as the tribe did not claim a re-distribution of the village allotments. The clearing of the woods and the breaking of the prairies being mostly done by the communities or, at least, by the joint work of several families—always with the consent of the community—the cleared plots were held by each family for a term of four, twelve, or twenty years, after which term they were treated as parts of the arable land owned in common. Private property, or possession "for ever," was as incompatible with the very principles and the religious conceptions of the village community as it was with the principles of the gens ; so that a long influence of the Roman law and the Christian Church, which soon accepted the Roman principles, were required to accustom the barbarians to the idea of private property in land being possible.[1] And yet, even when such property, or possession for an unlimited time, was recognized, the owner of a separate estate remained a co-proprietor in the waste lands, forests, and grazing-grounds. Moreover, we continually see, especially in the history of Russia, that when a few families, acting separately, had taken possession of some land belonging to tribes which were treated as strangers, they very soon united together, and constituted a village community which

---

[1] The few traces of private property in land which are met with in the early barbarian period are found with such stems (the Batavians, the Franks in Gaul) as have been for a time under the influence of Imperial Rome. See Inama-Sternegg's *Die Ausbildung der grossen Grundherrschaften in Deutschland*, Bd. i. 1878. Also, Besseler, *Neubruch nach dem älteren deutschen Recht*, pp. 11–12, quoted by Kovalevsky, *Modern Custom and Ancient Law*, Moscow, 1886, i. 134.

in the third or fourth generation began to profess a community of origin.

A whole series of institutions, partly inherited from the clan period, have developed from that basis of common ownership of land during the long succession of centuries which was required to bring the barbarians under the dominion of States organized upon the Roman or Byzantine pattern. The village community was not only a union for guaranteeing to each one his fair share in the common land, but also a union for common culture, for mutual support in all possible forms, for protection from violence, and for a further development of knowledge, national bonds, and moral conceptions ; and every change in the judicial, military, educational, or economical manners had to be decided at the folkmotes of the village, the tribe, or the confederation. The community being a continuation of the gens, it inherited all its functions. It was the *universitas*, the *mir*—a world in itself.

Common hunting, common fishing, and common culture of the orchards or the plantations of fruit trees was the rule with the old gentes. Common agriculture became the rule in the barbarian village communities. True, that direct testimony to this effect is scarce, and in the literature of antiquity we only have the passages of Diodorus and Julius Cæsar relating to the inhabitants of the Lipari Islands, one of the Celt-Iberian tribes, and the Sueves. But there is no lack of evidence to prove that common agriculture was practised among some Teuton tribes, the Franks, and the old Scotch, Irish, and Welsh.[1] As to the later survivals of the

---

[1] Maurer's *Markgenossenschaft ;* Lamprecht's " Wirthschaft und Recht der Franken zur Zeit der Volksrechte," in *Histor. Taschenbuch*, 1883 ; Seebohm's *The English Village Community*, ch. vi., vii., and ix.

same practice, they simply are countless. Even in perfectly Romanized France, common culture was habitual some five and twenty years ago in the Morbihan (Brittany).[1] The old Welsh *cyvar*, or joint team, as well as the common culture of the land allotted to the use of the village sanctuary are quite common among the tribes of Caucasus the least touched by civilization,[2] and like facts are of daily occurrence among the Russian peasants. Moreover, it is well known that many tribes of Brazil, Central America, and Mexico used to cultivate their fields in common, and that the same habit is widely spread among some Malayans, in New Caledonia, with several Negro stems, and so on.[3] In short, communal culture is so habitual with many Aryan, Ural-Altayan, Mongolian, Negro, Red Indian, Malayan, and Melanesian stems that we must consider it as a universal—though not as the only possible—form of primitive agriculture.[4]

Communal cultivation does not, however, imply by necessity communal consumption. Already under the clan organization we often see that when the boats laden with fruits or fish return to the village, the food they bring in is divided among the huts and the " long

---

[1] Letourneau, in *Bulletin de la Soc. d'Anthropologie*, 1888, vol. xi. p. 476.

[2] Walter, *Das alte Wallis*, p. 323; Dm. Bakradze and N. Khoudadoff in Russian *Zapiski* of the Caucasian Geogr. Society, xiv. Part I.

[3] Bancroft's *Native Races ;* Waitz, *Anthropologie*, iii. 423 ; Montrozier, in *Bull. Soc. d'Anthropologie*, 1870 ; Post's *Studien*, etc.

[4] A number of works, by Ory, Luro, Laudes, and Sylvestre, on the village community in Annam, proving that it has had there the same forms as in Germany or Russia, is mentioned in a review of these works by Jobbé-Duval, in *Nouvelle Revue historique de droit français et étranger*, October and December, 1896. A good study of the village community of Peru, before the establishment of the power of the Incas, has been brought out by Heinrich Cunow (*Die Soziale Verfassung des Inka-Reichs*, Stuttgart, 1896. The communal possession of land and communal culture are described in that work.

houses" inhabited by either several families or the youth, and is cooked separately at each separate hearth. The habit of taking meals in a narrower circle of relatives or associates thus prevails at an early period of clan life. It became the rule in the village community. Even the food grown in common was usually divided between the households after part of it had been laid in store for communal use. However, the tradition of communal meals was piously kept alive ; every available opportunity, such as the commemoration of the ancestors, the religious festivals, the beginning and the end of field work, the births, the marriages, and the funerals, being seized upon to bring the community to a common meal. Even now this habit, well known in this country as the "harvest supper," is the last to disappear. On the other hand, even when the fields had long since ceased to be tilled and sown in common, a variety of agricultural work continued, and continues still, to be performed by the community. Some part of the communal land is still cultivated in many cases in common, either for the use of the destitute, or for refilling the communal stores, or for using the produce at the religious festivals. The irrigation canals are digged and repaired in common. The communal meadows are mown by the community ; and the sight of a Russian commune mowing a meadow—the men rivalling each other in their advance with the scythe, while the women turn the grass over and throw it up into heaps—is one of the most inspiring sights ; it shows what human work might be and ought to be. The hay, in such case, is divided among the separate households, and it is evident that no one has the right of taking hay from a neighbour's stack without his permission ; but the

limitation of this last rule among the Caucasian Ossetes
is most noteworthy. When the cuckoo cries and an-
nounces that spring is coming, and that the meadows
will soon be clothed again with grass, every one in
need has the right of taking from a neighbour's stack
the hay he wants for his cattle.[1] The old communal
rights are thus re-asserted, as if to prove how contrary
unbridled individualism is to human nature.

When the European traveller lands in some small
island of the Pacific, and, seeing at a distance a grove
of palm trees, walks in that direction, he is astonished
to discover that the little villages are connected by
roads paved with big stones, quite comfortable for the
unshod natives, and very similar to the " old roads " of
the Swiss mountains. Such roads were traced by the
" barbarians " all over Europe, and one must have
travelled in wild, thinly-peopled countries, far away
from the chief lines of communication, to realize in full
the immense work that must have been performed by
the barbarian communities in order to conquer the
woody and marshy wilderness which Europe was some
two thousand years ago. Isolated families, having no
tools, and weak as they were, could not have conquered
it ; the wilderness would have overpowered them.
Village communities alone, working in common, could
master the wild forests, the sinking marshes, and the
endless steppes. The rough roads, the ferries, the
wooden bridges taken away in the winter and rebuilt
after the spring flood was over, the fences and the
palisaded walls of the villages, the earthen forts and
the small towers with which the territory was dotted—
all these were the work of the barbarian communities.
And when a community grew numerous it used to

[1] Kovalevsky, *Modern Custom and Ancient Law*, i. 115.

throw off a new bud.  A new community arose at a
distance, thus step by step bringing the woods and
the steppes under the dominion of man.  The whole
making of European nations was such a budding of
the village communities.  Even now-a-days the Russian
peasants, if they are not quite broken down by misery,
migrate in communities, and they till the soil and build
the houses in common when they settle on the banks
of the Amur, or in Manitoba.  And even the English,
when they first began to colonize America, used to
return to the old system ; they grouped into village
communities.[1]

The village community was the chief arm of the
barbarians in their hard struggle against a hostile
nature.  It also was the bond they opposed to oppression by the cunningest and the strongest which so
easily might have developed during those disturbed
times.  The imaginary barbarian—the man who fights
and kills at his mere caprice—existed no more than the
"bloodthirsty" savage.  The real barbarian was living,
on the contrary, under a wide series of institutions,
imbued with considerations as to what may be useful
or noxious to his tribe or confederation, and these
institutions were piously handed down from generation
to generation in verses and songs, in proverbs or triads,
in sentences and instructions.  The more we study
them the more we recognize the narrow bonds which
united men in their villages.  Every quarrel arising
between two individuals was treated as a communal
affair—even the offensive words that might have been
uttered during a quarrel being considered as an offence

[1] Palfrey, *History of New England*, ii. 13; quoted in Maine's
*Village Communities*, New York, 1876, p. 201.

to the community and its ancestors. They had to be repaired by amends made both to the individual and the community ;[1] and if a quarrel ended in a fight and wounds, the man who stood by and did not interpose was treated as if he himself had inflicted the wounds.[2]

The judicial procedure was imbued with the same spirit. Every dispute was brought first before mediators or arbiters, and it mostly ended with them, the arbiters playing a very important part in barbarian society. But if the case was too grave to be settled in this way, it came before the folkmote, which was bound " to find the sentence," and pronounced it in a conditional form ; that is, " such compensation was due, if the wrong be proved," and the wrong had to be proved or disclaimed by six or twelve persons confirming or denying the fact by oath ; ordeal being resorted to in case of contradiction between the two sets of jurors. Such procedure, which remained in force for more than two thousand years in succession, speaks volumes for itself ; it shows how close were the bonds between all members of the community. Moreover, there was no other authority to enforce the decisions of the folkmote besides its own moral authority. The only possible menace was that the community might declare the rebel an outlaw, but even this menace was reciprocal. A man discontented with the folkmote could declare that he would abandon the tribe and go over to another tribe—a most dreadful menace, as it was sure to bring all kinds of misfortunes upon a tribe that might have been unfair to one of its

---

[1] Königswarter, *Études sur le développement des sociétés humaines*, Paris, 1850.

[2] This is, at least, the law of the Kalmucks, whose customary law bears the closest resemblance to the laws of the Teutons, the old Slavonians, etc.

members.[1]  A rebellion against a right decision of
the customary law was simply "inconceivable," as
Henry Maine has so well said, because "law, morality,
and fact" could not be separated from each other
in those times.[2]  The moral authority of the commune was so great that even at a much later epoch,
when the village communities fell into submission to
the feudal lord, they maintained their judicial powers;
they only permitted the lord, or his deputy, to "find"
the above conditional sentence in accordance with the
customary law he had sworn to follow, and to levy for
himself the fine (the *fred*) due to the commune.  But
for a long time, the lord himself, if he remained a co-
proprietor in the waste land of the commune, submitted
in communal affairs to its decisions.  Noble or ecclesi-
astic, he had to submit to the folkmote—*Wer daselbst
Wasser und Weid genusst, muss gehorsam sein*—"Who
enjoys here the right of water and pasture must obey"
—was the old saying.  Even when the peasants became serfs under the lord, he was bound to appear
before the folkmote when they summoned him.[3]

In their conceptions of justice the barbarians evidently
did not much differ from the savages.  They also
maintained the idea that a murder must be followed by
putting the murderer to death ; that wounds had to be
punished by equal wounds, and that the wronged
family was bound to fulfil the sentence of the customary

---

[1] The habit is in force still with many African and other tribes.

[2] *Village Communities*, pp. 65–68 and 199.

[3] Maurer (*Gesch. der Markverfassung*, § 29, 97) is quite decisive
upon this subject.  He maintains that "All members of the community . . . . the laic and clerical lords as well, often also the partial
co-possessors (*Markberechtigte*), and even strangers to the Mark, were
submitted to its jurisdiction" (p. 312).  This conception remained
locally in force up to the fifteenth century.

law. This was a holy duty, a duty towards the ancestors, which had to be accomplished in broad daylight, never in secrecy, and rendered widely known. Therefore the most inspired passages of the sagas and epic poetry altogether are those which glorify what was supposed to be justice. The gods themselves joined in aiding it. However, the predominant feature of barbarian justice is, on the one hand, to limit the numbers of persons who may be involved in a feud, and, on the other hand, to extirpate the brutal idea of blood for blood and wounds for wounds, by substituting for it the system of compensation. The barbarian codes—which were collections of common law rules written down for the use of judges—"first permitted, then encouraged, and at last enforced," compensation instead of revenge.[1] The compensation has, however, been totally misunderstood by those who represented it as a fine, and as a sort of *carte blanche* given to the rich man to do whatever he liked. The compensation money (*wergeld*), which was quite different from the fine or *fred*,[2] was habitually so high for all kinds of active offences that it certainly was no encouragement for such offences. In case of a murder it usually exceeded all the possible fortune of the murderer. "Eighteen times eighteen cows" is the compensation with the Ossetes who do not know how to reckon above eighteen, while with the African tribes it attains 800 cows or 100 camels with their young, or 416 sheep

---

[1] Königswarter, *loc. cit.* p. 50; J. Thrupp, *Historical Law Tracts*, London, 1843, p. 106.

[2] Königswarter has shown that the *fred* originated from an offering which had to be made to appease the ancestors. Later on, it was paid to the community, for the breach of peace; and still later to the judge, or king, or lord, when they had appropriated to themselves the rights of the community.

in the poorer tribes.[1]  In the great majority of cases, the compensation money could not be paid at all, so that the murderer had no issue but to induce the wronged family, by repentance, to adopt him.  Even now, in the Caucasus, when feuds come to an end, the offender touches with his lips the breast of the oldest woman of the tribe, and becomes a " milk-brother " to all men of the wronged family.[2]  With several African tribes he must give his daughter, or sister, in marriage to some one of the family ; with other tribes he is bound to marry the woman whom he has made a widow ; and in all cases he becomes a member of the family, whose opinion is taken in all important family matters.[3]

Far from acting with disregard to human life, the barbarians, moreover, knew nothing of the horrid punishments introduced at a later epoch by the laic and canonic laws under Roman and Byzantine influence.  For, if the Saxon code admitted the death penalty rather freely, even in cases of incendiarism and armed robbery, the other barbarian codes pronounced it exclusively in cases of betrayal of one's kin; and sacrilege against the community's gods, as the only means to appease the gods.

All this, as seen, is very far from the supposed

---

[1] Post's *Bausteine* and *Afrikanische Jurisprudenz*, Oldenburg, 1887, vol. i. pp. 64 *seq.* ; Kovalevsky, *loc. cit.* ii. 164–189.

[2] O. Miller and M. Kovalevsky, " In the Mountaineer Communities of Kabardia," in *Vestnik Evropy*, April, 1884.  With the Shakhsevens of the Mugan Steppe, blood feuds always end by marriage between the two hostile sides (Markoff, in appendix to the *Zapiski* of the Caucasian Geogr. Soc., xiv. 1, 21).

[3] Post, in *Afrik. Jurisprudenz*, gives a series of facts illustrating the conceptions of equity inrooted among the African barbarians. The same may be said of all serious examinations into barbarian common law.

"moral dissoluteness" of the barbarians. On the
contrary, we cannot but admire the deeply moral principles elaborated within the early village communities
which found their expression in Welsh triads, in legends
about King Arthur, in Brehon commentaries,[1] in old
German legends and so on, or find still their expression
in the sayings of the modern barbarians. In his introduction to *The Story of Burnt Njal*, George Dasent
very justly sums up as follows the qualities of a North-
man, as they appear in the sagas :—

, To do what lay before him openly and like a man, without
fear of either foes, fiends, or fate ; . . . to be free and daring
in all his deeds; to be gentle and generous to his friends and
kinsmen ; to be stern and grim to his foes [those who are
under the *lex talionis*], but even towards them to fulfil all
bounden duties. . . . To be no truce-breaker, nor tale-bearer,
nor backbiter. To utter nothing against any man that he
would not dare to tell him to his face. To turn no man
from his door who sought food or shelter, even though he
were a foe.[2]

The same or still better principles permeate the Welsh
epic poetry and triads. To act " according to the
nature of mildness and the principles of equity," without regard to the foes or to the friends, and " to repair
the wrong," are the highest duties of man ; " evil is
death, good is life," exclaims the poet legislator.[3]
" The World would be fool, if agreements made on
lips were not honourable "—the Brehon law says.
And the humble Shamanist Mordovian, after having
praised the same qualities, will add, moreover, in his
principles of customary law, that " among neighbours

---

[1] See the excellent chapter, "Le droit de la Vieille Irlande," (also
"Le Haut Nord ") in *Études de droit international et de droit politique*,
by Prof. E. Nys, Bruxelles, 1896.

[2] Introduction, p. xxxv.

[3] *Das alte Wallis*, pp. 343-350.

the cow and the milking-jar are in common;" that "the cow must be milked for yourself and him who may ask milk;" that "the body of a child reddens from the stroke, but the face of him who strikes reddens from shame;"[1] and so on. Many pages might be filled with like principles expressed and followed by the "barbarians."

One feature more of the old village communities deserves a special mention. It is the gradual extension of the circle of men embraced by the feelings of solidarity. Not only the tribes federated into stems, but the stems as well, even though of different origin, joined together in confederations. Some unions were so close that, for instance, the Vandals, after part of their confederation had left for the Rhine, and thence went over to Spain and Africa, respected for forty consecutive years the landmarks and the abandoned villages of their confederates, and did not take possession of them until they had ascertained through envoys that their confederates did not intend to return. With other barbarians, the soil was cultivated by one part of the stem, while the other part fought on or beyond the frontiers of the common territory. As to the leagues between several stems, they were quite habitual. The Sicambers united with the Cherusques and the Sueves, the Quades with the Sarmates; the Sarmates with the Alans, the Carpes, and the Huns. Later on, we also see the conception of nations gradually developing in Europe, long before anything like a State had grown in any part of the continent occupied by the barbarians. These nations—for it is impossible

---

[1] Maynoff, "Sketches of the Judicial Practices of the Mordovians," in the ethnographical *Zapiski* of the Russian Geographical Society, 1885, pp. 236, 257.

to refuse the name of a nation to the Merovingian France, or to the Russia of the eleventh and twelfth century—were nevertheless kept together by nothing else but a community of language, and a tacit agreement of the small republics to take their dukes from none but one special family.

Wars were certainly unavoidable ; migration means war ; but Sir Henry Maine has already fully proved in his remarkable study of the tribal origin of International Law, that " Man has never been so ferocious or so stupid as to submit to such an evil as war without some kind of effort to prevent it," and he has shown how exceedingly great is "the number of ancient institutions which bear the marks of a design to stand in the way of war, or to provide an alternative to it." [1] In reality, man is so far from the warlike being he is supposed to be, that when the barbarians had once settled they so rapidly lost the very habits of warfare that very soon they were compelled to keep special dukes followed by special *scholæ* or bands of warriors, in order to protect them from possible intruders. They preferred peaceful toil to war, the very peacefulness of man being the cause of the specialization of the warrior's trade, which specialization resulted later on in serfdom and in all the wars of the " States period" of human history.

History finds great difficulties in restoring to life the institutions of the barbarians. At every step the historian meets with some faint indication which he is unable to explain with the aid of his own documents only. But a broad light is thrown on the past as soon

---

[1] Henry Maine, *International Law*, London, 1888, pp. 11–13. E. Nys, *Les origines du droit international*, Bruxelles, 1894.

as we refer to the institutions of the very numerous
tribes which are still living under a social organization
almost identical with that of our barbarian ancestors.
Here we simply have the difficulty of choice, because the
islands of the Pacific, the steppes of Asia, and the
tablelands of Africa are real historical museums containing specimens of all possible intermediate stages
which mankind has lived through, when passing from
the savage gentes up to the States' organization.    Let
us, then, examine a few of those specimens.

If    we    take    the    village    communities    of    the
Mongol Buryates, especially those of the Kudinsk
Steppe on the upper Lena which have better escaped
Russian influence, we have fair representatives of
barbarians in a transitional state, between cattle-breed-
ing and agriculture.[1]    These Buryates are still living
in "joint families;" that is, although each son, when he
is married, goes to live in a separate hut, the huts of
at least three generations remain within the same enclosure, and the joint family work in common in their
fields, and own in common their joint households and
their cattle, as well as their "calves' grounds" (small
fenced patches of soil kept under soft grass for the
rearing of calves).    As a rule, the meals are taken
separately in each hut; but when meat is roasted, all
the twenty to sixty members of the joint household
feast together.    Several joint households which live in
a cluster, as well as several smaller families settled in
the same village—mostly *débris* of joint households
accidentally broken up—make the *oulous*, or the village

[1] A Russian historian, the Kazan Professor Schapoff, who was
exiled in 1862 to Siberia, has given a good description of their
institutions in the *Izvestia* of the East-Siberian Geographical Society,
vol. v. 1874.

community ; several *oulouses* make a tribe ; and the forty-six tribes, or clans, of the Kudinsk Steppe are united into one confederation. Smaller and closer confederations are entered into, as necessity arises for special wants, by several tribes. They know no private property in land—the land being held in common by the *oulous*, or rather by the confederation, and if it becomes necessary, the territory is re-allotted between the different *oulouses* at a folkmote of the tribe, and between the forty-six tribes at a folkmote of the confederation. It is worthy of note that the same organization prevails among all the 250,000 Buryates of East Siberia, although they have been for three centuries under Russian rule, and are well acquainted with Russian institutions.

With all that, inequalities of fortune rapidly develop among the Buryates, especially since the Russian Government is giving an exaggerated importance to their elected *taishas* (princes), whom it considers as responsible tax-collectors and representatives of the confederations in their administrative and even commercial relations with the Russians. The channels for the enrichment of the few are thus many, while the impoverishment of the great number goes hand in hand, through the appropriation of the Buryate lands by the Russians. But it is a habit with the Buryates, especially those of Kudinsk—and habit is more than law—that if a family has lost its cattle, the richer families give it some cows and horses that it may recover. As to the destitute man who has no family, he takes his meals in the huts of his congeners ; he enters a hut, takes—by right, not for charity—his seat by the fire, and shares the meal which always is scrupulously divided into equal parts ; he sleeps where he has taken his evening meal. Altogether, the Russian conquerors of Siberia were so much struck by the communistic practices of the Buryates, that they gave them the name of *Bratskiye*—"the Brotherly Ones"—and reported to Moscow: "With them everything is in common; whatever they have is shared in common." Even now, when the Lena Buryates sell their wheat, or send some of their cattle to be sold to a Russian butcher, the families of the *oulous*, or the tribe, put their wheat and cattle together, and sell it as a whole. Each *oulous* has, moreover, its grain store for loans in case of need, its communal baking oven (the *four banal* of the old French communities), and its blacksmith, who, like the blacksmith of the Indian communities,[1] being a member of the community, is never paid for his work within the community. He must make it for nothing, and if he utilizes his spare time for fabricating the small plates of chiselled and silvered iron which are used in Buryate land for the decoration of dress, he may occasionally sell them to a woman from another clan, but to the women of his own clan the attire is presented as a gift. Selling and buying cannot take place within the community, and the rule is so severe that when a richer family hires a labourer the labourer must be taken from another clan or from among the Russians. This habit is evidently not specific to the Buryates; it is so widely spread among the modern barbarians, Aryan and Ural-Altayan, that it must have been universal among our ancestors.

The feeling of union within the confederation is kept alive by the common interests of the tribes, their folk-motes, and the festivities which are usually kept in

[1] Sir Henry Maine's *Village Communities*, New York, 1876, pp. 193–196.

connection with the folkmotes. The same feeling is, however, maintained by another institution, the *aba*, or common hunt, which is a reminiscence of a very remote past. Every autumn, the forty-six clans of Kudinsk come together for such a hunt, the produce of which is divided among all the families. Moreover, national *abas*, to assert the unity of the whole Buryate nation, are convoked from time to time. In such cases, all Buryate clans which are scattered for hundreds of miles west and east of Lake Baikal, are bound to send their delegate hunters. Thousands of men come together, each one bringing provisions for a whole month. Every one's share must be equal to all the others, and therefore, before being put together, they are weighed by an elected elder (always "with the hand": scales would be a profanation of the old custom). After that the hunters divide into bands of twenty, and the parties go hunting according to a well-settled plan. In such *abas* the entire Buryate nation revives its epic traditions of a time when it was united in a powerful league. Let me add that such communal hunts are quite usual with the Red Indians and the Chinese on the banks of the Usuri (the *kada*).[1]

With the Kabyles, whose manners of life have been so well described by two French explorers,[2] we have barbarians still more advanced in agriculture. Their fields, irrigated and manured, are well attended to, and in the hilly tracts every available plot of land is cultivated by the spade. The Kabyles have known many vicissitudes in their history; they have followed for

---

[1] Nazaroff, *The North Usuri Territory* (Russian), St. Petersburg, 1887, p. 65.
[2] Hanoteau et Letourneux, *La Kabylie*, 3 vols. Paris, 1883.

some time the Mussulman law of inheritance, but,
being adverse to it, they have returned, 150 years ago,
to the tribal customary law of old. Accordingly, their
land-tenure is of a mixed character, and private property
in land exists side by side with communal possession. Still, the basis of their present organization is
the village community, the *thaddart*, which usually
consists of several joint families (*kharoubas*), claiming
a community of origin, as well as of smaller families
of strangers. Several villages are grouped into clans
or tribes (*ârch*); several tribes make the confederation
(*thak'ebilt*); and several confederations may occasionally enter into a league, chiefly for purposes of armed
defence.

The Kabyles know no authority whatever besides
that of the *djemmâa*, or folkmote of the village
community. All men of age take part in it, in the
open air, or in a special building provided with stone
seats, and the decisions of the *djemmâa* are evidently
taken at unanimity: that is, the discussions continue
until all present agree to accept, or to submit to, some
decision. There being no authority in a village community to impose a decision, this system has been
practised by mankind wherever there have been village
communities, and it is practised still wherever they
continue to exist, *i. e.* by several hundred million men
all over the world. The *djemmâa* nominates its
executive—the elder, the scribe, and the treasurer;
it assesses its own taxes; and it manages the re-
partition of the common lands, as well as all kinds
of works of public utility. A great deal of work is
done in common: the roads, the mosques, the fountains,
the irrigation canals, the towers erected for protection
from robbers, the fences, and so on, are built by the

village community; while the high-roads, the larger
mosques, and the great market-places are the work of
the tribe. Many traces of common culture continue
to exist, and the houses continue to be built by, or with
the aid of, all men and women of the village. Altogether, the "aids" are of daily occurrence, and are
continually called in for the cultivation of the fields,
for harvesting, and so on. As to the skilled work,
each community has its blacksmith, who enjoys his
part of the communal land, and works for the community; when the tilling season approaches he visits
every house, and repairs the tools and the ploughs,
without expecting any pay, while the making of new
ploughs is considered as a pious work which can by no
means be recompensed in money, or by any other form
of salary.

As the Kabyles already have private property, they
evidently have both rich and poor among them. But
like all people who closely live together, and know
how poverty begins, they consider it as an accident
which may visit every one. " Don't say that you will
never wear the beggar's bag, nor go to prison," is a
proverb of the Russian peasants; the Kabyles practise
it, and no difference can be detected in the external
behaviour between rich and poor; when the poor
convokes an "aid," the rich man works in his field,
just as the poor man does it reciprocally in his turn.[1]
Moreover, the *djemmâas* set aside certain gardens and
fields, sometimes cultivated in common, for the use of

[1] To convoke an "aid, or "bee," some kind of meal must be
offered to the community. I am told by a Caucasian friend that in
Georgia, when the poor man wants an "aid," he borrows from the
rich man a sheep or two to prepare the meal, and the community
bring, in addition to their work, so many provisions that he may
repay the debt. A similar habit exists with the Mordovians.

the poorest members. Many like customs continue to
exist. As the poorer families would not be able to
buy meat, meat is regularly bought with the money of
the fines, or the gifts to the *djemmâa*, or the payments
for the use of the communal olive-oil basins, and it is
distributed in equal parts among those who cannot
afford buying meat themselves. And when a sheep
or a bullock is killed by a family for its own use on a
day which is not a market day, the fact is announced
in the streets by the village crier, in order that sick
people and pregnant women may take of it what they
want. Mutual support permeates the life of the
Kabyles, and if one of them, during a journey abroad,
meets with another Kabyle in need, he is bound to
come to his aid, even at the risk of his own fortune
and life; if this has not been done, the *djemmâa* of
the man who has suffered from such neglect may lodge
a complaint, and the *djemmâa* of the selfish man will
at once make good the loss. We thus come across a
custom which is familiar to the students of the mediæval merchant guilds. Every stranger who enters a
Kabyle village has right to housing in the winter, and
his horses can always graze on the communal lands for
twenty-four hours. But in case of need he can
reckon upon an almost unlimited support. Thus,
during the famine of 1867-68, the Kabyles received
and fed every one who sought refuge in their villages,
without distinction of origin. In the district of Dellys,
no less than 12,000 people who came from all parts of
Algeria, and even from Morocco, were fed in this
way. While people died from starvation all over
Algeria, there was not one single case of death due to
this cause on Kabylian soil. The *djemmâas*, depriving
themselves of necessaries, organized relief, without

ever asking any aid from the Government, or uttering the slightest complaint; they considered it as a natural duty. And while among the European settlers all kind of police measures were taken to prevent thefts and disorder resulting from such an influx of strangers, nothing of the kind was required on the Kabyles' territory: the *djemmâas* needed neither aid nor protection from without.[1]

I can only cursorily mention two other most interesting features of Kabyle life; namely, the *anaya*, or protection granted to wells, canals, mosques, marketplaces, some roads, and so on, in case of war, and the *çofs*. In the *anaya* we have a series of institutions both for diminishing the evils of war and for preventing conflicts. Thus the market-place is *anaya*, especially if it stands on a frontier and brings Kabyles and strangers together; no one dares disturb peace in the market, and if a disturbance arises, it is quelled at once by the strangers who have gathered in the market town. The road upon which the women go from the village to the fountain also is *anaya* in case of war; and so on. As to the *çof*, it is a widely-spread form of association, having some characters of the mediæval *Bürgschaften* or *Gegilden*, as well as of societies both for mutual protection and for various purposes—intellectual, political, and emotional—which cannot be satisfied by the territorial organization of the village, the clan, and the confederation. The *çof* knows no territorial limits; it recruits its members in various villages, even among strangers; and it pro-

---

[1] Hanoteau et Letourneux, *La Kabylie*, ii. 58. The same respect to strangers is the rule with the Mongols. The Mongol who has refused his roof to a stranger pays the full blood-compensation if the stranger has suffered therefrom (Bastian, *Der Mensch in der Geschichte*, iii. 231).

tects them in all possible eventualities of life. Altogether, it is an attempt at supplementing the territorial grouping by an extra-territorial grouping intended to give an expression to mutual affinities of all kinds across the frontiers. The free international association of individual tastes and ideas, which we consider as one of the best features of our own life, has thus its origin in barbarian antiquity.

The mountaineers of Caucasia offer another extremely instructive field for illustrations of the same kind. In studying the present customs of the Ossetes—their joint families and communes and their judiciary conceptions—Professor Kovalevsky, in a remarkable work on *Modern Custom and Ancient Law* was enabled step by step to trace the similar dispositions of the old barbarian codes and even to study the origins of feudalism. With other Caucasian stems we occasionally catch a glimpse into the origin of the village community in those cases where it was not tribal but originated from a voluntary union between families of distinct origin. Such was recently the case with some Khevsoure villages, the inhabitants of which took the oath of "community and fraternity."[1] In another part of Caucasus, Daghestan, we see the growth of feudal relations between two tribes, both maintaining at the same time their village communities (and even traces of the gentile "classes"), and thus giving a living illustration of the forms taken by the conquest of Italy and Gaul by the barbarians. The victorious race, the Lezghines, who have conquered several

---

[1] N. Khoudadoff, "Notes on the Khevsoures," in *Zapiski* of the Caucasian Geogr. Society, xiv. 1, Tiflis, 1890, p. 68. They also took the oath of not marrying girls from their own union, thus displaying a remarkable return to the old gentile rules.

Georgian and Tartar villages in the Zakataly district, did not bring them under the dominion of separate families ; they constituted a feudal clan which now includes 12,000 households in three villages, and owns in common no less than twenty Georgian and Tartar villages. The conquerors divided their own land among their clans, and the clans divided it in equal parts among the families ; but they did not interfere with the *djemmâas* of their tributaries which still practise the habit mentioned by Julius Cæsar ; namely, the *djemmáa* decides each year which part of the communal territory must be cultivated, and this land is divided into as many parts as there are families, and the parts are distributed by lot. It is worthy of note that although proletarians are of common occurrence among the Lezghines (who live under a system of private property in land, and common ownership of serfs [1]) they are rare among their Georgian serfs, who continue to hold their land in common. As to the customary law of the Caucasian mountaineers, it is much the same as that of the Longobards or Salic Franks, and several of its dispositions explain a good deal the judicial procedure of the barbarians of old. Being of a very impressionable character, they do their best to prevent quarrels from taking a fatal issue ; so, with the Khevsoures, the swords are very soon drawn when a quarrel breaks out ; but if a woman rushes out and throws among them the piece of linen which she wears on her head, the swords are at once returned to their sheaths, and the quarrel is appeased. The head-dress

---

[1] Dm. Bakradze, "Notes on the Zakataly District," in same *Zapiski*, xiv. 1, p. 264. The "joint team" is as common among the Lezghines as it is among the Ossetes.

of the women is *anaya*.   If a quarrel has not been
stopped in time and has ended in murder, the compensation money is so considerable that the aggressor is
entirely ruined for his life, unless he is adopted by the
wronged family ; and if he has resorted to his sword in a
trifling quarrel and has inflicted wounds, he loses for
ever the consideration of his kin.   In all disputes,
mediators take the matter in hand ; they select from
among the members of the clan the judges—six in
smaller affairs, and from ten to fifteen in more serious
matters—and Russian observers testify to the absolute
incorruptibility of the judges.   An oath has such a
significance that men enjoying general esteem are
dispensed from taking it : a simple affirmation is quite
sufficient, the more so as in grave affairs the Khevsoure never hesitates to recognize his guilt (I mean, of
course, the Khevsoure untouched yet by civilization).
The oath is chiefly reserved for such cases, like
disputes about property, which require some sort of
appreciation in addition to a simple statement of
facts ; and in such cases the men whose affirmation
will decide in the dispute, act with the greatest
circumspection.   Altogether it is certainly not a want
of honesty or of respect to the rights of the congeners
which characterizes the barbarian societies of Caucasus.

The stems of Africa offer such an immense variety
of extremely interesting societies standing at all intermediate stages from the early village community to
the despotic barbarian monarchies that I must abandon
the idea of giving here even the chief results of a
comparative study of their institutions.[1]   Suffice it to

[1] See Post, *Afrikanische Jurisprudenz*, Oldenburg, 1887 ; Mün-
zinger, *Ueber das Recht und Sitten der Bogos*, Winterthur, 1859 ;
Casalis, *Les Bassoutos*, Paris, 1859 ; Maclean, *Kafir Laws and
Customs*, Mount Coke, 1858, etc.

say, that, even under the most horrid despotism of kings, the folkmotes of the village communities and their customary law remain sovereign in a wide circle of affairs. The law of the State allows the king to take any one's life for a simple caprice, or even for simply satisfying his gluttony; but the customary law of the people continues to maintain the same network of institutions for mutual support which exist among other barbarians or have existed among our ancestors. And with some better-favoured stems (in Bornu, Uganda, Abyssinia), and especially the Bogos, some of the dispositions of the customary law are inspired with really graceful and delicate feelings.

The village communities of the natives of both Americas have the same character. The Tupi of Brazil were found living in "long houses" occupied by whole clans which used to cultivate their corn and manioc fields in common. The Arani, much more advanced in civilization, used to cultivate their fields in common; so also the Oucagas, who had learned under their system of primitive communism and "long houses" to build good roads and to carry on a variety of domestic industries,[1] not inferior to those of the early mediæval times in Europe. All of them were also living under the same customary law of which we have given specimens on the preceding pages. At another extremity of the world we find the Malayan feudalism, but this feudalism has been powerless to unroot the *negaria*, or village community, with its common ownership of at least part of the land, and the redistribution of land among the several *negarias* of the tribe.[2] With the Alfurus of Minahasa we find the

---

[1] Waitz, iii. 423 *seq.*

[2] Post's *Studien zur Entwicklungsgeschichte des Familien-Rechts.* Oldenburg, 1889, pp. 270 *seq.*

communal rotation of the crops; with the Indian
stem of the Wyandots we have the periodical redistribution of land within the tribe, and the clan-culture
of the soil; and in all those parts of Sumatra where
Moslem institutions have not yet totally destroyed
the old organization we find the joint family (*suka*)
and the village community (*kota*) which maintains its
right upon the land, even if part of it has been cleared
without its authorization.[1]    But to say this, is to say
that all customs for mutual protection and prevention
of feuds and wars, which have been briefly indicated
in the preceding pages as characteristic of the village
community, exist as well.    More than that : the more
fully the communal possession of land has been maintained, the better and the gentler are the habits.    De
Stuers positively affirms that wherever the institution
of the village community has been less encroached
upon by the conquerors, the inequalities of fortunes
are smaller, and the very prescriptions of the *lex
talionis* are less cruel ; while, on the contrary, wherever the village community has been totally broken
up, "the inhabitants suffer the most unbearable
oppression from their despotic rulers."[2]    This is
quite natural.    And when Waitz made the remark
that those stems which have maintained their ·tribal
confederations stand on a higher level of development
and have a richer literature than those stems which
have forfeited the old bonds of union, he only pointed
out what might have been foretold in advance.

More illustrations would simply involve me in tedious
repetitions—so strikingly similar are the barbarian

---

[1] Powell, *Annual Report of the Bureau of Ethnography*, Washington, 1881, quoted in Post's *Studien*, p. 290 ; Bastian's *Inselgruppen
in Oceanien*, 1883, p. 88.
[2] De Stuers, quoted by Waitz, v. 141.

societies under all climates and amidst all races. The
same process of evolution has been going on in mankind with a wonderful similarity. When the clan
organization, assailed as it was from within by the
separate family, and from without by the dismember-
ment of the migrating clans and the necessity of
taking in strangers of different descent—the village
community, based upon a territorial conception, came
into existence. This new institution, which had
naturally grown out of the preceding one—the clan—
permitted the barbarians to pass through a most
disturbed period of history without being broken into
isolated families which would have succumbed in the
struggle for life. New forms of culture developed
under the new organization; agriculture attained the
stage which it hardly has surpassed until now with the
great number; the domestic industries reached a high
degree of perfection. The wilderness was conquered,
it was intersected by roads, dotted with swarms thrown
off by the mother-communities. Markets and fortified
centres, as well as places of public worship, were
erected. The conceptions of a wider union, extended
to whole stems and to several stems of various origin,
were slowly elaborated. The old conceptions of justice
which were conceptions of mere revenge, slowly underwent a deep modification—the idea of amends for the
wrong done taking the place of revenge. The customary law which still makes the law of the daily life
for two-thirds or more of mankind, was elaborated
under that organization, as well as a system of habits
intended to prevent the oppression of the masses by
the minorities whose powers grew in proportion to the
growing facilities for private accumulation of wealth.
This was the new form taken by the tendencies of

the masses for mutual support. And the progress—economical, intellectual, and moral—which mankind accomplished under this new popular form of organization, was so great that the States, when they were called later on into existence, simply took possession, in the interest of the minorities, of all the judicial, economical, and administrative functions which the village community already had exercised in the interest of all.

# CHAPTER V

## MUTUAL AID IN THE MEDIÆVAL CITY

Growth of authority in Barbarian Society.—Serfdom in the villages.—Revolt of fortified towns: their liberation; their charts.—The guild.—Double origin of the free mediæval city.—Self-jurisdiction, self-administration.—Honourable position of labour.—Trade by the guild and by the city.

SOCIABILITY and need of mutual aid and support are such inherent parts of human nature that at no time of history can we discover men living in small isolated families, fighting each other for the means of subsistence. On the contrary, modern research, as we saw it in the two preceding chapters, proves that since the very beginning of their prehistoric life men used to agglomerate into *gentes*, clans, or tribes, maintained by an idea of common descent and by worship of common ancestors. For thousands and thousands of years this organization has kept men together, even though there was no authority whatever to impose it. It has deeply impressed all subsequent development of mankind; and when the bonds of common descent had been loosened by migrations on a grand scale, while the development of the separated family within the clan itself had destroyed the old unity of the clan, a new form of union, territorial in its principle—the village community—was called into existence by the social genius of man. This institution, again, kept men together for a number of centuries, permitting them to further develop their social institutions and to pass through some of the darkest periods of history, without being dissolved into loose aggregations of families and individuals, to make a further step in their evolution, and to work out a number of secondary social institutions, several of which have survived down to the present time. We have now to follow the further developments of the same ever-living tendency for mutual aid. Taking the village communities of the so-called barbarians at a time when they were making a new start of civilization after the fall of the Roman Empire, we have to study the new aspects taken by the sociable wants of the masses in the middle ages, and especially in the mediæval guilds and the mediæval city.

Far from being the fighting animals they have often been compared to, the barbarians of the first centuries of our era (like so many Mongolians, Africans, Arabs, and so on, who still continue in the same barbarian stage) invariably preferred peace to war. With the exception of a few tribes which had been driven during the great migrations into unproductive deserts or highlands, and were thus compelled period- ically to prey upon their better-favoured neighbours— apart from these, the great bulk of the Teutons, the Saxons, the Celts, the Slavonians, and so on, very soon after they had settled in their newly-conquered abodes, reverted to the spade or to their herds. The earliest barbarian codes already represent to us societies composed of peaceful agricultural communi- ties, not hordes of men at war with each other. These barbarians covered the country with villages and farmhouses;[1] they cleared the forests, bridged the torrents, and colonized the formerly quite un-inhabited wilderness; and they left the uncertain warlike pursuits to brotherhoods, *scholæ*, or "trusts" of unruly men, gathered round temporary chieftains, who wandered about, offering their adventurous spirit, their arms, and their knowledge of warfare for the protection of populations, only too anxious to be left in peace. The warrior bands came and went, prosecuting their family feuds; but the great mass continued to till the soil, taking but little notice of their would-be rulers, so long as they did not interfere with the independence of their village communities.[2] The new occupiers of Europe evolved the systems of land tenure and soil culture which are still in force with hundreds of millions of men; they worked out their systems of compensation for wrongs, instead of the old tribal blood-revenge; they learned the first rudiments of industry; and while they fortified their villages with palisaded walls, or erected towers and earthen forts whereto to repair in case of a new invasion, they soon abandoned the task of defending these towers and forts to those who made of war a speciality.

The very peacefulness of the barbarians, certainly not their supposed warlike instincts, thus became the source of their subsequent subjection to the military chieftains. It is evident that the very mode of life of

---

[1] W. Arnold, in his *Wanderungen und Ansiedelungen der deutschen Stämme*, p. 431, even maintains that one-half of the now arable area in middle Germany must have been reclaimed from the sixth to the ninth century. Nitzsch (*Geschichte des deutschen Volkes*, Leipzig, 1883, vol. i.) shares the same opinion.

[2] Leo and Botta, *Histoire d'Italie*, French edition, 1844, t. i., p. 37.

the armed brotherhoods offered them more facilities for enrichment than the tillers of the soil could find in their agricultural communities. Even now we see that armed men occasionally come together to shoot down Matabeles and to rob them of their droves of cattle, though the Matabeles only want peace and are ready to buy it at a high price. The *scholæ* of old certainly were not more scrupulous than the *scholæ* of our own time. Droves of cattle, iron (which was extremely costly at that time [1]), and slaves were appropriated in this way; and although most acquisitions were wasted on the spot in those glorious feasts of which epic poetry has so much to say—still some part of the robbed riches was used for further enrichment. There was plenty of waste land, and no lack of men ready to till it, if only they could obtain the necessary cattle and implements. Whole villages, ruined by murrains, pests, fires, or raids of new immigrants, were often abandoned by their inhabitants, who went anywhere in search of new abodes. They still do so in Russia in similar circumstances. And if one of the *hirdmen* of the armed brotherhoods offered the peasants some cattle for a fresh start, some iron to make a plough, if not the plough itself, his protection from further raids, and a number of years free from all obligations, before they should begin to repay the contracted debt, they settled upon the land. And when, after a hard fight with bad crops, inundations

---

[1] The composition for the stealing of a simple knife was 15 *solidi*, and of the iron parts of a mill, 45 *solidi*. (See on this subject Lamprecht's *Wirthschaft und Recht der Franken* in Raumer's *Historisches Taschenbuch*, 1883, p. 52.) According to the Riparian law, the sword, the spear, and the iron armour of a warrior attained the value of at least twenty-five cows, or two years of a freeman's labour. A cuirass alone was valued in the Salic law (Desmichels, quoted by Michelet) at as much as thirty-six bushels of wheat.

and pestilences, those pioneers began to repay their debts, they fell into servile obligations towards the protector of the territory. Wealth undoubtedly did accumulate in this way, and power always follows wealth.[1] And yet, the more we penetrate into the life of those times, the sixth and seventh centuries of our era, the more we see that another element, besides wealth and military force, was required to constitute the authority of the few. It was an element of law and right, a desire of the masses to maintain peace, and to establish what they considered to be justice, which gave to the chieftains of the *scholæ*— kings, dukes, *knyazes*, and the like—the force they acquired two or three hundred years later. That same idea of justice, conceived as an adequate revenge for the wrong done, which had grown in the tribal stage, now passed as a red thread through the history of subsequent institutions, and, much more even than military or economic causes, it became the basis upon which the authority of the kings and the feudal lords was founded.

In fact, one of the chief preoccupations of the barbarian village community always was, as it still is with our barbarian contemporaries, to put a speedy end to the feuds which arose from the then current conception of justice. When a quarrel took place, the community at once interfered, and after the folkmote

---

[1] The chief wealth of the chieftains, for a long time, was in their personal domains peopled partly with prisoner slaves, but mostly in the above way. On the origin of property see Inama Sternegg's *Die Ausbildung der grossen Grundherrschaften in Deutschland*, in Schmoller's *Forschungen*, Bd. I., 1878 ; F. Dahn's *Urgeschichte der germanischen und romanischen Völker*, Berlin, 1881 ; Maurer's *Dorfverfassung;* Guizot's *Essais sur l'histoire de France;* Maine's *Village Community;* Botta's *Histoire d'Italie;* F. Seebohm, Vinogradov, J. R. Green, etc.

had heard the case, it settled the amount of composition (*wergeld*) to be paid to the wronged person, or to his family, as well as the *fred*, or fine for breach of peace, which had to be paid to the community. Interior quarrels were easily appeased in this way. But when feuds broke out between two different tribes, or two confederations of tribes, notwithstanding all measures taken to prevent them,[1] the difficulty was to find an arbiter or sentence-finder whose decision should be accepted by both parties alike, both for his impartiality and for his knowledge of the oldest law. The difficulty was the greater as the customary laws of different tribes and confederations were at variance as to the compensation due in different cases. It therefore became habitual to take the sentence-finder from among such families, or such tribes, as were reputed for keeping the law of old in its purity; of being versed in the songs, triads, sagas, etc., by means of which law was perpetuated in memory; and to retain law in this way became a sort of art, a "mystery," carefully transmitted in certain families from generation to generation. Thus in Iceland, and in other Scandinavian lands, at every *Allthing*, or national folkmote, a *lövsögmathr* used to recite the whole law from memory for the enlightening of the assembly; and in Ireland there was, as is known, a special class of men reputed for the knowledge of the old traditions, and therefore enjoying a great authority as judges.[2] Again, when we are told by the Russian annals that some stems of North-West Russia, moved by the

---

[1] See Sir Henry Maine's *International Law*, London, 1888.

[2] *Ancient Laws of Ireland*, Introduction ; E. Nys, *Études de droit international*, t. i., 1896, pp. 86 *seq.*    Among the Ossetes the arbiters from three *oldest* villages enjoy a special reputation (M. Kovalevsky's *Modern Custom and Old Law*, Moscow, 1886, ii. 217, Russian).

growing disorder which resulted from "clans rising against clans," appealed to Norman *varingiar* to be their judges and commanders of warrior *scholæ ;* and when we see the *knyazes*, or dukes, elected for the next two hundred years always from the same Norman family, we cannot but recognize that the Slavonians trusted to the Normans for a better knowledge of the law which would be equally recognized as good by different Slavonian kins. In this case the possession of runes, used for the transmission of old customs, was a decided advantage in favour of the Normans ; but in other cases there are faint indications that the "eldest" branch of the stem, the supposed mother-branch, was appealed to to supply the judges, and its decisions were relied upon as just ;[1] while at a later epoch we see a distinct tendency towards taking the sentence-finders from the Christian clergy, which, at that time, kept still to the fundamental, now forgotten, principle of Christianity, that retaliation is no act of justice. At that time the Christian clergy opened the churches as places of asylum for those who fled from blood revenge, and they willingly acted as arbiters in criminal cases, always opposing the old tribal principle of life for life and wound for wound. In short, the deeper we penetrate into the history of early institutions, the less we find grounds for the military theory of origin of authority. Even that power which later on became such a source of oppression seems, on the contrary, to have found its origin in the peaceful inclinations of the masses.

In all these cases the *fred*, which often amounted

---

[1] It is permissible to think that this conception (related to the conception of tanistry) played an important part in the life of the period ; but research has not yet been directed that way.

to half the compensation, went to the folkmote, and
from times immemorial it used to be applied to works
of common utility and defence. It has still the same
destination (the erection of towers) among the Kabyles
and certain Mongolian stems; and we have direct
evidence that even several centuries later the judicial
fines, in Pskov and several French and German cities,
continued to be used for the repair of the city walls.[1]
It was thus quite natural that the fines should be
handed over to the sentence-finder, who was bound, in
return, both to maintain the *schola* of armed men to
whom the defence of the territory was trusted, and to
execute the sentences. This became a universal
custom in the eighth and ninth centuries, even when
the sentence-finder was an elected bishop. The germ
of a combination of what we should now call the
judicial power and the executive thus made its appearance. But to these two functions the attributions of
the duke or king were strictly limited. He was no
ruler of the people—the supreme power still belonging
to the folkmote—not even a commander of the popular
militia; when the folk took to arms, it marched under
a separate, also elected, commander, who was not a
subordinate, but an equal to the king.[2] The king
was a lord on his personal domain only. In fact, in
barbarian language, the word *konung*, *koning*, or
*cyning*, synonymous with the Latin *rex*, had no other
meaning than that of a temporary leader or chieftain

---

[1] It was distinctly stated in the charter of St. Quentin of the year
1002 that the ransom for houses which had to be demolished for
crimes went for the city walls. The same destination was given to
the *Ungeld* in German cities. At Pskov the cathedral was the bank
for the fines, and from this fund money was taken for the walls.
[2] Sohm, *Fränkische Rechts- und Gerichtsverfassung*, p. 23; also
Nitzsch, *Geschichte des deutschen Volkes*, i. 78.

of a band of men. The commander of a flotilla of boats, or even of a single pirate boat, was also a *konung*, and till the present day the commander of fishing in Norway is named *Not-kong*—"the king of the nets."[1] The veneration attached later on to the personality of a king did not yet exist, and while treason to the kin was punished by death, the slaying of a king could be recouped by the payment of compensation: a king simply was valued so much more than a freeman.[2] And when King Knu (or Canute) had killed one man of his own *schola*, the saga represents him convoking his comrades to a *thing* where he knelt down imploring pardon. He was pardoned, but not till he had agreed to pay nine times the regular composition, of which one-third went to himself for the loss of one of his men, one-third to the relatives of the slain man, and one-third (the *fred*) to the *schola*.[3] In reality, a complete change had to be accomplished in the current conceptions, under the double influence of the Church and the students of Roman law, before an idea of sanctity began to be attached to the personality of the king.

However, it lies beyond the scope of these essays

[1] See the excellent remarks on this subject in Augustin Thierry's *Lettres sur l'histoire de France*, 7th letter. The barbarian translations of parts of the Bible are extremely instructive on this point.

[2] Thirty-six times more than a noble, according to the Anglo-Saxon law. In the code of Rothari the slaying of a king is, however, punished by death; but (apart from Roman influence) this new disposition was introduced (in 646) in the Lombardian law—as remarked by Leo and Botta—to cover the king from blood revenge. The king being at that time the executioner of his own sentences (as the tribe formerly was of its own sentences), he had to be protected by a special disposition, the more so as several Lombardian kings before Rothari had been slain in succession (Leo and Botta, *l. c.*, i. 66–90).

[3] Kaufmann, *Deutsche Geschichte*, Bd. I. "Die Germanen der Urzeit," p. 133.

M

to follow the gradual development of authority out of
the elements just indicated. Historians, such as Mr.
and Mrs. Green for this country, Augustin Thierry,
Michelet, and Luchaire for France, Kaufmann, Janssen,
W. Arnold, and even Nitzsch, for Germany, Leo and
Botta for Italy, Byelaeff, Kostomaroff, and their
followers for Russia, and many others, have fully told
that tale. They have shown how populations, once
free, and simply agreeing "to feed" a certain portion
of their military defenders, gradually became the serfs
of these protectors; how "commendation" to the
Church, or to a lord, became a hard necessity for the
freeman; how each lord's and bishop's castle became
a robber's nest—how feudalism was imposed, in a
word—and how the crusades, by freeing the serfs who
wore the cross, gave the first impulse to popular
emancipation. All this need not be retold in this
place, our chief aim being to follow the *constructive*
genius of the masses in their mutual-aid institutions.

At a time when the last vestiges of barbarian
freedom seemed to disappear, and Europe, fallen
under the dominion of thousands of petty rulers, was
marching towards the constitution of such theocracies
and despotic States as had followed the barbarian
stage during the previous starts of civilization, or of
barbarian monarchies, such as we see now in Africa,
life in Europe took another direction. It went on on
lines similar to those it had once taken in the cities
of antique Greece. With a unanimity which seems
almost incomprehensible, and for a long time was not
understood by historians, the urban agglomerations,
down to the smallest burgs, began to shake off the
yoke of their worldly and clerical lords. The fortified

village rose against the lord's castle, defied it first, attacked it next, and finally destroyed it. The movement spread from spot to spot, involving every town on the surface of Europe, and in less than a hundred years free cities had been called into existence on the coasts of the Mediterranean, the North Sea, the Baltic, the Atlantic Ocean, down to the fjords of Scandinavia; at the feet of the Apennines, the Alps, the Black Forest, the Grampians, and the Carpathians; in the plains of Russia, Hungary, France and Spain. Everywhere the same revolt took place, with the same features, passing through the same phases, leading to the same results. Wherever men had found, or expected to find, some protection behind their town walls, they instituted their "co-jurations," their "fraternities," their "friendships," united in one common idea, and boldly marching towards a new life of mutual support and liberty. And they succeeded so well that in three or four hundred years they had changed the very face of Europe. They had covered the country with beautiful sumptuous buildings, expressing the genius of free unions of free men, unrivalled since for their beauty and expressiveness; and they bequeathed to the following generations all the arts, all the industries, of which our present civilization, with all its achievements and promises for the future, is only a further development. And when we now look to the forces which have produced these grand results, we find them—not in the genius of individual heroes, not in the mighty organization of huge States or the political capacities of their rulers, but in the very same current of mutual aid and support which we saw at work in the village community, and which was vivified and reinforced in the Middle Ages by a new form of unions, inspired by the very same spirit but shaped on a new model—the guilds.

It is well known by this time that feudalism did not imply a dissolution of the village community. Although the lord had succeeded in imposing servile labour upon the peasants, and had appropriated for himself such rights as were formerly vested in the village community alone (taxes, mortmain, duties on inheritances and marriages), the peasants had, nevertheless, maintained the two fundamental rights of their communities : the common possession of the land, and self-jurisdiction.   In olden times, when a king sent his vogt to a village, the peasants received him with flowers in one hand and arms in the other, and asked him—which law he intended to apply: the one he found in the village, or the one he brought with him? And, in the first case, they handed him the flowers and accepted him ; while in the second case they fought him.[1]  Now, they accepted the king's or the lord's official whom they could not refuse ; but they maintained the folkmote's jurisdiction, and themselves nominated six, seven, or twelve judges, who acted with the lord's judge, in the presence of the folkmote, as arbiters and sentence-finders.  In most cases the official had nothing left to him but to confirm the sentence and to levy the customary *fred.* This precious right of self-jurisdiction, which, at that time, meant self-administration and self-legislation, had been maintained through all the struggles ; and even the lawyers by whom Karl the Great was surrounded could not abolish it ; they were bound to confirm it. At the same time, in all matters concerning the com-

[1] Dr. F. Dahn, *Urgeschichte der germanischen und romanischen Völker*, Berlin, 1881, Bd. I. 96.

# DO NOT BIND
SUBMIT TO QUALITY STATION FOR MEASUREMENT

# DO NOT BIND
SUBMIT TO QUALITY STATION FOR MEASUREMENT

# DO NOT BIND
SUBMIT TO QUALITY STATION FOR MEASUREMENT

# DO NOT BIND
SUBMIT TO QUALITY STATION FOR MEASUREMENT

# DO NOT BIND
SUBMIT TO QUALITY STATION FOR MEASUREMENT

# DO NOT BIND
SUBMIT TO QUALITY STATION FOR MEASUREMENT

# DO NOT BIND
SUBMIT TO QUALITY STATION FOR MEASUREMENT

1

**CHECK FOR BANDING OR STRIPES IN THE PRINT AREA**

Primary Side

**CHECK FOR BANDING OR STRIPES IN THE PRINT AREA**

**CHECK FOR BANDING OR STRIPES IN THE PRINT AREA**

Primary Side

**CHECK FOR BANDING OR STRIPES IN THE PRINT AREA**

Secondary Side

**CHECK FOR BANDING OR STRIPES IN THE PRINT AREA**

# The Culture of Incompetence

## The Mind-Set That Destroys Inner-City Schools

John Cartaina

iUniverse, Inc.
New York   Bloomington

# The Culture of Incompetence
## The Mind-Set That Destroys Inner-City Schools

iUniverse books may be ordered through booksellers or by contacting:

iUniverse
1663 Liberty Drive
Bloomington, IN 47403
www.iuniverse.com
1-800-Authors (1-800-288-4677)

ISBN: 978-1-4401-6413-2 (pbk)
ISBN: 978-1-4401-6414-9 (ebook)

Printed in the United States of America

iUniverse rev. date: 11/12/09

This book is dedicated to
my wife, Gloria, for all her love and support;
my wonderful children, Diane and John; my daughter-
in-law, Helene;
God's gift, our grandchildren, Zoe and Jack;
and the children of Paterson, New Jersey.

# CONTENTS

Preface................................................................ix

The Culture of Incompetence.........................1

The Teacher as a Person ...............................13

Teacher-Student Rapport: ...........................24

The Twinkle in the Eye................................33

The Living School .......................................43

Classroom Instruction.................................55

Messages to the Stakeholders......................69

Epilogue......................................................95

## PREFACE

The nun told my mother I could stay if I didn't cry. I was four years and four months old, and that is my lasting image of day one in kindergarten. Just how tall was that nun? She looked like a mountain hovering over me. She scared me to death.

No child should ever feel afraid, lonely, or threatened in school. Today's classroom should be an oasis of safety and security in our chaotic world. Many years later as a teacher myself, I was always drawn to the students who seemed left behind or left out. Maybe I saw that scared kindergarten child in them. I empathized with the shy eighth grade girl from India who read novels all day in class. She did her assignments and was well behaved. In a tough inner-city classroom, she could easily have flown below the radar. But she never looked comfortable to me, and of course, that just couldn't be. Why was she shy, and why did she not want to mingle? I had to get beyond the novel that she used as a protective barrier. I would deliberately call on her, and she would push down the novel with an expression that said, "Can't you see I'm busy?" We eventually became such good friends that her mom invited me to their home. I never liked to see sad or isolated children in my class.

Teaching is a noble profession. It rests on a pledge between an adult and a child that is as sacred as the oath taken by doctors and ministers. A bad doctor kills one person at a time, but a bad teacher can kill the spirits of twenty children at a time. Teachers must inspire. They must exude dignity and integrity. If you don't want to be a role model, then find another profession. Teaching is a profession loaded with humanity. We have lost a lot of the opportunity for humanity in today's classroom of high-stakes state testing. Ironically, the emphasis on standardized testing can actually leave some children behind. We are supposed to be teaching children—not treating them like political pawns and overwhelming them with tests created by people who have not been in a classroom in years. We are their surrogate parents—not their testing coordinators. Quality instruction delivered by caring teachers can raise scores higher than the drill and kill strategies employed in some inner city schools.

Teaching can be exhilarating and frustrating at the same time. Only people who strive to touch the lives of children every day can understand the mountains of joy and the valleys of sadness that good teachers traverse.

This book is written for those people who see teaching as a mission to improve the lives of children who, through no fault of their own, do not receive the quality education that other children receive. It is for those concerned parents who drag themselves to school to visit a teacher after working the second or third shift in a factory. It is for those people who see education as a human and civil right whose quality should not be based on socioeconomic status or geographic location. It is for those teachers and administrators who bang their heads against the bureaucratic wall with occasional success.

I spent thirty-two years in the Paterson, New Jersey, school district. It is a poor, urban community in that very dense area of New Jersey between Newark and New York City. It has always been a port of entry city where Irish, Italian, Jewish, and other ethnic groups raised their children to become part of the American Dream. Today the groups are African-American, Latino, Arabic, and many others. The groups may be different, but the dreams are still alive, although housing and schools are older and the silk factories have become condos or crack houses.

I am proud of my time in the Paterson school district. I am prouder of the many parents, teachers, and administrators who strove to improve the lives of children. I love the children of that city—always have and always will. This book is in no way a condemnation of their efforts; rather, it is a salute to their dream and a call for help, support, and assistance.

Urban education today is either invisible to mainstream America or the pawn of the ten-second sound bite of politicians. Many Americans take the same view of urban education as they do of the homeless: out of sight, out of mind. They have segregated themselves in beautiful homes along the suburban sprawl and don't relate to the problems of the inner city. After all, if those people worked harder and didn't have so many children, their lives would be better.

No politician can get elected unless he is committed to improving education. We've had so many education presidents, governors, and senators that we could start a museum. They create standards, demand accountability, and blame parents for not preparing their children properly for school. Unfortunately, the funding for programs shuts down as soon as the camera lights go off. Politicians need to look in the mirror carefully. Do they truly want to improve the lives of children, or are their programs ineffective nonsense proposed just to get elected. Some politicians look at schools as employment opportunities for friends regardless of whether ability and a position really exist—"We'll just make up a title to suite the patronage." Some inner-city school districts should have red lights next to their signs—"We sell positions to any politician who can advance our agenda."

When I was a child a beer company's slogan consisted of three rings: purity, body, and flavor. Today, three rings deny inner-city children a quality education: incompetence, racism and politics. We accept the culture of incompetence in inner-city school districts because it reinforces our perceptions of minority children as nice but unable to learn as well as wealthy suburban children.

Racism is the elephant sitting in the corner of every classroom, administrator's office, and faculty room in the inner city. It permeates our relationships and clouds our judgment. Racism is the overt and covert use of race in decision-making. Racism can be found in the contractors who build inferior schools because "they" won't know any better.

Once, I organized a dinner for our eighth grade graduates. The menu was supposed to include soup. When I saw no soup on the table, I questioned the owner, who nonchalantly said, "'they' don't eat soup."

Racism permeates the employment process in many inner-city school districts. We need to replace a black male principal with a black male principal. This school is predominately Latino, so we need a Latino principal. Racism exists when a minority parent refuses to support teachers because they are white and live in the suburbs; quality not color should determine parental support of teachers. Racism clouds the judgment of teachers who expect less and less of their students as the number of minority students increases. Racism clouds the judgment of some experts and community members who believe that white teachers can't understand and teach black children. It is the quality of the teacher not the color of his/her skin that matters. As we become more and more of a diverse society, racism bounces back and forth among many races and ethnicities.

Every person who directly or indirectly affects the education of inner-city children is a stakeholder in the future of our country. Our democracy can not thrive when only children living in "good" neighborhoods get quality education. Political and business leaders; education experts at the university, state, and district level; administrators and teachers; and parents can continue to accept incompetence, racism, and politics as the norm, or they can realize that those factors are an affront to what our country is supposed to stand for: equality of opportunity. The culture of incompetence is the antithesis to the noble words of the Declaration of Independence. If we ignore the message of our founding document, then this can not be the country our forefathers envisioned. This is a country of hope and promise. All stakeholders in the education of our children must make choices to perpetuate the legacy of the Declaration of Independence.

I write this book in anger and hope. I have strong opinions on how to improve the quality of education in inner-city schools, and I hope my suggestions help those stakeholders who really care. As I sometimes rant about the need for change, many diverse memories flood my consciousness. As I add them to the story, I hope those memories uplift the spirits of the stakeholders.

Finally, I write this book because it just isn't fair. We parade some inner-city children as success stories in front of the cameras and shout, "She made it. Why can't everyone else?" Sure, there are superhuman children who can survive any environment and succeed. Two of my favorite people in Paterson today are success stories who overcame every type of obstacle the streets could throw in their way. I love them and am proud of them. But do we expect superhuman efforts from all wealthy children in excellent schools? No. So why must a child born in poverty be superhuman to succeed? Why can't he just be a kid and get a good education regardless of where he was born? Why can't a twelve-year-old city kid just be a twelve-year-old kid?

God bless the administrators, teachers, parents, and students who strive for excellence in a climate of racism, politics, and incompetence.

# The Culture of Incompetence

## Oh Well, That's an Inner-city School

Incompetence is accepted as the norm in poorly-run, inner-city schools. "Oh well, that's Paterson, Chicago, or New York" is a common expression used by teachers and administrators throughout the cities of America. Expectations for success are low due to a history of failure. Failure breeds more and more failure. It feeds upon itself. We assume failure and never expect success.

Visualize an inner-city school with a majority of students from low socioeconomic and sometimes broken families. Some of the students are African-American, and others do not speak English at home. Is your vision similar to one of the Hollywood movies about schools in the inner city? Why is the *Lean on Me* movie our image rather than an image of students working productively? Some schools are blackboard jungles, but when we accept that image as the norm, we perpetuate the problem and produce more incompetent schools.

Low expectations are compounded by our culture's historic confusion about race and poverty. Are poor

1

people lazy, and do they really want to be poor? Can black children really learn as well as white children? Those questions have haunted our history, and they lay among the muck and silt of any river of change. Try to initiate change, and those sometimes dormant sediments rise from the bottom and cloud the chances of success.

If any initiative is introduced to improve the quality of education, some people simply say, "Be patient, this too shall pass." They are veterans of the inner-city wars, and they know the history of previous efforts. The effort to change is stymied by a failure to recognize the complexity and scope of the problem, which is centuries old and intertwined like a spider's web.

All problems must be attacked simultaneously. Sure, we need new school buildings, but if we don't improve the quality of teachers and administrators, what good will the new science lab be? Yes, we need smaller class sizes, but how much more successful will we be with twenty children from dysfunctional families instead of thirty? If a child is up all night because of an abusive mother and an absent father, will she learn more because there are fewer children in the classroom? Money alone will not solve the problem. Smaller class sizes alone will not solve the problem. Higher pay for teachers alone will not solve the problem. To break the cycle of unequal educational opportunity for inner-city children, we must break the backs of many problems simultaneously. We must attack racism among and between whites, blacks, and Hispanics. We must unchain the grip of sleazy politics from the halls of boards of education. If only one aspect of the problem is addressed, the winds of change are destroyed by the remaining overwhelming problems.  Only then can we change the culture of failure and incompetence. Willy-nilly stopgap measures will always fail. The problems of inner-city schooling have historic implications, and there is no magic pill. If we attempt one solution without considering its impact on the other problems, the ignored weeds will strangle the seeds of change. Racism, incompetence, politics, and money must be addressed at the same time.

The core of the struggle to break the cycle of incompetence must include a belief system that all children can learn. Community leaders, administrators, teachers, students, and responsible government agencies must believe in the value and ability of poor children to learn, leave their egos at the entrances of the cities, and concentrate all their energies to redeem this national disgrace. All stakeholders are responsible for the quality of education in inner-city schools and therefore must make choices. Those choices can be responsible, or they could be to bury one's head in the sand and blame someone else for the problems. Let's be responsible for the future generations of American citizens.

The culture of incompetence incestuously breeds itself. It gives grossly inefficient teachers an excuse to continue working the same way. Why? Initiatives for change are either ill conceived or inadequately implemented or funded. So when they fail, incompetent teachers can smugly say, "I told you so." They continue to teach poorly, give everyone a passing grade for substandard work, and the community has a new pool of illiterate adults who help perpetuate a culture of crime in the city—"It's not me, it's the system"; "It's downtown's fault."

I witnessed so many new literacy programs and testing strategies that were heralded as the latest panacea. They were to be the best inventions since sliced bread. Most failed because they were poorly funded, were not supported by the staff or the community, or were devoured by other problems. The incompetent teacher simply bided his time, waiting for the program to fail. His incompetence was rewarded and perpetuated.

Every faculty room has a "Donut Dan" who sits in the corner and complains about parents, students, administrators, and the board of education. His constant complaining masks his incompetence. Every time incompetence is rewarded, Donut Dan wins and teaches another day. Every time we fail, we provide Dan with more ammo for his negativity. When an uncertified, incompetent administrator is appointed, Dan simply throws up his hands and says "it" is the new administrator's fault. When we accept inferior teaching as the norm, we make Dan look like a prophet. The acceptance of the culture of incompetence validates his existence. We justify his sitting in the corner and getting another donut stain on his tie. He is the mascot for the culture of incompetence. When we slay the dragon of incompetence, we will eliminate the necessity for his existence.

Dedicated teachers and administrators fight the pervasive culture of incompetence every day. Instead of rewarding those dedicated teachers, we give them more work: "Please be on an extra curriculum committee"; "Please take this extra troubled child"; "You're the one who can do this report well." How many years can we expect our excellent teachers to bang their heads against the wall? Thousands of young, quality teachers leave the poorer systems, disenchanted and disgusted. They are the people we should nurture and support not the winners in the system of incompetence. The future success of any school system lies in the hands of new, young, energetic teachers. We can't afford to lose them.

I lost an outstanding, young teacher who was not supported or appreciated at a large, comprehensive, inner-city high school. According to an administrator in the school, she wasn't very good. She was intelligent, bold, and displayed initiative to make changes. She was exactly what that administrator didn't want. He wanted someone who would say "Let's keep quiet, maintain the status quo, and just get to 3:00 PM." She moved on to an excellent suburban school where she immediately began teaching an advanced placement course. She wasn't good enough to teach in the failing school, but she was good enough to teach an honors class in an excellent school. Maybe if she had just kept her mouth shut, hadn't expected any change, she would have been evaluated glowingly. Part of the culture of incompetence is the acceptance of inferior teaching as the norm—"What the heck, these kids can't learn anyway."

When I hired that excellent teacher, I had also hired three other young, white women for the same department. Silly me, I hadn't checked their race or gender. When I came to visit the four of them, the security guard told me that some people in the school were guessing how long the four white women would stay. The implied message for most schools is always the same: "Give us some big men, preferably dark in color and skilled in Spanish. Only they can handle the kids. We don't care if they can teach; can they control 'them'?" We must ensure that great teachers are not like salmon swimming upstream, struggling to reach the top of the waterfall, and then too exhausted to continue.

Some poor teachers who stay become administrators. They learn the system well—keep your mouth shut, play the game, fill out the paperwork. They align themselves with the correct political group in power, and get appointed to positions in order to get out of the classroom. They run a school and are supposed to exhibit educational leadership. However, they are followers who wouldn't recognize quality instruction if it bit them on the behind. Some incompetent supervisors haven't left their offices "downtown" for so long, they would need a map to find the schools in their districts. No one would know they were alive if they didn't pass around a meaningless memo several times a year. They never cause "trouble," and remain in their jobs forever.

No supervisor or principal observed me during my first ten years of teaching. At that time, many inner-city districts were just looking for warm bodies to fill positions; they were not concerned about the quality of instruction, just with the body count. Some administrators raised the level of incompetence to an art form.

One administrator collected plan books and asked bilingual students to open the books to the last page with writing on it. Then the young girl or boy used the administrator's stamp on the current week's lesson plans. Why choose a bilingual student? Because she didn't know what she was reading. The administrator lasted for a long time in the district. She was politically connected, and when the school and parents finally got disgusted, the superintendent simply transferred her to another school. Her incompetence was rewarded.

Communities must hire visionary, charismatic superintendents for inner-city school districts. Children need superintendents who believe in their abilities to learn and succeed. People need to rally behind this leader and trust that he has a mission along with the ability and stamina to see it through. Arrogance should be checked at the board of education's door. The superintendent must be willing to work with the existing staff and community. He must be   humble enough to recognize that one person does not have all the answers. Troubled districts don't need bullies or egomaniacs with the my-way-or-the-highway mentality. A good superintendent does not want the job to pad his resume; he wants to make positive, sustainable change. Such a candidate is difficult to find for boards of education, because the pool of superintendents shrinks every year. However, if the political culture strangling districts is eliminated, more and more people will be attracted to the position. Currently, many superintendents are products of the political system where back-scratching and arranging quid pro quos replace vision and integrity.

A great superintendent or principal in a district that is predominately black or Hispanic doesn't have to be black or Hispanic. That's racism at its insidious worst. We need the late Dr. Martin Luther King, Jr., to remind boards of education that they need to hire based on the content of a man's character not the color of his skin. If the mentality is that a black school must have a black principal, then the community is cutting off its nose to spite its face. The small pool of qualified candidates for the position drastically shrinks when we limit the list of hires to one ethnic or racial group. Students need quality not color; they need to expand their horizons not limit them. Do communities that have that mentality really have the success of its children first and foremost on its agenda? Or do politics, racism, and hatred of anything white cloud its vision. It's about the children and no one or nothing else.

To break away from the culture of incompetence, we must take the appointment of teachers, administrators, and supervisors out of the hands of politicians. They do not have the right to dole out educational jobs as patronage to loyal supporters; I don't care how far back in our history this practice has been active. Because it has been done for so long doesn't make it right. It was wrong "back in the day," and it is wrong today. In addition to examining potential principals' resumes, trace their ties to local, state, or national politicians.

We must limit tenure for principals, administrators of schools, and all central office supervisors. These people affect the lives of too many children every day. They are not justices of the Supreme Court of the United States; they should not receive life appointments. Principals should be groomed in the same mold as superintendents. They should be instructional leaders and organized managers. They must believe and trust that all students can learn. Again, arrogance and ego need to be left at the schoolhouse door. Principals need to nurture and support teachers rather than conduct power play exercises. The message must be simple: "I demand quality instruction, will support sound educational projects, and will praise every positive act."

The principal must be a highly visible person in the school. I've been in schools where the main office counter is like a Berlin Wall. Administrators live happily on one side and believe that the school on the other side has no problems. Teachers and students have as much success seeing an administrator as East Germans had climbing over the wall during the Cold War.

Additionally, there should be written and oral tests for all administrative positions. The results of the tests should be made public, so everyone knows which candidate should get the next available opening. The written test must be created by a committee of trained professionals from the local district, university staff, and educational experts from the community. Oral tests should be conducted by a committee of teachers, administrators, and parents.

Unions and district officials should establish merit pay for teachers and administrators. The criteria could include, but should not be limited to, evaluations and observations, attendance, punctuality, parental contact, curricular and instructional initiatives, and extra curricular activities. Merit pay should be above and beyond normal contractual raises. If we want quality people, we must recognize their efforts and accomplishments.

We must fight to keep excellent teachers. Give new teachers buddies from within the school who can see them through difficult times. Professional development aimed at practical classroom experiences should be provided for all new teachers and should provide a support system for all teachers in their first three years of service. They should meet regularly with new teachers from other grades in the school and even from other schools in the district. They need to vent their frustrations, exult in their successes, and listen to fellow colleagues who are in a similar situation. New teachers should not feel alone and isolated in the classroom. They must feel that despite socioeconomics, decrepit housing, and unsafe streets, they have a chance to reach children who really need help.

I was very wet behind the ears as a first-year teacher. My enthusiasm was certainly tested by my lack of both preparation and supplies and an abundance of students. Many new teachers struggled along with me during that year. We went to a local bar/restaurant every Friday after school with some veteran teachers for food, drink, and conversation. Those venting sessions really helped me through that first difficult year. Like me, new teachers need support from all quarters.

University professors from local institutions should be hired to conduct some support sessions, because this places accountability at the appropriate levels. Education professors who prepared students for teaching must then support them in the field. If a professor's original message was accurate and effective, then encouragement, support, and refinement of the message are all that are needed.

What should we do about tenure for teachers? How do we get Donut Dan out of the profession? How do we ensure that quality teachers are not victims of incompetent, politically connected administrators? Tenure must be examined on a state and national level. There must be consistent, reasonable criteria to determine if a teacher deserves to be rehired. Tenure is granted after three years and one day of continuous teaching. A qualified administrator conducts several observations and evaluations each year for non-tenured teachers. If an administrator can't determine if a teacher is effective after three years, then something is wrong with the system of supervision or the administrator. Is the administrator given enough time to observe teachers properly, or is he called to boring, useless meetings at the central office?

Attainable, practical criteria to determine teacher effectiveness must be established on the state level at least, and each teacher must meet those criteria at regular intervals. A qualified teacher should not fear those criteria or retribution from incompetent administrators. If the standards are professional and accurate, it should be a "piece of cake" for the excellent teacher. The criteria should be a challenge and a goal for the average to good teacher and an exit sign for the incompetent teacher. Groups of qualified professionals with no political or personal agenda should be given the responsibility to develop those criteria.

Change can be difficult and disturbing. Inertia and acceptance are easy to maintain. However, we are Americans. We are from a nation founded on a philosophy that will always be an experiment and a challenge to accomplish. Not many countries are founded on the concept that all men are created equal; that we derive our rights from our creator and not from a governmental body.

The Declaration of Independence and the acceptance of incompetence in education as the norm are diametrically opposed to each other. Our founding document established the bar of equality of opportunity for all, regardless of race, religion, or wealth. We must not accept the belief that poor children can not learn as well as other Americans. When we follow that philosophy, we throw our very integrity as a nation into question.

Abraham Lincoln, the Civil War, and the ensuing amendments to the Constitution redefined our identity as a nation. That purpose must be redefined again as all Americans fight to break away from the culture that has framed inner-city schools far too long. All children must believe in their right to a quality education. Every American adult has a responsibility to address, vilify, and eradicate the culture of incompetence in and from our schools. Every child in America is our child. Let's make the right choices for our children. We must reward excellence, not incompetence.

# The Teacher as a Person

You Believed in Us When We Didn't Believe in Ourselves

The role of teacher as a person is critical in inner-city schools. A teacher's humanity to students can be as important as the skills she learned in education classes; the teacher must connect to children in a personal way. All students can evaluate a teacher's humanity very quickly. If the teacher is not "the real deal," students will turn him off immediately. She's just another fake adult in their lives.

Generally, ineffective inner-city schools have many children from low socioeconomic neighborhoods. Some poor children have been ignored, abused, or neglected. If they are to learn, they must respect and relate to their teachers as people first. How many adults can inner-city children trust in their lives? One or both of their parents may be absent from the home. A tired grandmother may be the only force holding the family together; I met many warm-hearted grandmas who fought desperately to save grandchildren from the streets that had taken their own children.

Children watch television and notice the material comforts that other people their age enjoy. They know their books are outdated. They can see the ceilings that are chipping and falling in their classrooms. Some of their teachers never consider them important, but the teacher as a person may be the only adult the child can trust. A male teacher may have an even greater responsibility to be a role model. How many successful males are present in the lives of inner-city kids? There is a tremendous opportunity for all teachers to step up and make a difference in the lives of children at a time when children may not have anyone else. The connection must be real, genuine, and continuous. This point cannot be taken lightly at all.

There is a very easy measuring stick that can be used to judge how a teacher should treat students. She should treat them as she would want another teacher to treat her own biological children. She shouldn't ignore or isolate herself from her class and then complain about her own child's uncaring teachers. If she expects her own children to be treated with respect and dignity in school, then she needs to do the same for the children in her class. I remember some middle class teachers of various ethnic and racial backgrounds telling me that their own children were enjoying projects and activities in social studies in their schools. Yet when I visited those teachers' classes they taught children from poor neighborhoods with "dittos" and useless busywork. If interesting projects are good for their own children, then why aren't they good for students in their classes? Shame on them.

I particularly remember an eighth grade student of mine, who was an immigrant from Portugal. Because of family and language issues, she was turning sixteen in the eighth grade. She was a wonderful student and a pleasant young lady. When that young lady entered the class on her sixteenth birthday, she was greeted to the music of "Sixteen Candles," a cake, and her adoring classmates. I encouraged her to go to college, but she was concerned because she would be graduating from high school at twenty. She needed to help her family before thinking about college. Whatever successes or failures awaited her in the future, at least she knew on her sixteenth birthday that her classmates and teacher loved her as a person first and a student/peer second. She was recognized as a unique individual and treated as such.

When I was an administrator in a performing arts high school, I had the luxury of knowing many students by their first names. The school was very small, and students remained after school for rehearsals and performances. We developed overlapping schedules to free up classrooms for practices and electives. Each morning, I opened the school door at three different times and greeted less than one hundred students as they entered the building. I had the time and opportunity to talk to many of them about their homes, friends, and aspirations.

Talking to students on an equal, human level has a great impact on them. A teacher must really care about them if he asks questions about their lives. Their experiences must matter to that teacher. If he thinks they are worthy, maybe they are worthy. As a student said in her graduation address, "Teachers believed in us when we didn't believe in ourselves." Caring teachers helped her overcome a lack of self-esteem caused by a violent inner-city society that tried to rob her of her humanity and childhood. A teacher should make the attempt to connect to his students in a human, sincere way. It will be well worth it. He shouldn't be a poster child for incompetence; he should be a face of concern and caring.

How do teachers make a personal connection with their students? First and foremost, they must respect the children as equal human beings. Sure, the children do not have the knowledge, wisdom, or life experiences of the teacher. However, their equality is God-given and cemented in the philosophy of our Declaration of Independence. We often hear the complaint that children in school do not respect their elders, but I believe that respect should originate in the teacher. We are the experts in child psychology and teaching pedagogy. Shouldn't we recognize the importance of respect for humanity before the young, inexperienced child understands its importance? I hate when teachers tell students that they must earn the teacher's respect. How sanctimonious and egotistical is that? They are the children, and we are the professional adults. We must respect them as equals.

On the first day of most new school years, I wrote the words "respect" and "expect" on the blackboard. I respected them as fellow human beings, and that respect was guaranteed before I even knew them personally. Respect was a given from day one. I also expected a lot from day one, because I respected my students as wonderful human beings. My philosophy of education and practical classroom management strategies were built on those two words.

Excellent teachers have similar human characteristics regardless of where they teach. Those characteristics must include respect for children and enthusiasm for the profession. Excellent teachers obviously enjoy the company of children and have a limitless supply of energy. They are passionate and patient and have the courage to persevere. Excellent teachers are tolerant, compassionate leaders who exhibit a sense of humor. They must use the paradox of flexibility and firmness in their instruction and classroom management. They must ooze self-confidence and be able to motivate groups of people. Of course, I have probably described someone whose name is Jesus, Mohammed, or Moses. However, quality, functional school districts attract, demand, and hire people with some or most of those characteristics. Should inner-city districts settle for less just because the children are poor, abused, or neglected? There are inner-city teachers who display many of the traits mentioned above. Some maintain those traits throughout their careers. Some become disenchanted because they bang their heads against the culture of incompetence too often. We still must remember the crucial need for the teacher as a human being. Inner-city children need as many quality human beings to touch their lives as possible. Districts must attract, demand, hire, and support those teachers. To settle for less is to accept the culture of incompetence.

Poor school districts perpetuate incompetence by steering excellent teachers toward the few good schools in the district. Inferior teachers get transferred to the inferior schools. Do we accept inferior teachers because of our perceptions of students in the inferior schools: "Let's not waste an excellent teacher in that school. They can't learn anyway." We don't believe that poor children—many of whom are black and Hispanic—can learn. Teachers think, "Just keep moving them right along. When they are sixteen, seventeen, or eighteen, they will be someone else's problem." Sometimes I think large inner-city high schools are just holding cells for the city and state jails.

I vividly remember my first year as a teacher. One of my scariest moments was when the principal asked to see me after passing my classroom door. Of course, he never entered my room, but he still asked to see me. He had been the principal of the building for over twenty years and had a large, veteran, female staff. He brusquely said, "Cartaina, it looked like a good lesson you were teaching, but why did you take off your suit jacket?" Jump ahead thirty years to when I was a supervisor in the same district. Every time I visited that school, I took extra time walking around to stir up memories. One day, I attended an assembly program in the building. As all the classes streamed into the old auditorium, I was struck by the fact that most of the male teachers did not have a jacket or tie; some did not even have a collared shirt. If a teacher wants to dress like a night custodian, then grab a broom. If a person wants to be a teacher, then look, act, and talk like a professional.

The term "burnout" is overused and mistaken for incompetence. The flames of many burned-out teachers were never lit in the first place. Teachers should be passionate, excited, and alive for children who have little to cheer about. You can purchase e-tickets for a plane; I want "e-people" as my inner-city teachers. The *E*s stand for "enthusiasm," "energy," "enjoyment," and "expectations."

I had the privilege of working with great teachers during my two years as a traveling science teacher. They defined the term "e-teacher." One teacher conducted a staff development workshop during her last week before retirement. She was the most energetic, excited person in the room. Her flame never went out. We operated a traveling planetarium from school to school. We carefully led a small class and its teacher into the dark canvas cave to tell them famous constellation stories. I was the youngest of the four, and I was inspired by the dedication, enthusiasm for learning, and love of children the others had. Can you imagine how much the children enjoyed the presentations?

Teachers must have enthusiasm for their jobs and their professions. They should stop complaining. If things are so bad, they should change jobs or professions. Are teachers constantly complaining because they need a smoke screen to hide their incompetence? Is the best defense a good offense? Teachers should try not to be Donut Dan. They shouldn't do minimum work and complain constantly. They shouldn't add to the culture of incompetence. They should strive for excellence and become part of the solution.

Being an excellent teacher is a difficult job. Teachers are responsible for raising the intellectual, emotional, and social growth of twenty to thirty totally different children who come from totally different socioeconomic backgrounds. Why should students want to learn if their teacher doesn't want to be in the classroom with them? How can they have a positive outlook on their future when the person leading them is negative, disgusted, or depressed? Students easily read a teacher's body language and facial expressions.

Entering my last year as a public school employee, I had over three hundred sick days in the bank. Why did I keep so many days? How did I accumulate them? From my mother and father, I learned that you get up every morning and go to work. You don't make excuses. Just do the best you can. If I, as a teacher, took off ten to fifteen days per year just because I could, why should students come to school when they have to combat far greater hurdles than I have to to get to school?

Teachers in inner-city school districts must manifest leadership in their schools and classrooms at all times. They are in charge and must be take-charge people. By no means does that mean that they should be a dictator. I am thinking of the kind of leadership that is active and proactive in the students' personal and school lives. They must display self-confidence even though they are probably scared to death. A teacher needs to be firm but fair; these are two model adjectives for quality classroom management. Teachers should be motivators using many different methods. Some children are high maintenance and need constant encouragement. Other children are very independent and just need a wink of reassurance. Teachers must inspire them to believe that their potentials are limitless when they seem so limited. Teachers must believe in them when they don't believe in themselves. Teachers should take a stand and demonstrate their beliefs, not learn the system and how it works so they can move up the ladder. Teachers should learn how to beat the system for their students. They should learn which battles to fight and which to save for other days.

I worked for Dr. Morris Waldstein during my time as a science resource teacher. He was the greatest person, teacher, and administrator I have ever met. He told me that every teacher needs to find his or her niche. I'm sure he didn't mean a niche of comfort. Some teachers find a comfort zone, hide, and waste away. Other teachers decide that the classroom is not for them. They weasel their way through the system and find or create a position "downtown." Once there, they move irrelevant memos around central office to justify the fake position. Some become assistant superintendents. These are some of the people who make decisions that affect our children.

Teachers should challenge themselves as people and professionals to improve every year. The big picture may be depressing, but they should remain positive, refresh themselves often, and keep looking for the bright spots. They shouldn't beat themselves up after a bad day; they should get up off the mat and enjoy their students the next day. They should always look for new programs or strategies to improve their teaching skills. Enthusiasm for the profession will keep them fresh and alert. A wonderful woman in charge of staff development at the end of my teaching career introduced the district to the 4MAT model of teaching, which was based on the latest brain theories, and I thought it was great. I enjoyed the staff development and volunteered to be a trainer. I utilized the new strategies in my classroom, even though I had already been teaching for almost twenty-five years. It felt great being with other teachers who still had the flame to learn new methods to improve their quality of instruction. I gave up a week of my summer, but the recharging of my batteries was worth every minute.

Living in an inner-city community robs children of their childhood and humanity very early. There are reasons why twelve year olds can murder and show no remorse. They may have lost their souls to the street a long time before. Children absolutely need to see a human being in front of the class every day. That human being should be passionate, tolerant, understanding, and compassionate. Those adjectives are not synonyms for "pushover." The challenge for the teacher is to constantly display those human qualities and develop an orderly community of learners at the same time.

Some veteran teachers will advise new teachers not to get too personal with their needy children—"They will break your heart. You will not be able to function as a teacher if you constantly wear your heart on your sleeve." I totally disagree with that advice. A teacher's heart will be broken many times if she is an excellent teacher. Frustration lives with the great teacher daily. It's an occupational hazard. However, it's how she bounces back from frustration that will determine how long she can excel. She needs to clear her mind, do something good for herself, relax, and come back the next day ready to affect a child's life.

It is unfair to expect teachers to get up off the mat on many days and continue to be positive influences unless they receive the support of all appropriate stakeholders. We are not looking for martyrs to educate children. We are asking people with quality personality traits to work hard, because the rewards are great. However, teachers should not be scaling the mountain alone. There is no excuse for school leadership, the business community, the media, and the politicians not to provide the necessary support. If a person makes a choice to try to be an excellent teacher, then the entire American society must make the same choice.

There is no coincidence that "The Teacher as a Person" chapter is near the beginning of this book. Humanity is vital to the success of a teacher and to the lives of children, who—through no fault of their own—have no one to guide them. The teacher should be a real person, be a real teacher. He should help change the culture of his school. He shouldn't sell his soul as a teacher for laziness and disinterest. He needs to teach children not statistics, ethnicities, or colors. He shouldn't use children to further his personal agenda. He shouldn't perpetuate the culture of incompetence, rather he should be a leader and advocate for our most precious resource: children.

# TEACHER-STUDENT RAPPORT:

## The Heart of Classroom Management

You teach in an inner-city school? How can you control "those kids"? Aren't you afraid? You can't do anything for those kids; it's their parents' fault.

Those are just some of the questions and comments that I have heard, over the course of my career, whenever I told someone I taught in an inner-city school district. Of course, "controlling" children was never the reason I became a teacher. I was hoping to educate children not to control them. Control is not the goal of any teacher. It is the process to reach the goal of educating children.

How many times have I heard a principal tell me to visit the third floor of his school? "It's so quiet; I've got good teachers up there." "Is anyone learning?" should have been my response. Quiet and learning are not necessarily synonymous. Some of my class discussions in social studies on sensitive issues became very animated. Try talking about the eighth amendment and capital punishment in a class full of African-American males. A math teacher jokingly stopped me in the hallway with a complaint.

She had a great lesson planned for the day, but when my students entered her room still debating the death penalty, she was so intrigued by their passion that she let them continue. Social studies should be controversial and relevant, and excited children may not be misbehaving; they may be learning. Some administrators would have been better suited being wardens in jails. Control and silence is all that matters to them.

The teacher as a person must bring his humanity to the classroom every day. That humanity, as discussed previously, is integral to any personal or professional success. No classroom management system in an inner-city environment can succeed without quality teacher-student rapport. That relationship can not flourish unless the teacher consistently displays a humanity, which may be missing in the child's personal life.

Classroom management is crucial for educating children rather than just controlling them. It can be defined as the system of organized procedures necessary for a nurturing climate of learning. That system is created and implemented by caring, qualified teachers who have developed a warm, genuine relationship with their students. Classroom management is preventive and proactive; discipline is corrective. The better the classroom management, the less discipline a teacher needs. If teachers walk into school in the morning at the same time as the students and have poor or no lesson plans, they will need a lot of discipline. The more discipline a person needs, the more control becomes the main reason for her existence as a teacher.

I remember many teachers who had to run to the sign in book in the morning so they wouldn't be marked late.

Then they ran to their classes or out to the playground to meet their students. It took the teacher ten or fifteen minutes to get settled; all the while, the students sat with nothing worthwhile to do. Someone would enter the room, and the teacher would complain about how badly the students were behaving. She exclaims, "These students are not the same as they used to be." with a condescending tone and look.

Good classroom management incorporates both skill and humanity. Classroom management is a function of many factors. I would stress these essential components: teacher-student rapport, quality lesson plans, and preparation/organization.

I often used the phrase, "Respect a lot, Expect a lot," as the theme for my classroom management style. It was always plastered on one of my very inartistic bulletin boards in class. To me, it symbolized my philosophy about the relationship between teacher and students. The heart of classroom management is the relationship between the teacher and students. No list of rules, point systems, or games can compensate for the failure of teachers as human beings to develop a genuine rapport with the fellow human beings they will be working with for a year. The teacher must be genuine and proactive in the development of this relationship.

Elaborate and complicated point systems or games never work for effective classroom management. The Bluebirds have accumulated 10,000 points this week, but the Jaybirds are close behind with 9,900. If I raise my left hand instead of my right is that a demerit? I don't want to lose points for my team. Now what did the teacher just say about adding fractions? The point system becomes the focus of attention instead of the means to focus attention.

How does a new teacher develop a relationship with a group of children he has just met? He should be himself, be human. He should read, understand, and internalize the previous chapter of this book. If that chapter does not make sense to him, then he should choose another profession. It is a teacher's responsibility to treat his students like fellow human beings, equal to him in every way except the expertise of teaching and the experience of adulthood. Respect their humanity.

Inner-city children are often robbed of their childhood or forced to act tough or streetwise because of their environment. Many days they come to school and do all the silly things that any child would do in school. Other days those same children wear their street masks for defense to cover up the hurt and violence they felt the night before. Never forget that they are human beings first and foremost.

I chaperoned several three-night class trips to places like Washington DC and Boston. We always had a great time, and the children behaved magnificently. A group of thirty students and three teachers spent four days and three nights in the Virginia area. The bus broke down on our way home. Needless to say, we were all exhausted and didn't need the extra excitement. As the bus pulled up to our school at two or three in the morning, instead of early the evening before, I noticed that all the students were sleeping. The bus driver wanted to fully illuminate the bus. The children were all asleep, and many had a favorite stuffed animal, blanket, or some other sleeping prop from home. At that point, they all looked like teenage children. They were innocent creatures safe in their dreams. The anger and resentment they carried with them because of where they lived was replaced by the quiet of a good sleep. I gently touched each of the children to wake them up. It was one of the most precious moments of my career. I might have touched them to wake them, but they touched me with their innocence. Never forget they are human beings first and foremost.

Too often teachers in inner-city schools forget that their students are children. Sometimes students represent the teacher's "cause" to change the world, or they are lumped together as a "them," as almost inanimate objects that just don't want to learn. To politicians and their lackeys, students can become pawns in the crazy world of raising test scores.

They are children first before they are enrollment data, a tenth of an increase in a test score, or a member of a racial or ethnic group. I argued with a teacher once who insisted that a child is a Latina first and a child second. I could not believe the stupidity and cruelty of that belief. The child was merely a vehicle for that teacher to push her own political agenda and spread her hatred for white men in power.

There is very little that one can learn in a textbook about establishing classroom rapport; we must celebrate and respect the individuality of the child. This is definitely a case where the "science" of teaching must take a back row to the personal humanity of the teacher. The teacher must either love them, and want to be with them every day or get out. He shouldn't waste the children's time with any other excuse.

Children must know their teacher cares for them. There is no gray area in a teacher's relationship with students. Children can smell a phony just as easy as they can smell the bad pizza for lunch.

Who respects a student? It is the teacher who takes the time to listen to a child, even though the child tells the same story every Monday morning. It is the teacher who wants to know about their students' baby sisters, ill grandmas, and pets that get them into trouble. The teacher respects the student when she shows his parents respect when they visit the class. The child knows that his parent doesn't speak English well or is not dressed as nicely as other parents or maybe combs her hair in an uncool manner. The teacher will win the hearts and souls of her students by just showing respect to their family.

Teachers who water down the curriculum because "they" just won't get it have no respect for the right of all children to a quality education. Teachers who pander to inner-city students by setting expectations so low and excuses so high have no respect for them. We must respect and trust a student's ability to learn. That sounds like a simple statement, but the unspoken belief that "these kids can't learn as well as other children" permeates the culture of mediocrity, incompetence, and racism in inner-city schools.

However, school teachers are not miracle workers. If a student comes to kindergarten in an inner-city school already several years behind, then all stakeholders need to work together to fix the reasons why she is several years behind. We have the resources and intelligence to fix the problems; we just don't have the will and conviction to do so. It is easier to allow the problems to support and perpetuate our racist beliefs that "those kids" can't learn.

Teachers who respect every child's right and ability to learn and who develop high expectations for success must be the cornerstones of an effective, humane classroom management system. Use "respect" and "expect" as an umbrella to cover all rules in the classroom: "Because we respect each other, we don't swear or curse at each other." Teachers cannot accept crude, street language in the classroom. If they accept it, they sanction it. Do they sanction it because they do not believe the students can behave better?

As a student, I would expect to come to a safe, nurturing classroom each day; I would expect to find my teacher prepared, organized, and anxious to greet me. When I taught grades six and eight in middle school, we were still in the Stone Age, and students were placed in self-contained classes. Students only left the room for lunch and special subjects like gym, art, and music. Keep them under control. Classes moving through the halls had to be orderly so as to not disturb other classes. Instead of threatening the class, I told them I expected them to walk quietly because they respected both themselves and their teacher. I expected much of them, and they needed to respect themselves enough to walk like students with a purpose.

I like the word "choices" instead of "rules" for creating the expectations for class behavior. Students choose to respect each other, because that is the correct thing to do. They don't choose to respect each other because they are afraid of the teacher. The teacher must be firm but fair in implementing the respect and expect guidelines.

As the year progresses, the class can revisit the terms respect and expect. The students can actually gain ownership of classroom management by discussing and expanding the two terms. Empower the students to define and outline behaviors that exhibit respect and high expectations. This can grow into a plan of action for the class. Students will definitely follow a plan that they have helped create.

There are many non-instructional activities in the daily life of a classroom that can destroy the teaching process if not accomplished correctly. Collecting homework, going to the nurse or bathroom, raising one's hand, and passing out papers are just some of those activities that can cause chaos if not planned well in advance. Once the teacher-student rapport has been allowed to germinate and grow the teacher can proceed with the structure necessary for successful classroom instruction. Without the rapport in place, the above activities can only be accomplished through threats, screaming, or sheer cruelty.

All students want and need structure in the classroom. Students who come from broken homes and have little structure in their family life particularly need and want it. Teachers should develop structure and routines in those classroom management activities mentioned before that could doom instruction if mishandled.

The great fallacy in education is the belief that a teacher must be cruel and mean until Christmas and then lighten up a little. The teacher can develop firm but fair structures in the classroom and still be the human person mentioned in the previous chapter. A teacher does not need to be Hitler to succeed. Again, he may have a quiet class, but who is learning?

If teachers want a genuine rapport with their students to maintain quality classroom management, then those teachers must create an atmosphere of learning and purpose through respect and high expectations. Before reading a textbook on classroom management, a teacher should look into his heart and soul and his reason for becoming a teacher in the first place. If those reasons are sound, then he can build a successful classroom management system.

# The Twinkle in the Eye

## Learning the Skills of Teaching

Teaching is a very difficult job. We expect teachers to be knowledgeable in their subject matters and to love and understand children. In our diverse culture, we also expect them to be social workers and surrogate parents. It is the most difficult, yet most rewarding profession a person can choose. It is also a profession that chooses you, as I noticed quite frequently in the passion of student teachers.

Once people realize they have the heart to teach, how do we ensure that they will learn and use the skills of teaching? Teacher preparation and support must be a lifelong process that begins in the early stages of college life and continues through retirement. Teachers in the field must feel like they have a lifeline of support from education experts from the university level to central office personnel in their district to the community itself.

Training may begin as soon as an undergraduate student chooses education as a major at a certified college or university. Prospective teachers choose a core group of academic courses in subjects like language arts, math, science and social studies. Those courses must mirror the standards of syllabi from courses that majors in those subjects should master. There must never be a difference between a United States history course for history majors and one for education majors. How can someone teach the content if he doesn't master it? We expect children to meet world-class standards on state assessments, yet we produce education majors who have not received world-class training. Content for education majors must never be watered down.

When I was a supervisor of social studies, I was astonished at the transcripts of graduating college students looking for employment as teachers. Courses that I considered electives passed as core subject requirements. People were hired to teach world history who had never taken that course at the college level. Geography is a skill infused in all grades, yet some graduating seniors never had a geography class in college. Some colleges and universities have addressed the issue, creating specific core course selections. As I stated in the previous paragraph, the level and complexity of content for courses taken by education majors must be the same as courses taken by history, English, math, and science majors. Teachers must be historians, mathematicians, scientists, and linguists who also have the heart and skill to reach children. I absolutely do not believe in the phrase, "Those who can do, and those who can't teach." We owe our students the most qualified teachers possible rather than just people who "couldn't make it in the real world." If we are to change the negative climate of incompetence, we must staff our schools with great teachers who have warm hearts and are masters of content.

Many new teachers enter the profession as a second career. Thirty- or forty-year-old moms who have registered their youngest children in kindergarten now want to follow their dreams of becoming a teacher. Fathers who are tired of the rat race at work and want to be at home for their children at night change careers to become teachers.

I have had the privilege of working with many passionate people who made the life-altering decision to educate young children. They were nervous and sometimes unsure of themselves, but they amazed me with their determination to absorb every bit of advice I gave them. I have been humbled by their desires to change the lives of children. To this day, some of them still call or email me with questions and concerns. It has been one of the highlights of my career to work with such dedicated people. I framed a letter of thanks from one group, and I read it regularly to remind me of the hope for the future. They are definitely the type of people that the universities, communities, and districts should nurture for many years. They have made great personal choices for themselves and for the future of our society.

Many Americans are quick to point fingers at teachers as the cause of the failure of education in our inner-city schools. Most politicians can't use the word "teacher" in a sentence without using the word "accountability." Teachers must respect their students and treat them with dignity and humanity. But who respects teachers? Who treats teachers with respect and dignity? It is a bit hypocritical to ask teachers to act in a professional, humane manner when we don't treat teachers the same way.

All new teachers must feel connected to a university, a school mentor, and a district supervisor for at least the first few years of their careers. I have truly enjoyed observing and evaluating student teachers. But I believe the connection to the university professors should not end upon graduation or certification. That valuable relationship should continue for the first few years of the intern's teaching career. Think of the powerful connections that could be developed. A supervisor of student teachers could become one of their mentors until the novice teachers gain tenure. The university could choose regional sites to provide masters level education courses taught by the same mentors. The novice teachers could have weekly contact with the same education professors who taught them in college or supervised them in the field.

Colleges and school systems must become partners in the lifelong process of preparing and supporting teachers. Inner-city schools, especially, should open their doors to colleges to be used as laboratories for educational training. College classes in urban education should spend some of their time in the public school systems. Several class periods should be conducted right in the schools. College students should see the daily operations of the schools—victories and failures. When I was an undergraduate, the university actually had a high school on the campus. We observed our college professors teaching high school students and were able to discuss those classes in our next college session. Colleges and district training must mirror the real world. What better way than to have the professor teach in a real high school?

Once those new teachers become quality veterans, they can become mentors to new teachers. Both the mentors and new teachers should have a lifetime of support and training. We cannot afford to throw teachers into the deep water of their first classroom and then forget them. "Sink or swim" should be replaced with "swim and we'll be there to help you." We don't want Donut Dan to become a hero and mentor when he says, "Forget everything you learned in college, this is the real world."

One of the first people a new hire should meet is the supervisor or director of staff development. That district person is a valuable support person for the novice teacher. Just as the professor is the link to the university, the staff development supervisor is the link to subject supervisors, district office personnel, and all new opportunities for creative programs and strategies. The teachers should always feel like they are in a supportive cocoon. Then they will never free-fall after a bad day, because someone with a great deal of experience is there to catch them. That is the only way we can recruit and keep excellent new teachers. They must feel they are part of a great mission to improve instruction.

The supervisor of staff development has the responsibility of creating interesting, relevant training sessions for teachers. There have been new, exciting breakthroughs in brain theories regarding learning and behavior. New teachers need constant support with classroom management skills and conceptualizing quality lesson plans. A myriad of great topics should be addressed at training sessions for teachers. The skill of teaching always needs fine-tuning.

However, district teacher training sessions can be boring and useless. Incompetent district supervisors who don't make frequent connections with their staff often provide horrific training sessions. Since they don't spend the time getting to know their staff, they don't know the staff's needs. It is the same parallel with a teacher and a class of students. Teachers must know their students and make real, personal connections with them. Then they learn what and how their students need to learn. Supervisors who spend all day in their offices writing meaningless memos will create meaningless workshop sessions. They will pontificate about some topic totally alien to a teacher's real-life situation. The teachers sit in the sessions and count the minutes on the clock on the wall. Often these sessions make teachers realize what it must be like for a student in a boring class. What about the student who experiences chronic failure in traditional classroom settings? What hell we put our failing students through.

It was with that mentality that I attended a teacher training workshop in one of our very old, hot buildings on a spring day in the early seventies. I was sitting in an old one-piece wooden student desk hoping not to catch my pants on one of its many splinters. It was an afternoon session, so I had already taught four hours of class in the morning. The room was very warm, and I tried to look out through the dilapidated, stained windows for some sign of spring.

As I glanced away from the window, I noticed a tall elderly man with a full head of gray hair enter the room. He carried a small cardboard box that had, at one time, contained Girl Scout cookies. What caught my eye, however, were his eyes. He had a twinkle in his eyes like a small boy getting his first chemistry set. He was excited to be with us, and the smile on his face told us that he was having fun.

Dr. Morris Waldstein took a few pieces of electrical wire, sockets, bells, and batteries out of his box. It was playtime and show–and-tell all rolled into one. He began to demonstrate how science should be hands on and fun. I was amazed by the man's entire persona. How could I, as a young teacher, complain, when this veteran of many teaching wars was obviously having a ball playing science with us?

That day began a relationship that changed my professional life. He was the greatest man I ever met. He was as rare a teacher and supervisor as he was a great man. Academic brilliance and humility lived within the same person. He expected great things from students and teachers, yet he was extremely compassionate and kind. Our relationship lasted until his death several years later, and there wasn't one time that I was in his presence that I wasn't in awe of him. He had the ability to connect to his staff in a real, personal manner.

Dr. Waldstein was the science supervisor, and he asked me to participate in several districtwide projects. Students in my sixth grade class received weather equipment from Doc, and each morning, they measured indicators such as temperature, rainfall, barometric pressure, wind speed, and direction. Then we called Eastside High School with the recordings. No matter how many times I made a mistake, Doc was patient and kind. He explained cloud formations over and over until I could recognize them easily.

Doc wanted me to leave the classroom and join him as a traveling Mr. Wizard. Officially, I was a science resource teacher, but we were really his Boy Scouts. I say "we" because three sensational science teachers joined me: Vicki Madden, Josie Culmone, and Catharine Whitaker. We lived out of the trunks of our cars for one full year. My wife was actually a Girl Scout leader at the time, so I had wires, batteries, bells, and sockets in old Thin Mints boxes, animal specimens in Tagalong boxes, and so on. We traveled from school to school, performing demonstration lessons for teachers in front of their own classes. It was a hectic, busy year, but I would have walked through a wall for that man.

One of our favorite activities of the year was a portable planetarium. It was really a large inflatable canvas bag with a state of the art projector that displayed the constellations on its cave-like ceiling. We had so much fun telling stories about the constellations. Bootes is a constellation shaped like an ice cream cone. Of course, its flavor was heavenly hash. Orion and Leo roared through the heavens as we moved the sky across the seasons. Recycled air was pumped into the planetarium, but after several lessons per day, we were as groggy as a drunk walking out of a bar at 3 AM.

One of the first times we performed a lesson for students was on a stage in an elementary school. The four of us were very nervous, but the session went very well, and the students learned astronomy in a fun-filled hour. As we left the planetarium, we were horrified to see that Dr. Waldstein had been sitting in the first row of the auditorium listening to us. Would this great man think we had performed adequately? We walked off the stage, praying for his approval. When we reached him, there were tears in his eyes. Over and over again, he told us how proud he was of us. I have never had a private audience with the pope, but Doc's pride in us that day must be close to that feeling.

Dr. Morris Waldstein passed away as a result of asbestos cancer very soon after retiring. I hope to God that I have been some source of pride to him as he has sat in heaven with his torn cookie box of wires, bells, and batteries.

Where are the Morris Waldsteins today? How many administrators, supervisors, and teachers have the twinkle in their eyes? I believe there are thousands of dedicated teachers who recognize that the profession is more than an eight-to-three job; it is an opportunity to change a child's life. Sadly, too many dedicated teachers leave the profession within a few years. Often, the most incompetent teachers find a little niche of comfort and just waste away.

We must build a network of support for new teachers to sustain them throughout their careers. Every professional that the new teacher comes in contact with must be part of the support system. Think of the power of this network. Undergraduates, student teachers, new and veteran teachers, mentors, administrators, and district and university personnel must all work together to smash the status quo. The culture of incompetence is insidious, systemic, and pervasive just like the problems that attack inner-city schools. A new climate of professionalism, support, and enthusiasm must become the new culture, the new status quo. That new culture will support the heart and skills of new teachers. University officials and school district personnel must make practical choices to support and nourish new teachers so those teachers can do the same for their own students.

# THE LIVING SCHOOL

## Experienced Educators Can Determine If a School Is Effective within an Hour of The First Visit

I do believe that experienced educators can identify an effective school in a matter of hours. The extent of the effectiveness may need more time, but obvious clues jump out at us as we walk down the halls of most schools. That's because a school is not a building; it is a living organism that performs a vital social function within the walls of a physical plant. It can be vibrant with the exhilarating sounds of students learning and growing, or it can be stifling with dysfunction and despair. A school develops the social, emotional, and academic lives of children, or it destroys natural curiosity and creates illiterate, dysfunctional citizens who depend on of participating in that society.

What are the obvious clues that tell us if a school is an effective institution? Walk into any inner-city high school. Are the hallways dark and foreboding? Are the bulletin boards and display cases bright, positive, and most of all, up-to-date? When the bell rings for the change of

classes, do the students move with a purpose? Do the security guards have to shepherd reluctant students to class and away from known hiding places for those who cut school? Go to the cafeteria during lunch. Does it resemble a jail with teachers, security, and maybe even police officers acting like guards? Do the students clean up after themselves, or must lunchroom monitors clean all the tables? Is there a buzz of happy students talking, or is there a rumble of discontent? Is it obvious that one small altercation will cause chaos?

While the students are in classes, walk down the halls. Look at the students' faces as you peek into the rooms. Do they seem engaged in the lessons? Are they active learners? Are the teachers facilitating and inspiring learning or simply chalking and talking and passing out meaningless dittos? Is there a positive energy in the class? Great administrators can feel the rapport between students and teachers as they observe classes. There are positive vibes in the room. All of the answers to the above questions can be found within an hour or two at most schools. If the answers are all negative, then the school acts like a different social institution: it is simply is a holding pen for the county jail. I've seen high school classroom doors that are metal and have tiny slits for light. What kind of building comes to mind?

A child enters school as a curious, impressionable creature, anxious to grow and learn. A living school can provide the nurturing environment for that child's growth. What is a living school? I'll define it with these words: hope, purpose, joy, and learning. It is an environment and a state of mind where adults and children believe in each other, work together, and grow together. Adults grow as much as the children in a living school. They both develop and feed off each other's enthusiasm. Children bring their innate curiosity, and adults bring life experiences. The symbiotic relationship nurtures everyone. It can be a place an inner-city child runs to in the morning to escape the chaos of life in the streets.

Departments of education attempt to lower the truancy rate in some schools. If the school itself is chaotic, then life at school mirrors life at home. Why should a child attend a chaotic school? He might as well "chill in the street." I'm amazed that the rate is not higher considering some of the dysfunctional schools I have seen.

Some of my favorite schools were definitely living organisms. Students and teachers were happy as they moved about the building. There were always pleasant hellos from administrators, teachers, and students. Every bit of wall space was covered with students' work.

One particular third- or fourth-grade class drew templates of the fifty states. Parents agreed to bake a small sheet cake and cut it into the shape of their child's state. On a given day, all the cakes came to school, were pieced and iced together, and then were eaten by everyone. The superintendent of the district, community leaders, and the local media attended the official "cutting of the country," giving the event a star-studded scene. Quality education is fun and engaging. Quality education helps students pass state-mandated tests without stress or drill. Students in that particular school easily passed their state-mandated tests, because quality projects enhance learning and improve scores.

Unless districts, administrators, and teachers accept the concept of the school as a living organism, the focus of the school remains the next state report, meaningless meetings, and other bureaucratic trivia. Instead of the school being alive, the bureaucracy flourishes. Trivial forms and meetings become ends in themselves. We serve a dysfunctional system rather than for the children. The organism grows diseases like illiteracy and complacency instead of nurturing children to become responsible adults.

A living school must continue to work on the health of its body. The health of the body becomes better or worse every day. A school doesn't remain static. It needs to grow, or it will fester and die. Just as a healthy body needs nutrition and exercise, a healthy school needs massive doses of humanity, skill, and caring. Those massive doses are the lifeline for the living school. Principals are head doctors for the living school. They provide intravenous feeding in the form of food and drink for the staff on long staff development or teacher conference days. They nourish their staff by defending them. They are heroes for the staff when they model professionalism, enthusiasm, and passion for learning. They exemplify leadership when they attend a difficult seminar at an inconvenient time or place so they can transmit the information to the staff. They never send their staff where they are not willing to go first.

The living school operates within a physical plant, so the quality and maintenance of the facility should be a district priority. Yes, the buildings may be old, and some are decrepit and falling apart. However, we must never accept certain scenarios. First, the teachers, parents, and community must fight at every board meeting for proper maintenance and repairs of school buildings. I have seen and heard of many cases where plumbing, electrical, and carpentry work done in schools was so deficient that if it had been done in my house, I would not pay the contractor and would probably sue for damages. Just like we treat students like our own children, we must treat our schools and classrooms like our second home. Stop accepting incompetence. I think many people would actually be surprised to see quality workmanship when new windows or doors get installed in our schools. We would be amazed if an asbestos cleanup went well and stayed on schedule. Why? Because inner-city schools are draped in the culture of incompetence, and we are conditioned to accept failure.

Teachers and students deserve to work and learn in clean buildings. There may be graffiti on the outer walls and a leaky roof, but we should expect clean and healthy rooms—on a daily basis. I worked with one principal who walked around the school with a can of Comet an hour before the opening bell. He demanded a clean building but was willing to do his part. Teachers and administrators shouldn't accept floors that haven't been swept in a few days or overflowing garbage cans that are not emptied on a regular basis. Floors should be washed, scraped, and waxed during breaks. Desks, bulletin boards, and chalkboards should be cleaned and repaired. Children should be able to use bathrooms that have stall doors that work, toilets that flush, and enough toilet tissue. We should expect the same sanitary conditions at work as they would at home. We shouldn't accept anything less.

While I was teaching at the performing arts school, the board and superintendent decided to privatize the custodial staff. Our outstanding head custodian greeted his new team one afternoon. They were all very nice ladies who didn't speak a word of English. He couldn't tell them how, what, or where to clean.

That same school had a power unit that provided heat or air conditioning. To operate the system, one just had to close and secure the doors. One day I decided to open the doors of the unit very wide to see if there was a filter inside. I don't know if there was a filter, but there sure was a science experiment growing in there. I asked an administrator if there was any scheduled maintenance program. I was assured that people came to inspect the units, and no bad air circulated in the rooms. So we continued to work with runny noses and congested sinuses. The physical discomfort was nothing compared to the debilitating spirits caused by disinterest.

New construction in inner-city school districts can be a comedy of errors. One of the high schools in my district was bulging at the seams in the late sixties and early seventies. Instead of building an additional school, the board of education attached an annex to the original building. The connecting hallways did not mesh into the old to look like one big school. The open area for sports had to be eliminated, so students had to walk several blocks for gym classes. The heating and cooling systems never worked properly in the new patchwork school. Some rooms were very hot, and some very cold.

Who thinks of these things? Better yet, how do they get away with it?

One of the best inner-city schools I ever worked in was burned to the ground. The new facility had a play area on the roof that the children never used because of shoddy workmanship. The building shook for several years because the roof was not attached properly. A national scandal would have erupted if that had happened in a wealthy school district. The building doesn't shake any longer, but the students still do not play on the roof.

Unfortunately, stories like that are all too common in inner-city schools. The culture of incompetence accepts such behavior as the norm. People think, "inner cities are corrupt, so why shouldn't I get my piece of the action?"

If teachers work in dilapidated buildings, they must make every effort to make the building livable. Again, it is a question of school culture. The outside of the building may look like a dungeon, but the inside can contain a vibrant learning community. It is the teacher's responsibility to make the classroom bright, cheery, and an environment conducive to learning.

How can someone maximize the existing situation? He can find free posters at a travel agency to cover peeling plaster or paint. National Geographic distributed four-by-six-foot laminated maps of the world to each classroom in the country one year. That map covered a lot of ugly walls in Paterson that year. He can cover demoralizing spaces with timelines and science projects. I had so many timelines in my high school history classes that my caricature in the yearbook had a timeline on my tie. What can possibly boost children's self-esteem more than to see their work hanging up on a wall? Wow! A child who gets no praise or encouragement in a dysfunctional family has a teacher who actually thinks her work is so good that it should be hung in the hall. Nothing exemplifies the climate of a school like active, engaged children doing projects and proudly showing their work.

Some schools do not have an available classroom for art instruction. Teachers use a wagon or cart and travel from room to room. Art-on-a-cart can be a negative or a positive experience. Again, it depends on where your spirit lies. Have you accepted the culture of incompetence, or are you a member of a living school?

One art teacher in a small school in Paterson was a spectacular member of that living school. Each year, he checked with the social studies teachers in the upper grades for curriculum input. Students then created a mural that complemented their social studies learning. The hallways of that third floor became an extension of the classroom. Not only was it beautiful and bright, but it was also evidence that children were learning—in one of the oldest school buildings in the state.

One of the most memorable moments of my teaching career came when we hatched eggs in the classroom. A fellow teacher had provided us with a small incubator and fertilized eggs. The children actually heard the chicks break through the shells and saw them fight for life. The windows in our classroom were old, and the button latches did not work well, so I wired the windows closed that night, because it was winter, and the baby chicks were still wet. As I entered the room early the next morning, I noticed immediately that the room was frigid. One of the wires had snapped. God sent my very good friend— the district's science supervisor—to my class soon after

I arrived. We grabbed some aquarium tubing and fake fur used for static electricity experiments. We warmed the chicks with the fur and gave them "mouth-to-beak" resuscitation. Much to the children's glee, the chicks survived. Those decaying windows did not kill the chicks or defeat the spirit of my students.

Teachers shouldn't ever accept the terrible conditions that await you in an inner-city classroom. If they accept corruption, incompetence, and complacency, then they have become part of the problem instead of part of the solution.

A living school can function well, even in an old facility, if the head of that organism is an outstanding leader. Principals must be the instructional leaders of the school, and it is their responsibility to create a climate that fosters learning. Positive attitudes and vibrations must begin with the principal. It is the principal's responsibility to combat the culture of incompetence with the culture of learning. Mission statements are ignored sheets of paper in most schools, but quality teachers can recognize the real or de facto mission of the principal within the first month of school. The principal's mission must clear and simple: "Everyone who enters this building is responsible for the education, safety, and growth of children. People will act in a professional manner between and among staff members and as surrogate parents toward children." The principal is the leader, lightning rod, and shoulder for the entire staff. It is a daunting, task.

A staff can be a forest of many different trees. Sometimes a principal needs to address the forest and sometimes the individual trees. Let's look at the trees first. Greet all teachers and staff members as they enter the office every morning—"Hi, how's your daughter? Is she over her flu?" Personal comments of concern build a mutual feeling of trust and humanity that can carry a school staff through rough times. If the goal is to have children come to school eagerly each day, shouldn't it begin with treating teachers personally and professionally so they come to school eagerly? School leaders should walk around the building and comment positively on activities and projects that actively engage students in learning. They should be highly visible but positive forces throughout the school. I have had the privilege of working with and for several outstanding school leaders. They smile in spite of enormous stress and strain. They are human beings who understand that other caring human beings share the goal of quality education for all students.

In addition to treating the staff individually, the principal must look at the teachers as a cohesive team—the forest. The team must develop a chemistry that reflects the personality and professionalism of the principal, who then must have the ability to manage and lead the increasingly diverse group of teachers in schools. The challenge is to meld various styles, cultures, and abilities into a team that loves to work hard for children. Teachers need to trust the leadership of their principal, and the principal must establish a zone of comfort for the staff. In a very real sense, teaching in an inner-city school is like going to battle every day. The teachers face enemies in the streets, the bureaucracy, and the culture of incompetence. They need to feel like the principal is on their side, willing to fight with them and for them. There is a strong parallel between the trust and respect between teacher and student and the trust and respect between principal and teacher.

Instructional leadership must complement personal leadership. Teachers must not only trust the principal as a person but also her ability to create an environment for learning. It is not good enough to simply house children like a jail does. Once the students and teachers are safe in their classes, then what happens? The goal is to educate them not just to keep them safe from harm. The principal must prepare staff development workshops that emphasize learning rather than simply how to enter and exit the cafeteria. Teachers must leave the building every day, thinking, "This is a place where we teach children." If dedicated teachers don't have that feeling in their souls, then the incompetent teacher wins again. That teacher can walk out the door looking like they just walked out of a salon, thinking, "I fooled them again." The living school needs spirit to flow through its veins. Enthusiasm for one's job is almost as important as respect for the children. Children need to see happy and excited teachers and administrators enter the building each day. It's so important that it bears repeating: why should children want to come to school each day when the teachers don't seem to be happy to come to school each day?

School spirit games and activities are vital to the inner-city school. School leaders should let the students and teachers combine ideas to create a colorful physical education uniform. They should empower the students to select certain days of the month for spirit days. They need to bring life to the school. The goal is clear. We want students to see their school as an oasis from the realities of the street—"I like to go to school. It's fun, and we learn there."

Some veterans of the teaching wars may look upon my living school as a corny, quixotic dream. It is not a dream. I have seen real, living schools where teachers hug each other like lifelong friends and treat children like family members. Adults and children look forward to going to school each day. It is not easy, but building a living school can be a stimulating experience. There is a little negative man sitting on everyone's shoulders that will tell you it is impossible. But if teachers refuse to accept the culture of incompetence, they have overcome the first major hurdle. The rest is positive energy, stress, and fun. Our children are worth it.

A school will always have a culture of its own, whether teachers work at it or not. It can be negative, self-defeating, and demoralizing, or it can be beacon of positive vibrations, hope, and excitement. The living school can be the centerpiece of the assault on the culture of incompetence. All stakeholders must work to make their schools living environments for teachers and students to thrive and learn. Choose to live rather than to exist in the status quo of incompetence, racism, and politics.

## CLASSROOM INSTRUCTION

### If It's Not Tested, It's Not Taught

Many inner-city principals are overwhelmed by security issues inside and outside of the building—violence, guns, and overcrowded classrooms. It takes an incredible leader to remember the true purpose for that large building: the instruction of children. I have found that most principals are concerned for their students, but the tsunami of negative factors can destroy their beachhead of instructional plans. Long-range planning can fall by the wayside as principals try to just get through the day. The difficult job of being a quality principal got exponentially harder with the advent of state testing.

The Standards and Assessment Movement of the last few decades has added tremendous pressure to the already challenging job of the school leader. The overt or covert message from the central office is always "raise the test scores." Bureaucrats at the local, national, and federal level often confuse instruction with assessment and standards with assessment. Instruction is teaching; assessment is the measurement of that teaching. Standards are the bar, and assessment is the measurement of success in reaching that bar. Prepping students for a state test for a significant period of time is not in the child's best interest. It is in the interests of superintendents, state education officials, and local real estate agents who want to read a headline in the newspaper that scores rose in the city. The tragedy lies in the fact that all our eggs are in one basket. Our children's future rests on tests whose validity is often challenged. Instead of mandating a curriculum that includes quality works of literature in language arts classes, district officials purchase consumable books with such titles as *Strategies and Techniques* in order to pass a state test. The assessment bureaucrats working at the state level distribute practice test questions to prepare students for the test. Some districts then hire private companies to produce more practice questions so we can drill the students to death. Maybe scores will rise a few percentages because students have practiced similar questions. But what have they learned?

I totally understand and agree with the idea of taking time on a regular basis to review concepts for a state test and even to develop test-taking skills. However, if a school or district must spend months prepping for a test, then their original instructional plan was either flawed or nonexistent.

My boss at one time was the director of curriculum. When she left, her position was not filled for over a year. The importance of curriculum at that time for that district was clear: "If it is not tested, don't teach it, and don't fund it." The state appointed a superintendent, and the district sold its soul to the pressures of state bureaucrats. They stopped teaching children and prepped them, like trained seals, to pass a questionable test. Leaders only thought of making themselves look good instead of putting the children first.

A primordial, significant question must be asked. Why can't children in inner-city schools pass the state tests at the same rate as schools in wealthier, suburban areas? In all my years as teacher, administrator, district supervisor, and state committee member, I've never heard that question asked. All stakeholders should have asked themselves that question, and then inner-city schools could have faced the issues confronting city children together. Was the question never asked because the stakeholders felt they knew the answer? Did ignorance and bigotry stop them from an intelligent discussion of the question?

I have always believed that all children can learn as long as the playing field is made fair and level. All races and ethnicities can learn equally, if all members of society work hard to ensure equity. However, if we assume that poor children can't learn anyway, then it is easy to accept the drill-and-kill test prep mentality. We need to have faith in our children and believe that they can learn and succeed if given world-class curriculum guides, excellent teachers, and diverse, engaging instruction. We need to believe that parental responsibility, reduction of poverty, and students working hard are the path to success. If we accept the culture of incompetence as the norm, then we perpetuate the existence of illiteracy, crime, and overcrowded jails.

The focus of any important, state testing meeting must be the refusal to accept the culture of incompetence as the norm. Such meetings should demand the social, political, and economic changes necessary to give all children a fair chance at passing the tests. When all children are given the opportunity of quality instruction, then we can compare districts. Until that time comes, all we do is cover up the real problems with wasteful preparations for unreliable tests. I am absolutely convinced that, when all stakeholders work together to provide equality of opportunity in all schools, all students can easily pass any state-mandated tests.

All inner-city boards of education must take a stand and create challenging curriculum guides that become the marching orders for quality instruction. Quality instruction is impossible without a stimulating, thorough curriculum. Classroom management is impossible without quality instruction. All the best practices in education are related. The strategies for success are as interwoven as the negative factors that destroy schools.

Students in an inner-city school district need an intense, thorough curriculum equal to the best suburban schools. If we accept the concept that inner-city children come to school in kindergarten further behind their counterparts in suburbia because of poverty and other socioeconomic conditions, then why aren't we trying to catch up instead of dumbing lessons down? Shouldn't the school be the vehicle to drive students to overcome the debilitating effects of life in the inner city?

Let's find the best schools in our state, and create a curriculum that equals them in quality and excellence. I want one superintendent in an inner-city school district to have the guts to call the most successful school district in the state to borrow its curriculum guides. I want that superintendent to slam those guides down on a table during the next district meeting and ask, "Why doesn't ours look like this?" I never witnessed anyone in the higher echelons of power in the Paterson School District contact the Mountain Lakes or Milburns of the state to compare curricula guides. I did see a carousel of people in authority move from Newark to Paterson to Jersey City, padding their resumes with new titles. All three districts were operated by the state at one time because of gross incompetence, yet each district hired each others' brain trust.

Curriculum committees must include the best teachers, administrators, and community members possible. Too often, the same people reply to postings for curriculum committees because they think it is an easy way to make some extra money. Leaders should make it a competitive process where people earn places on the committee because of their abilities, experience, and vision.

That committee must then create a guide so thorough that it looks like a teacher resource guide. A new teacher should be able to pick up the guide and feel assured that he knows what content must be taught for the entire year and what the best practices are to implement that content. A national expert on curriculum should be able to pick up the district guide and believe that, if implemented properly, a child will master national standards. Do we owe our children anything less? Wouldn't if be wonderful if a child graduating from high school in an inner-city school looked at national standards during one of his college classes and said, "Yeah, I learned that."

Curriculum guides must be the bibles for all teachers, especially new ones. It provides their marching orders as they try to touch the lives of many children. It is almost a history of best practices that have succeeded in the past. No army can succeed without a great battle plan. The curriculum guide is the educational plan. Caring teachers who have developed a warm rapport with their students can implement the plan and know that children will learn. Children who are learning have no time to misbehave. The brains behind classroom management must be a thorough, creative, engaging curriculum.

The only way schools will succeed is if children succeed. Children succeed when they are taught well, challenged with a quality curriculum, and pushed by caring, competent teachers. Instruction is the key. All stakeholders in the educational process have the responsibility to create the cocoon for teachers and students to thrive.

Curriculum guides throughout the country are very similar. Is United States history different in Kansas, Texas, or New York? Are physics or chemistry different subjects in different states? Why are children in affluent communities taught a more challenging, comprehensive version of subjects? Isn't it the grossest form of racism to dumb down a curriculum for inner-city children? Do we allow our preconceived racial perceptions to form our level of expectations?

The quality of instruction in inner-city classrooms can be as diverse as the student body itself. It ranges from creative and brilliant to incompetent and destructive. Many inner-city teachers create quality plans and interesting activities in spite of the people who are supposed to help them—not because of them.

I have personally witnessed excellence that brought tears to my eyes, because I knew what the teacher had to do to successfully implement her plan. I have seen mock trials from grades two to twelve, community redevelopment plans created by students in all grades, and of course, the amazing high school students who won the state National History Day contest almost every year. All those teachers should be the heroes of the district; they should be praised and supported by all stakeholders. They should not feel isolated or as though they are struggling against a tide of incompetence. They should be riding on the wave of change.

The quality of instruction can be abysmal in many inner-city classrooms. Old textbook and ditto activities are followed by textbook and ditto activities. Children are bored and fail. The more they fail, the more textbook and ditto activities they get—at a lower level. Incompetence is a self-perpetuating phenomenon. The more the students fail, the less and less is taught and learned. It is a vicious spiral. Poor teaching is followed by poor teaching, as teachers blame everyone but themselves for the failure of students to learn. Teachers give close-minded lessons assessed by open book tests. If a student is quiet, his lowest grade is a C. If some incompetent teachers had to sit through their own lessons, they would jump out of the first window they could open.

Great instruction begins with excellent teachers preparing interesting, diverse, and challenging lessons. Students in excellent schools have a rich, stable structure at home for support. They have a stake in the system, and they know that learning and passing are key to future success. Whether they think school is a game or not, they are players, because they know it is important and that they can succeed if they pass. Many students in quality schools can survive a mediocre or poor teacher because they have support at home or in the community.

Inner-city students need years of quality teachers in a row. They need quality teachers who inspire, challenge, and believe in them. Inner-city students need to be convinced that they can succeed. They don't believe in themselves or in the system. Put a textbook on their desks and a teacher in front of them who lectures or reads it aloud all day, and they tune out immediately. "I am a failure in school, and I have a failure as a teacher." As I have often repeated, failure breeds failure, and the spiral is always downward. So the failures get greater and more severe each year. That failure can lead to a life in jail, or multiple unwanted pregnancies for females.

"Student-centered learning" and "active learners" are terms we hear often in education. They are strategies that are essential in the inner-city school. Students must be engaged and active learners. They must be convinced that learning is important and that they have a chance to be successful in school and the community.

One of the most tragic fallacies in urban education is that less planning is necessary because the students learn less or more slowly than students in effective schools. Students in failing schools need the highest quality lesson plans just like they need the highest quality teachers. The further behind the student is academically, the more thoughtful and interesting the lesson plans need to be. Usually a "slow" group of students in an ineffective school gets basic, boring, drill-and-kill types of lessons. The materials are much lower than the grade level, but sometimes the tease is that the materials are "high interest." If students successfully complete a text that is several years behind their grade level, what have they accomplished? They are seventh graders who received an A on a fifth-grade test.

What type of lesson plan is most effective in failing schools? Lesson plans must engage the students in the learning process. The lessons must empower the students to assume responsibility for their learning. Students must be active learners in the process. Instruction must include learning that begins with the student's experiences in life but expands their knowledge to places they have never seen. Teachers can't limit instruction to the students' experiences. That is a great place to start, but teaching Hispanic, Korean, or African-American students only their own culture and history confines them to what they already know. They must experience the language, history, and science of the mainstream culture so they can develop dreams and goals of success.

"Engage" and "empower" are two chic verbs in education. People with titles often need to invent new terms to justify their jobs. However, those two verbs are very much on target in the instructional process. Think about what an inner-city child owns or has the authority to control. He has very little ownership over material things and very little control over other issues in his life. If he can claim ownership over a school project, do you think he will remember the experience longer than the practice worksheet to a test? I remember an ethnic festival that the students and teachers planned and carried out together like it was yesterday—I was in seventh grade.

There was a nonprofit foundation in the city of Paterson that provided teachers and students with grants for innovative projects. My eighth grade class decided to build an environmental center on our school grounds. We won the grant and created the plans. A local concrete company donated cement and poured our slab for nothing—everyone is a stakeholder. My students carried lumber, sawing wood, and nailing the building together. They were boys and girls from a housing project that had a horrible record of drugs and crime, and they were working together in harmony. For years after the project was completed, there was never any graffiti on our center, yet there was plenty of graffiti on the school building. Why? It was their center.

I once found a design to build all the planets to scale in size and distance from the sun, that would work perfectly for my eighth grade class, which was at the intersection of two long corridors. The children loved the idea and quickly made each planet out of balloons and papier mâché. We painted the planets and accurately hung them along the longer of the two corridors. The children were very proud of their project. A student from another class knocked the earth down and stuffed it into the boys' commode. At the end of the day, one of my students ran into class, frantically yelling, "Mr. C, the earth is going down the toilet!" Looking around at my old classroom that needed new desks and a paint job, I said, "I know." It was a magical moment, but I was struck by how sincere and upset the student was. It was his solar system.

Can we think of other school assignments that would get students so upset? I have seen countless other quality projects that are much better than the ones I have described. They all have the same aspects in common. Students are given responsibility and authority over their work. The assignments are fun and meaningful. Everyone remembers them for a long time.

Students in ineffective schools in the inner cities have experienced a lifetime of failure, lies, and false hope. They have been told over and over again that they don't count and that they are just biding their time in school. As I have said, some schools are simply holding cells for the county jail. The teacher chalks and talks, hands out dittos—mainly for control and discipline—and the day ends slowly with little or no instruction. Children recognize a teacher who wants to include them in the instructional process; they will be a friend and learner for life—"Mrs. Smith thinks enough of me to make me take responsibility for my learning and includes me in planning how the lesson is taught. Wow." The affective result to an instructional activity will be as great as or greater than the cognitive knowledge and skills learned. The residual effect will be a student who looks forward to coming to class, saying, "I need to do well because Mrs. Smith thinks I'm important." The student learns pride in her work and success. She thinks, "I can do this!"

I worked part time for the New Jersey Center for Civic and Law-Related Education. That center encourages teachers to utilize dynamic activities that engage students in the process of learning civics. One of my favorite activities is called a continuum. It is really a spectrum of opinions. The teacher presents a controversial issue or an interesting question with a yes or no answer that relates to the students' life experiences. "Should a curfew for teenagers be established in a town due to the increased level of crime and drug activity late at night?" was one of my favorite continuum questions. Students who support the question wholeheartedly stand on one side of the continuum line, and students who oppose it strongly stand on the other end. Students judge where on the line they should stand. This is a kinesthetic, thinking activity that forces students to listen and evaluate their friends' opinions. Students love it and remember it for weeks.

When I taught high school social studies, students knew that we would form a circle to discuss a controversial issue, several times a month. I always chose a topic related to the curriculum, looking for one that might affect their lives personally. They needed to think and discuss fairly, and respect their classmates' opinions. The discussion was always followed by a writing assignment. It sure did beat a boring worksheet.

Boring busywork centered on worksheet after worksheet only encourages discipline problems. All children of the twenty-first century have been raised in a culture of instant gratification and the entertain-me-now mind-set. Computers and video games over stimulate children with fantastic imagery and nonstop action. Inner-city children are surrounded by more stimuli with loud, violent, crowded streets and sometimes broken, dysfunctional families. Boring, watered down curricula dull the senses and can be very frustrating. When frustration reaches a boiling point, major discipline problems occur. We create our own monsters with a dull curriculum, which leads to ineffective instruction. If qualified, sensitive teachers are the soul of classroom management, then a great curriculum guide implemented by those great teachers is certainly the brains behind effective classroom management.

I have observed many classrooms in my career, and I have noticed a disturbing trend in classroom instruction. Too many teachers pander to their children instead of educating them—"We are having a quiz tomorrow, but don't worry. I didn't make it too hard. In fact, here are the two essay questions that you should study tonight. Tomorrow, you will have twenty minutes to study before you take the quiz. If you are having trouble, you can open your textbooks for the last few minutes of the test to check your answers." If I were a student in that class, I would never study. Why should I?

Life can be tough. Work can be hard. Start studying.

In addition to pandering to students, we have substituted quantity for quality in the work we accept from students. Some say, "I did all the worksheets you gave us. Why didn't I get an A on my report card?" Class worksheets carry the same weight on a report card as tests. How can that be possible? The quality of a student's work must be valued higher than the quantity of worksheets he submits. Will their bosses give him worksheets when he is an adult, or will he be expected to think, solve problems, and create? Let us choose quality as our yardstick for success.

Parents play a major role in reviewing their children's schoolwork. If someone's children receive an A on a report card without seeing substantial work, then something is wrong. Your children should not receive an A just because they are well behaved and complete their schoolwork. They should earn an A because they have mastered the material and have demonstrated that mastery in projects or tests. If parents are only interested in their children getting A's so they can get into a good college, then that mentality will only damage their child in the future. Sooner or later—in college or a career position—they will need to demonstrate their abilities, which does not mean completing ditto number fifteen. Parents are stakeholders, and they have a choice. Parents need to fight for quality of instruction.

If we are to improve instruction in inner-city schools, then we must accept certain prerequisites. Poverty and racism have combined to destroy many families in inner-city neighborhoods. Those broken families are faced with the challenges of guns, violence, drugs, and abuse. If we accept those facts as unfortunate truths, then we must admit that those families may be less stable, structured, and disciplined.

Children from those families spend most of their days at an institution we call school. That school must be more stable, structured, and disciplined than the home. Stability, structure, and discipline mean functionally strong schools with quality curriculum and instruction. Students need to look at the clock and say, "Wow, where did the day go?"

# Messages to the Stakeholders

## Choices

A beautiful, intelligent Hispanic young woman prepares to go to school on a very cold morning. She is a member of a warm, loving family that came to America to pursue the dream that millions before them have chased. Her doting father goes into the garage to warm the car before taking his children to school. Moments later, she enters the garage to find that her father has been fatally wounded by criminals trying to steal the family car.

The many problems facing inner-city schools are as interwoven as a spider's web. No one problem can be fixed without addressing them all simultaneously. Therefore, all stakeholders in the process of educating children must do their appropriate parts to solve the problems. Those stakeholders include federal and state officials, all educators, parents, students, the media, and community members. No group can escape part of the blame, and all groups must work together to solve the mess. No one can point fingers unless they point at the mirror. All stakeholders have choices to make concerning the education of our next generation of American citizens. These are complex choices when one discusses the problems of inner-city education.

The essence of a democracy is opportunity. All citizens, regardless of socioeconomics, race, or ethnicity must have the opportunity to achieve the American Dream with hard work and determination. That is one of the cornerstone beliefs of our country. Thomas Jefferson told us that the pursuit of happiness is an inalienable right bestowed by our Creator. It is the mandate of all stakeholders to ensure that the playing field where youngsters struggle to reach the American Dream is level and fair. There is the profound belief that all men are created equal under God and implicit with that belief is the concept of equality of opportunity. What people do with that opportunity is their choice. All stakeholders must work together to make sure that all children have a fair and equal choice.

The battle for the American Dream can be found in the daily struggles of individuals and families who strive to make a living so that their children can have a better life than them. It is the game of American life. All of us must believe that we are viable members of the game with a legitimate chance at success if we work hard and obey the rules. American children from functioning families who attend effective schools know the rules and work hard to succeed in order to please themselves and their parents.

As I have previously stated, there are children in middle and high school in the inner city who know that they are not in the lineup of life. They are not a member of the team striving for the American Dream. If a child in the inner city doesn't feel like he is part of the system growing up, he will become an apathetic adult, a career criminal, or a dysfunctional ward of the state. It is the responsibility of all the stakeholders to put every child on a team that is fighting for the American Dream. Schools, communities, politicians, parents, and the media must do their parts, and likewise the students must respond with responsible behavior of their own. People who choose not to join the struggle for the American Dream have no one to blame but themselves. Opportunity is implicit in the birth of every American, and both the powerful and the neglected have choices to make when opportunity knocks.

More importantly, if all citizens don't believe in the opportunity to succeed, then the democracy is in danger. If the public school system fails to prepare young adults for that challenge, then we will have a country of the few rich, a struggling middle class, and the permanently poor. The gap between rich and poor is widening in this country and that is partly a result of our failing public school system. Some children in inner-city schools can barely read by twelfth grade, and they have very few skills in a competitive, information-age world. They have no stake in the system. They get passed from one grade level to the next without the skills necessary for success. Whatever fancy term we give to the pass-them-along system, it is still social promotion. Social promotion moves students through and out of the system and keeps the retention statistics on the school report cards palatable.

Some wealthy and some middle-class students know they must succeed in school so they can go to college and get a good job. It's what their parents want, and it is the way to live well. If I am thirteen years old in the seventh grade in an inner-city school and can't read or write, what are my chances of success? I am not in the game of life. I'm a spectator in the cheap seats. Why should I stay? Why should I behave? There were days when I would sadly watch my classes work silently. Are there students sitting in front of me who are already doomed to a life of dependency and failure? Who has the best chance to survive in the real, adult world?

It is long past due for our stakeholders to make some choices. It is time our country made the education of all children its top priority. Nothing short of national security should bump educating children from the top of our priorities. The compass of our country must always point to *C* for child. If the country doesn't want to take the time or reserve the resources to prepare the next generation of leaders, then what kind of country will we have in the near future?

I believe our country has lost its moral purpose. We have become slaves to materialism and sensationalism. Drug- or alcohol-addicted celebrities are our teenagers' heroes. We need documentary reports on television that tell parents that it is good for them to prepare a family meal that is attended by all members. Our children demand the latest in fashion, and we buy it regardless of appropriateness of the outfit for the age of the child. As the Paterson principal who became famous for his bat and bullhorn said, "If you have Calvin Klein on your behind, you have nothing on your mind." (I disagreed with everything else he said.)

Several years ago, I saw a picture of the Washington Monument with scaffolding surrounding all its sides. If our inner-city children are to succeed, they must be like that monument—strong and with a solid foundation of scaffolding stakeholders to support them. If our society truly wants inner-city children to succeed, then it must be the scaffold and grow with the child as she struggles through life. It is our responsibility and duty. Students must respond to the stakeholders and utilize the resources provided by the scaffold to be successful. It is their responsibility and duty. Everyone must make choices.

No one should expect inner-city students to be superheroes in order to succeed in life. That is totally unfair. Students must however, recognize that education is the vehicle to escape the horrors of poverty, drugs, and crime. They must also recognize the dangers of the streets and the lure of quick money or superficial happiness. All stakeholders must support students as they try to recognize both facts, but the students themselves must bear their share of the responsibility. If there are community, religious, or governmental programs to support students, they should go for it. They shouldn't wait for the programs or blame the lack of those programs for their decisions to not work hard in school, stay out all night, or participate in activities and scenarios that are dangerous to your health and chances at success. It takes more than one person or entity to form a relationship. Students must respond to the network of people and groups who are helping build their support scaffold. When the network provides students with the advice and materials for success, they must not throw them away.

Students in successful schools pay attention in class and participate in lessons, activities, and programs. They do that because they want to be in the game of life and they want future financial and personal success. They don't do it because they are white. For students to choose not to take notes, pass tests, or do homework because they might "act white" is absurd and self-destructive. Some people will deliberately keep children out of the game of life for selfish or racist motives, but for them to choose not to enter the game gives those racist actions validity—"See. Even they don't care about themselves."

It doesn't take a grown white man like me to tell any minority student that life will difficult for them. Socioeconomics and racism hinder many minority children from birth. That child always seems to play catch-up throughout his or her life. As the census figures illustrate very clearly, in this country, poverty is centered on single women and children. If a young, inner-city teenage male engages in casual, unprotected sex, then his actions are irresponsible and criminal. You are committing a crime against society, your female partner, and your ethnic group when you impregnate a woman and then walk away from your responsibility. Yes, it is much worse for an inner-city male to perform that deed than for a middle- or upper-class male who commits the same act of irresponsibility. Your actions have deliberately put the child further behind at the start of life and actually helped to keep that child out of the game of life. It is immorality at its lowest point. The immorality has nothing to do with the morals of having sex at an early age. That is a private choice for you. However, when you know how difficult it is to succeed in the inner city and carelessly place your child even further behind where you started, that is even more egregious. Shame on you and the choice you made.

Our country is a country of immigration. The origin and language of the immigrants have changed throughout history, but they all come to America looking for a better life. My grandfather came to America from Italy at age thirteen. He barely had enough money for a cheap ticket on the ship of dreams. His voyage was so bad, he never returned to visit his native country. He began working soon after entering the country and knew that he had to learn some English in order to work in the country. How else could he communicate with his bosses?

Immigrants today have the luxury of quicker travel to and from their native country due to the invention of the airplane. They can also purchase phone cards to call family that was left behind. Unfortunately, that leads many new immigrants to speak of their native country as "my country." Immigrant children who attend American schools and whose families plan to live and work in this country must consider the United States their country. How else can students succeed educationally, socially, and economically if they do not even consider the place they live to be their country? They need to choose America and become American so they and the schools in the inner city have a chance for success.

The streets are the enemy of any inner-city family trying to succeed in life. There are many stories of families who have been hardworking citizens and still lost a child to the street. I never encountered a better family than the one mentioned in the opening narrative of this chapter, and yet they lost their father. The streets are predatory and relentless.

All stakeholders must assist young children to choose work, family, and school over the lure of the streets. Again, young people need to assume some of the responsibility. They need to learn the difference between right and wrong at an early age from a parent, grandparent, guardian, or any role model. Some young people in the inner city do not have one or two parents but do have a responsible adult to teach them right from wrong. That is why the stakeholders must be as relentless as the streets in fighting for and with young people. As I have said many times, the problems of inner-city schools are complex and interwoven like a spider's web. All problems must be addressed simultaneously by all stakeholders if we have any chance of success.

Additionally, health problems plague many young people living in the inner city. Environmental factors such as dilapidated housing can cause diseases, such as asthma. Poor eating habits can lead to hypertension and obesity. Alcohol, drugs, and guns can be found as easily as a candy bar. Those factors are all part of the stranglehold of the streets. Students can't assist the streets in their own destruction. They need to clean their rooms and their clothes. They need to learn about proper diet, even if it means cooking for themselves. They need to participate in gym classes and other exercise programs found in places like a Boys and Girls Club or a YMCA. They need to be proactive against the streets and not be willing volunteers. There is a fine line between being a victim and a volunteer. Students shouldn't point a finger at other stakeholders for their problems, if they have not done their share.

It is totally un-American for any young person to be kept out of the game of life. No young person should willingly drop out of the game for the short-term enjoyment of the streets. If he wants to be successful, he must be in the mainstream game of life. Where are the successful jobs, the good schools, and the safe streets? Where does he want to live? He needs to dress and talk like he wants to be successful. Why have young people modeled themselves after criminals by wearing their pants below their butts? Why would a business owner want to hire a young person after he has seen his underwear? The child needs to comb his hair, take the rag off, and hold his hat when he enters a building.

There is one English language. Rules for grammar are found in English textbooks in all high schools in America. There aren't different rules in New York, Chicago, or Los Angeles. If children expect to work in a successful, mainstream business, they need to speak mainstream American English, not some illiterate nonsense some people pass off as a cultural language. We must promote success, and success means speaking in an educated manner.

Students do not need to give up any of their cultural enjoyments because they speak English correctly, dress appropriately, and comb their hair. No one is asking them to give up their roots or heritage. One of the definitions of success is the ability to function in the majority, mainstream culture. Dressing and talking "ghetto" defines a student as a ghetto resident for life. Why limit himself to the environment that may be destroying him? He needs to expand his horizons and join the game of life. Students have difficult choices to make in life. There are people willing to support their struggles, but they have to bear their share of the work and responsibility. Students need to choose life, not instant gratification.

What do good parents do for their children? A father is not someone who simply makes babies. That is the easy part. There is a major difference between a male who makes babies and a dad. Most women, likewise, can reproduce. Not all women who reproduce are mothers. Adults who choose to become parents must know and understand the lifetime responsibilities of parenthood. A student who chooses to become a parent is a fool.

Real parents read to their children from birth, set parameters for their children's behavior, and fight for them and with them on all important issues. Real parents know where their children are at all times of the day. They watch what their children eat, what they watch on television, and what video games they play.

I will repeat that all stakeholders must help and be parents, if necessary, in their fight for children. The phrase "It takes a village" is never more meaningful than in the inner city. Parents in or on the cusp of poverty may have to work two jobs or work second or third shifts at a factory. Who watches their children? We all must watch their children. That means school, church, and community programs. It means the neighbor next door, or an aunt or uncle.

My parents were raised during the Great Depression and had many siblings. Yet if they misbehaved, a neighbor was sure to either cuff them or tell their parents. Police walking the beat knew all the children in the neighborhood and recognized when a child was near danger or in danger.

The streets have so terrorized people in poverty that they live in isolation and fear their neighbors. We need to regain and reform our neighborhoods so neighbors can act like extended family members. Neighbors and friends do not have a choice. They must be that surrogate parent when a child needs one. The spiritual and economic growth of our country depends on the success of our inner cities. We can not continue to have pockets of poverty that feed upon themselves generation after generation. The loss of talent and the draining of funds will destroy our country. We are all in the same national boat. Every American must be a parent to a child who needs one. We can not continue to sit on our porches and watch children go astray. Community police and the clergy must emphasize this point to all citizens.

It is a dangerous, scary world for all children to live in today. When I was ten, I could go to a park, play all day, and just make sure I was home before the streetlights went on. Today, parents can send their children out for the day and never see them again. Some parents are trapped in their own worlds and do not know what is going on in their children's lives. Other parents become helicopter parents, hovering over their children at all times and swooping down at the first sign of trouble. Helicopter parents have great motives, but may inadvertently hurt their children in the end. All children must be taught self-reliance. That is essential for future success. Parents need to make sure their children can solve appropriate problems at a young age. They need projects or chores to build their sense of responsibility. Parents should give them situations in which they must learn self-reliance. Children must assume some of the responsibility for their success in school, but that ability to be responsible doesn't grow overnight. Parents need to raise their children to be winners who know how to walk, talk, and dress like winners.

Everyone must fight for our children and fight with them when necessary. I once had a high school student who drove me nuts over assignments and grades. I would not give up on him, but he sure pushed me to the limit. We never really had many friendly conversations, because I was always in his face when necessary and he was always complaining or making excuses. At the end of the year, we all signed each others' yearbooks. The comment he left on his picture for me was very simple: "Don't change." That comment hit me like a bolt of lightning.

We are all parents to any children who need one.

As a grandparent, it is very easy for me to give into every one of my grandchildren's requests. I do not have the same responsibility as their parents. Parents must say no to some of their children's requests. They must fight for their children and fight with their children at certain times. Parents are the adults.

It is not right for children to be in the streets late at night, especially on a school night. I have driven home from board of education or PTA meetings and watched ten year olds in the streets after ten o'clock at night. That makes no sense.

It is harder for parents in the inner city to be good parents because the dangers are greater. Their children will complain and scream, but they will love their parents more because they stood up to them and the streets.

I worked for a state-appointed superintendent of schools who decried the fact that inner-city children did not have computers at home and had to do their homework at the kitchen table. Where did he think I did my homework as a child? Is there any better visual than a parent sitting with a child at the table doing homework? Isn't that the time for conversations about the day?

Parents must talk to and with their children. "Our neighbor told me he saw you with that boy who always gets into trouble. Why are you playing with him? What did you do after school?" Remember, it is a parent's responsibility to know where her children are at all times. If she does not have the time, ability, or desire to be a great parent, then why be a parent at all? If she is so young that she does not even know who she is, then how can she be that great parent?

Parents are the primary stakeholders in the education of their children. Their choices are first and paramount in their children's lives. Their support is the first floor of the scaffold. There is no manual that guarantees success as a parent. However, all adults who conceive children must make the choice to work hard to raise those children to the best of their abilities and to fight the forces that will tear down the support scaffold around a child's life. People who conceive children with no thought at all have already made their choices.

Parents need to make the right choice. They need to be a fighting role model for their children. They need to be great parents.

Are all forms of media out of control and only interested in money, ratings and sales? Have sensationalism and greed overtaken truth and integrity as motivation for the media? The quick story to beat the competitors gets the front page rather than the truthful, insightful story. The merit of the story is secondary to its immediate impact.

The common good is a term that has been abandoned in the race for ratings.

The common good is an essential theme for a democracy. We give up some of our personal freedoms to join a society that works for the good of all. I am a stout believer in the first amendment to our Bill of Rights, but selfish people must not use that amendment to advance their own greed or interest. All media have a responsibility to the common good. Does the story have merit, or does it just have shock value?

Media moguls are stakeholders in the education of children in the inner city, and they have many choices to make for the common good. Do they want to see a new generation of qualified, informed citizens purchasing their products, or do they want to make the quick buck regardless of ethics? We are asking children in the inner city not to be influenced by the lure of easy, quick money from drugs and violence. Should we not expect the same of the multimillionaires who run our media?

Everyone needs to make choices to break the culture of incompetence in our schools. All stakeholders must attack that culture at the same time and with the same vigor. If teachers try to improve quality instruction but do not get support from the media, businesses, and parents, then the scaffold of support crumbles.

Television and video games can be babysitters for inner-city children before and after school. Adults believe that, as long as the children are off the streets, they are safe. What are the children watching and playing? They are watching shows and commercials that glorify the drug and alcohol lifestyles of the heroes of pop culture. They watch "reality" shows that create heroes out of the most dysfunctional celebrities. Reality must show hardworking, caring parents who strive to provide their children with better lives. Reality must be children working hard in school, believing in the system, and willing to forego easy money. Yet there are many so-called reality shows where self-destructive behavior is championed. No child should watch that nonsense, especially not children who may not have adult role models. Should the latest celebrity who ditches rehab be their hero?

Sports heroes have a major responsibility in our society. Whether they like it or not, they are role models for children who fantasize their future lives to be the lives of sports heroes. The number one sport for children in the inner city is basketball. How can we expect young adults to dress, talk, and walk like successful people in mainstream society, when they watch professional basketball players dress and talk like members of the ghetto culture? Can children in the inner city interview for a job dressed like a "gangsta"? Can we please stop the after game press conferences with rags on our heads and our hats on sideways? Where in the business world do you see people dressed like that? Are we deliberately cultivating and glorifying a culture that can be harmful to young adults in the inner city just to make money? One in a million young boys will ever play professional sports. Yet how many young boys will talk, dress, and act like their sports heroes?

Million-dollar athletes have a responsibility and a choice. If they influence young children to look and act "ghetto," then they condemn those children to a life in the ghetto. Only one in a million will get out to play professional basketball. What will happen to the rest?

We must encourage young adults to dress properly and not to wear their pants so low that their underwear is visible for everyone to see. I walked into a clothing store during a recent Christmas holiday season, and there was a mannequin with its pants deliberately low enough for the mannequin's crack to be showing. Crack kills in more ways than one. How irresponsible was that company for that type of marketing? What choices have they made for the good of the country?

Business owners complain about the young people who apply for a job with little or no skills. They also complain about the number of young people who drop out of school, loiter around, and shoplift in their stores. Those owners have a responsibility to not pander to the latest fad just to make a quick buck. If they want responsible citizens who can work and shop in their stores, then they must be active stakeholders who support the efforts of people who want to break the culture of incompetence in our schools. No person or group can be left on the sidelines if we are to break the cycle of poverty and despair.

In the past, I have called newspapers to cover an event that illustrated hardworking, young adults doing positive things for their communities. I called, but reporters rarely showed up. For many years, one high school in Paterson regularly won the state National History Day competition. Because of one victory, they traveled to Washington DC for the national competition. While I sat with thousands of other high school students at the awards ceremony, I was struck by the lack of diversity in the crowd. Our inner-city students went to bat against the best schools in the country. Do you think that would make a good story?

I introduced the Academic Decathlon to the Paterson district. Again, this was a competition where young inner-city students matched their abilities to read, research, and think about a common topic against the best schools in the state. Don't you think the common good would be served with a lengthy story about their struggles to study, and earn money for the competition?

Media and businesses shouldn't complain about the quality of young adult that leaves an inner-city school if they shirk their responsibilities as stakeholders in the process. They are willing volunteers in the culture of incompetence rather than a positive member of society, trying to destroy it.

Let that same newspaper get a call about guns in schools, and they will send a team of reporters and cameramen. Are they choosing what news stories are good for the community, or are they sensationalizing and creating the news to sell newspapers? Who makes choices like that? Do they really care if children succeed in school and in life? They have a choice to de-emphasize cheap, quick stories that just sell more newspapers. Again, I say that if we are asking our young people in the inner cities not to seek the cheap thrill, we should expect the same from our media moguls.

Newspapers need to follow a positive story that illustrates productive young citizens. The children need the boost, the community needs to know, and it is the newspapers' responsibility as stakeholders to build the scaffold for students to climb out of the poverty of the inner city. We don't need a simple picture with one minority student holding a plaque. We are talking about quality journalism that tells a meaningful story for the betterment of the common good. Remember those types of stories? Let's have stories that glorify the struggles children overcame in order to succeed. Let's have stories in which young adults relish the joy of hard work.

The media has a choice and a responsibility. If it does not want to provide positive stories about quality individuals, then it has chosen to abandon children and create a new generation of young adults who will commit crimes. Then it can perpetuate the stories it really wants to talk about. Shame on the media.

African-Americans have been the one minority group that has suffered the most poverty and discrimination in history. That is no secret or historic discovery. Their history is a story of the middle passage, of slavery, of Jim Crow, and of migrations to Northern cities. Poverty translates into life in the inner city, poor health, and inferior schools.

There have been some major strides in the struggle for economic, political, and social equality. There is a significant black middle class, and it has the opportunity to reach out and pull some family and friends out of the cycle of poverty. That cycle breeds some of the negative factors that have crushed our schools. Breaking the cycle of poverty is one of the main strategies to destroy the culture of incompetence in our poor schools.

African-American students, parents, and business leaders can read about the choices that have already been discussed in previous pages. They must also recognize the uniqueness of their people's historic dilemma and make important choices. There is still systemic racism in this country, but there is a larger group of people who want to see African-American children succeed. They want all children, regardless of color or ethnicity, to have equality of opportunity in housing, jobs, and schools. One hundred percent of all our energy should focus on providing the new generation of young African-American children with a quality education.

There are only a certain number of hours in the day; there is only a certain amount of energy in our lives. African-Americans can use that time and energy to work together with other concerned groups toward the same goal, or they can use their time hating white people for what happened in the past. The culture of incompetence in our schools will not be destroyed if they spend their precious time hating everything white. It is their choice.

I volunteered to be a member of a community group to work on strategies to improve education. It was after school, and as a volunteer, I was not financially compensated. That fact never bothered me, because many teachers donate their time on after-school committees. I attended the first meeting, and a fellow member—an African-American female—challenged me, saying that I could not possibly understand the problems of the community because I lived in a white suburb. Her anger and resentment were palpable. She made a choice to use her time and energy for that tirade instead of working positively for change. I will never feel or understand the pain that person may have encountered during her life in the city. It would have been be presumptuous for me to say, "I understand." However, her body and soul were facing backward not forward.

I also tried to help an excellent, white female teacher who was being accosted by an African-American parent, again claiming that she could not teach her child well because she lived in a white suburb.

Is it time in our history for all Americans to eliminate the word "race" from our national consciousness? Race is a physical description that is rather obsolete. If we discard race, will racism as a term be next for the trash? Children who need quality education are human beings first, and deserve great teachers just because they are human. We sometimes forget that teaching is about one wiser human being influencing the life of an equal, though younger, human being.

Aren't the words "culture" and "ethnicity" more accurate and positive ways to celebrate our differences and similarities? We will soon find out that we are more alike than we are different. How many times do we walk into an upscale restaurant to find a "new" item on the menu that momma made for us many years ago because families were large and money was short? Whether it is pasta *e fagiole* for Italians, chitterlings for African-Americans, or *arroz* and beans for Hispanics, all groups have shared and still share the same miseries and glories of history. Parents—all families fight the same fight. They should join together with other people who want their child to succeed. Great teachers have no particular skin color, religion, ethnicity, or gender. They have soul, dedication, and passion. Every child needs great teachers.

Politicians are the scapegoats for many of society's problems. We blame politicians for not doing enough, and then we blame them when government gets too big, because they have tried to solve problems. Money is the only solution for many politicians. It looks good for reelection when politicians can show their constituents that they brought this program or that project to their district. Money is also the most expedient solution used by many politicians. Cable news has reduced many interviews with politicians to thirty-second sound bites. Therefore, politicians who almost always are running for something must get their message out quickly, and quickly usually means something that is not thought out in depth.

Money is an important solution to the crisis in inner-city education, but it is one of the many solutions that must be initiated at the same time. The term "Good money after bad" fits well if you send lots of money to a district without changes in leadership, with commitment from the community, and with improvements in curriculum and instruction. The problems have deep roots; therefore, the solutions must be implemented with vision and depth.

Depth is a difficult word for politicians who need to run for reelection every two years. Yet depth of knowledge, insight, and purpose are necessary to attack the culture of incompetence that has festered for generations. A program here or a program there will not solve the problems. Money here or money there will not solve the problems. Politicians need to be statesmen or stateswomen and listen to all stakeholders, examine all options, and act like all children in America are their own children.

The purpose of public service for a politician should be to make the country a better place for all Americans. What better way to change the country than through improvements in education? Schools are the vehicle for a better tomorrow, because today, we are training the people who will run the country in the future. Politicians need to have some foresight to examine the issue of education in depth and over the long haul. There are no quick fixes, but there are some difficult, significant choices that all politicians can make.

Teachers are held "accountable" by many stakeholders for the failure of education in inner-city America. Since the problems are so deep and intertwined, everyone should only point a finger of blame at themselves. Often, people who want to blame teachers for all problems have fuzzy but fond memories of sitting in a classroom as a child. Everyone sat quietly in a class of forty-five children back in 1945. They never dared to talk back to their teachers, because, if a note went home, their parents would smack them before even reading it. They did their homework, and they passed. The older we are, the clearer those dreams seem to be.

The structure and types of families have changed; teachers don't receive the respect they should; and those wonderful schools are now sixty years older. That does not absolve schools from their responsibilities for the successes and failures of our education system. They have their fair share of the blame for the failures and of responsibility for the change.

All stakeholders must work together to provide every young child in America today with a fair, equal educational opportunity. The schools definitely can not do it alone in today's world. If politicians must be the public's bully pulpit for reform, then the schools must absolutely be the vehicle to drive the reforms from the drawing room to the classroom. Schools are the foot soldiers for change.

My daughter is one of the best teachers I have ever met. She and her friends face the debilitating effects of negative social forces every day. They see children coming to school in the winter without a coat. They know that some inner-city children come to school on Monday morning hungry because they had little to eat all weekend.

Schools can also be the laboratory for positive change. New curricular programs are piloted in an inner-city classroom, not in a politician's office, a business luncheon, or a university conference room. Schools must be the focal point for change, because they are also the lightning rod of the argument. All schools have choices. They can fight for their children and strive to be living schools. If they remain stagnant and static, then that is a choice as well.

All inner-city schools must analyze and prioritize the needs of their buildings, students, and staff. I am not talking about a silly, long, bureaucratic form provided by a governmental agency in order to seek state funds. I am talking about a down-to-earth document created by the people who live it every day—teachers, administrators, parents, and students. "Classroom 105 has a leak in the ceiling"; "the seventh grade needs new math textbooks"; "We need to have better curricular dialogue in the building"; and "How can we help the increasing number of Pakistani children entering the school"—these are the types of topics and concerns each school must address on their own. Administrators must be the leaders of these discussions, but if an administrator is an incompetent, political hack, then teachers on the same floor or at the same grade level could develop the dialogue. It can be done, because I have seen it done. Schools shouldn't make the choice to do nothing.

The entire school community must determine which factors it can control and which ones it must fight with the government, business, and community agencies for its fair share. The school community must maximize its positive features. An inner-city school may be located near a museum, a park, or a university. Reach out to those institutions for curricular, instructional, and volunteer assistance. Use the park as a science lab. Can the students identify the trees and bugs living right next to them? Paterson is famous for Great Falls, the second largest waterfall east of the Mississippi River. How many students have visited Great Falls? The age of the school may be an asset. Some old schools still have sliding walls, which means that teachers can team teach with large and small rooms as needed for instruction.

Teachers and administrators need to keep their rooms and schools clean. I have seen schools where children are allowed to eat, drink, and chew gum all day long. Then at the end of the day, the last class just throws all the junk out of their desks onto the floor for the janitors to sweep and clean. Yuck! When those children work as adults someday, will they be allowed to act that way with their office desks? Maybe that is one reason why so many students can't keep good jobs when they are older. It is the parents' responsibility to make sure their children keep their rooms clean at home, so they have some experience to keep their classroom clean and neat.

If I walk into any classroom, there should be evidence that children are working in that room. Teachers should hang students' work prominently throughout the room. Bright colors will help erase dingy corridors in the building or in the neighborhood. Students should create their own bulletin boards. Ownership and empowerment will carry over into their work at home and at school.

Teachers should demand that their rooms be cleaned every night. There is no excuse for everyone in the building not to be doing their assigned jobs. If any stakeholder accepts the culture of incompetence, then we have made the choice to perpetuate atrocious conditions in the school. Rooms should be swept, garbage emptied, and boards washed. Can we expect children to work hard in a room that has not been cleaned in days? Again, it is an example of "They don't care about me, why should I care about me?"

Newspapers and paperwork can make a classroom and school look chaotic and sloppy if they are allowed to be left on floors or windowsills. Every school should have a recycling program for paper and bottles. Is there a better way to teach responsibility and citizenship? Is there any better way for a school to prove that it wants to be a living school than with examples of building a better environment for tomorrow? Maybe schools can use the money from the recycling program to plant flowers or plants in or outside the school. Teachers and administrators need to make the choice to have their school live.

# Epilogue

## Is It Brauhway or Broadway?

Education is simple yet profound. Parents literally or figuratively walk their prize children to the schoolhouse door every September, and say essentially, "Here. They are yours for the next year. Treat them with the same kindness and respect that I do, and teach them to be better people and citizens. Teach them about our great country and its promise of equality and opportunity. Teach them to get along with one another and to judge people by the content of their character and not the color of their skin. Teach them to be responsible yet compassionate adults."

I was blessed to become a grandpa a few years ago, and the feeling is indescribable. If you are a grandpa, or hope to be one someday, think about the children in dilapidated schools with old books. Think about them walking home through even more dilapidated streets to get to a home where someone may or may not be there to care for and love them. If we are all our brothers' keepers, then, in a real sense, we are all someone's grandpa or grandma. We do have a social, moral, and ethical responsibility to one another, especially to those who can't fend for themselves. Be a grandpa or grandma to someone who needs you. Be a great American.

All Americans can use the bully pulpit to demand responsibility from all the stakeholders toward the education of children. It is wrong for children to make babies, and we need to tell teenagers that. It is wrong for parents not to know where their children are or for parents to allow them to walk the streets at all hours of the day and night. Politicians, clergy, parents, and teachers need to make those speeches.

Teachers must be professionals at all times and reform themselves when necessary. They must revile the Donut Dans in their teacher lounges as pariahs rather than as prophets. Teachers need to be compensated for quality work, but they must recognize that, if salaries do increase substantially, it is their responsibility to keep their professional houses clean. No stakeholder should accept the culture of incompetence, but teachers must be at the front of the line of protest. They work with the children every day and see the waste of human lives each year.

Politicians must be the vehicle that demands accountability and cooperation from all stakeholders responsible for the education of our future adult citizens. They are representatives of Americans and are, therefore, are also responsible to all stakeholders. If inner-city education is to be changed, then all stakeholders must be aligned and fighting the same fight.

Teachers need to stand up for their students. They must not accept the status quo. They need to be an advocate for their profession—it is a noble, righteous one. They should not prostitute their teaching certificates to politicians, the streets, or expediency. Teachers need to choose the road less traveled and fight for the children who do not have the power to fight for themselves.

Teaching is a unique profession filled with hope and despair. There were many stories of frustration in this book, but I believe I balanced them with examples of hope and joy. If someone loves children and believes in their abilities, that person should become a teacher and feel the ecstasy. Hope does spring eternal as long as there are new generations of young teachers willing to put their courage and enthusiasm on the line for the neediest of children.

When I was a supervisor of social studies classes, I moved around the city on a daily basis. I met many former students who still lived and worked in the community. I almost ran into two lovely former students, who were driving their car in the downtown area. They stopped short and yelled, "Mr. C, we need directions." They had always been hardworking, pleasant, young ladies in school, and I was pleased to see that they had retained that charm. I gave them directions, which included a turn on Broadway. They roared hysterically, and I asked them what was wrong?

Still laughing, they exclaimed, "Mr. C, there's no *D* in Brauhway."

I hope and pray that those two women will maintain their innocence and charm for their entire lives. Teachers just have to love children and want to work with them to improve their lives. That is the joy and hope of education.

When I was still teaching in the performing arts high school, there was a move toward site-based management and budgeting. I was a member of a committee to plan for the following school year. One teacher wondered how many students in our performing arts school had ever been to Broadway to see a hit play. We decided to raise money and send our children to Broadway. We chose our play, *Bring in 'Da Noise, Bring in 'Da Funk*, did some additional begging for funds, and one day, sent three hundred students, faculty, and community members to Broadway on seven buses.

Since it was an expensive trip, we wanted to make a full day out of it. That morning each major teacher— there were seven majors that students could audition in for acceptance into the school—selected a school or landmark in Manhattan that was relevant to the students' future professions. All seven buses were to descend on Times Square for lunch and a walk to the theater.

Needless to say, I was very nervous. Would the students all get to eat and arrive at the theater on time? But as my bus entered Times Square, I could count the other six arriving from different directions. We all had had time to eat and form a long line outside the theater.

It was a spectacular day, and one that those aspiring artists and performers would never forget. There was enough joy, hope, and enthusiasm for the world coming from those buses on the way home from the theater. That's what makes the broken school windows, bad heat, and old books bearable.

I believe a task force on education should be established by the president of the United States and funded by Congress. That task force should include representatives of all stakeholders in the future of our educational system, specifically the inner cities. A plan with the significance and scope of the Marshall Plan after World War II must be created and funded. The plan must be supported by all Americans with the same enthusiasm as when we backed the trips to the moon in the sixties. Poverty, poor health and nutrition, low expectations, old facilities, single-parent families, racism, politics, and many more issues must all be open to discussion.

Let's not think about the quick fix, because there is none. Let's not decide on singular choices of money or accountability, responsibility or support. Tackle all problems and solutions simultaneously.

As I have said repeatedly, the problems facing inner-city schools today are interwoven like a spider's web. Throw money or time at just one, and the other problems will entangle and destroy any positive gains. We must address every problem that faces the education of young, poor children with equal vigor.

Let's not think about which group likes or distrusts what group. Let's leave our personal issues, prejudices, and hatred at home and think about this generation of young American children entering school. Everything needs to be pulled out from under the rug, and everything must be put on the table. Let's stop pointing fingers at one another and join those fingers with other people who believe in the promise of America as stated in our Declaration of Independence. If we can't prepare our children for the future, what is our reason for existence as a nation? Let's break away from the culture of incompetence by making intelligent choices aimed at educating all children in America equally and fairly.

God bless all parents who still believe that the American Dream can be achieved through a quality education. God bless all teachers who work long hours and buy chalk and markers out of their own meager salaries, because they believe that all children can learn. God bless community and business members who believe that they have a responsibility to the common good. Finally, God bless all children who fight monsters in the streets every day to learn algebra, physics, and the history of this great country. Especially, God bless the children of Paterson, New Jersey.

Secondary Side

**CHECK FOR BANDING OR STRIPES IN THE PRINT AREA**

Secondary Side

**CHECK FOR BANDING OR STRIPES IN THE PRINT AREA**

Secondary Side

CHECK FOR BANDING OR STRIPES IN THE PRINT AREA

**CHECK FOR BANDING OR STRIPES IN THE PRINT AREA**

Secondary Side

**CHECK FOR BANDING OR STRIPES IN THE PRINT AREA**

Primary Side

Made in the USA
Lexington, KY
02 May 2010

munity's domain, the folkmote retained its supremacy and (as shown by Maurer) often claimed submission from the lord himself in land tenure matters. No growth of feudalism could break this resistance; the village community kept its ground; and when, in the ninth and tenth centuries, the invasions of the Normans, the Arabs, and the Ugrians had demonstrated that military *scholæ* were of little value for protecting the land, a general movement began all over Europe for fortifying the villages with stone walls and citadels. Thousands of fortified centres were then built by the energies of the village communities; and, once they had built their walls, once a common interest had been created in this new sanctuary—the town walls—they soon understood that they could henceforward resist the encroachments of the inner enemies, the lords, as well as the invasions of foreigners. A new life of freedom began to develop within the fortified enclosures. The mediæval city was born.[1]

[1] If I thus follow the views long since advocated by Maurer (*Geschichte der Städteverfassung in Deutschland*, Erlangen, 1869), it is because he has fully proved the uninterrupted evolution from the village community to the mediæval city, and that his views alone can explain the universality of the communal movement. Savigny and Eichhorn and their followers have certainly proved that the traditions of the Roman *municipia* had never totally disappeared. But they took no account of the village-community period which the barbarians lived through before they had any cities. The fact is, that whenever mankind made a new start in civilization, in Greece, Rome, or middle Europe, it passed through the same stages—the tribe, the village community, the free city, the state—each one naturally evolving out of the preceding stage. Of course, the experience of each preceding civilization was never lost. Greece (itself influenced by Eastern civilizations) influenced Rome, and Rome influenced our civilization; but each of them began from the same beginning—the tribe. And just as we cannot say that our states are *continuations* of the Roman state, so also can we not say that the mediæval cities of Europe (including Scandinavia and

No period of history could better illustrate the constructive powers of the popular masses than the tenth and eleventh centuries, when the fortified villages and market-places, representing so many "oases amidst the feudal forest," began to free themselves from their lord's yoke, and slowly elaborated the future city organization ; but, unhappily, this is a period about which historical information is especially scarce : we know the results, but little has reached us about the means by which they were achieved. Under the protection of their walls the cities' folkmotes— either quite independent, or led by the chief noble or merchant families—conquered and maintained the right of electing the military *defensor* and supreme judge of the town, or at least of choosing between those who pretended to occupy this position. In Italy the young communes were continually sending away their *defensors* or *domini*, fighting those who refused to go. The same went on in the East. In Bohemia, rich and poor alike (*Bohemicæ gentis magni et parvi, nobiles et ignobiles*) took part in the election ;[1] while the *vyeches* (folkmotes) of the Russian cities regularly elected their dukes—always from the same Rurik family—covenanted with them, and sent the *knyaz* away if he had provoked discontent.[2] At the same

---

Russia) were a continuation of the Roman cities. They were a continuation of the barbarian village community, influenced to a certain extent by the traditions of the Roman towns.

[1] M. Kovalevsky, *Modern Customs and Ancient Laws of Russia* (Ilchester Lectures, London, 1891, lecture 4).

[2] A considerable amount of research had to be done before this character of the so-called *udyelnyi period* was properly established by the works of Byelaeff (*Tales from Russian History*), Kostomaroff (*The Beginnings of Autocracy in Russia*), and especially Professor Sergievich (*The Vyeche and the Prince*). The English reader may find some information about this period in the just-named work of

time in most cities of Western and Southern Europe, the tendency was to take for *defensor* a bishop whom the city had elected itself; and so many bishops took the lead in protecting the "immunities" of the towns and in defending their liberties, that numbers of them were considered, after their death, as saints and special patrons of different cities. St. Uthelred of Winchester, St. Ulrik of Augsburg, St. Wolfgang of Ratisbon, St. Heribert of Cologne, St. Adalbert of Prague, and so on, as well as many abbots and monks, became so many cities' saints for having acted in defence of popular rights.[1] And under the new *defensors*, whether laic or clerical, the citizens conquered full self-jurisdiction and self-administration for their folkmotes.[2]

The whole process of liberation progressed by a series of imperceptible acts of devotion to the common cause, accomplished by men who came out of the masses—by unknown heroes whose very names have not been preserved by history. The wonderful movement of the God's peace (*treuga Dei*) by which the popular masses endeavoured to put a limit to the endless family feuds of the noble families, was born in the young towns, the bishops and the citizens trying to extend to the nobles the peace they had established

M. Kovalevsky, in Rambaud's *History of Russia*, and, in a short summary, in the article "Russia" of the last edition of *Chambers's Encyclopædia*.

[1] Ferrari, *Histoire des révolutions d'Italie*, i. 257; Kallsen, *Die deutschen Städte im Mittelalter*, Bd. I. (Halle, 1891).

[2] See the excellent remarks of Mr. G. L. Gomme as regards the folkmote of London (*The Literature of Local Institutions*, London, 1886, p. 76). It must, however, be remarked that in royal cities the folkmote never attained the independence which it assumed elsewhere. It is even certain that Moscow and Paris were chosen by the kings and the Church as the cradles of the future royal authority in the State, because they did not possess the tradition of folkmotes accustomed to act as sovereign in all matters.

within their town walls.[1]   Already at that period,
the commercial cities of Italy, and especially Amalfi
(which had its elected consuls since 844, and frequently
changed its doges in the tenth century)[2] worked out
the customary maritime and commercial law which
later on became a model for all Europe; Ravenna
elaborated its craft organization, and Milan, which had
made its first revolution in 980, became a great centre
of commerce, its trades enjoying a full independence
since the eleventh century.[3]   So also Bruges and
Ghent; so also several cities of France in which the
*Mahl* or *forum* had become a quite independent
institution.[4]   And already during that period began
the work of artistic decoration of the towns by works
of architecture, which we still admire and which
loudly testify of the intellectual movement of the
times.   "The basilicæ were then renewed in almost
all the universe," Raoul Glaber wrote in his chronicle,
and some of the finest monuments of mediæval architecture date from that period: the wonderful old
church of Bremen was built in the ninth century,
Saint Marc of Venice was finished in 1071, and the
beautiful dome of Pisa in 1063.   In fact, the intel-

---

[1] A. Luchaire, *Les Communes françaises;* also Kluckohn, *Geschichte des Gottesfrieden,* 1857.  L. Sémichon (*La paix et la trève
de Dieu,* 2 vols., Paris, 1869) has tried to represent the communal
movement as issued from that institution. In reality, the *treuga Dei,*
like the league started under Louis le Gros for the defence against
both the robberies of the nobles and the Norman invasions, was a
thoroughly *popular* movement. The only historian who mentions
this last league—that is, Vitalis—describes it as a "popular community" ("Considérations sur l'histoire de France," in vol. iv. of
Aug. Thierry's *Œuvres,* Paris, 1868, p. 191 and *note*).

[2] Ferrari, i. 152, 263, etc.

[3] Perrens, *Histoire de Florence,* i. 188; Ferrari, *l. c.,* i. 283.

[4] Aug. Thierry, *Essai sur l'histoire du Tiers État,* Paris, 1875, p.
414, *note.*

lectual movement which has been described as the
Twelfth Century Renaissance[1] and the Twelfth Century Rationalism—the precursor of the Reform[2]—
date from that period, when most cities were still
simple agglomerations of small village communities
enclosed by walls.

However, another element, besides the village-
community principle, was required to give to these
growing centres of liberty and enlightenment the
unity of thought and action, and the powers of
initiative, which made their force in the twelfth and
thirteenth centuries.   With the growing diversity of
occupations, crafts and arts, and with the growing
commerce in distant lands, some new form of union
was required, and this necessary new element was
supplied by the *guilds*.   Volumes and volumes have
been written about these unions which, under the
name of guilds, brotherhoods, friendships and *druzhestva, minne, artels* in Russia, *esnaifs* in Servia and
Turkey, *amkari* in Georgia, and so on, took such a
formidable development in mediæval times and played
such an important part in the emancipation of the
cities.   But it took historians more than sixty years
before the universality of this institution and its true
characters were understood.   Only now, when hundreds of guild statutes have been published and studied,
and their relationship to the Roman *collegia*, and the
earlier unions in Greece and in India,[3] is known, can

[1] F. Rocquain, "La Renaissance au XII᷎ siècle," in *Études sur
l'histoire de France*, Paris, 1875, pp. 55–117.

[2] N. Kostomaroff, "The Rationalists of the Twelfth Century," in
his *Monographies and Researches* (Russian).

[3] Very interesting facts relative to the universality of guilds will be
found in "Two Thousand Years of Guild Life," by Rev. J. M.

we maintain with full confidence that these brotherhoods were but a further development of the same
principles which we saw at work in the *gens* and the
village community.

Nothing illustrates better these mediæval brotherhoods than those temporary guilds which were formed
on board ships. When a ship of the Hansa had
accomplished her first half-day passage after having
left the port, the captain (*Schiffer*) gathered all crew
and passengers on the deck, and held the following
language, as reported by a contemporary :—

"'As we are now at the mercy of God and the waves,' he
said, 'each one must be equal to each other. And as we are
surrounded by storms, high waves, pirates and other dangers,
we must keep a strict order that we may bring our voyage to
a good end. That is why we shall pronounce the prayer for
a good wind and good success, and, according to marine law,
we shall name the occupiers of the judges' seats (*Schöffen-
stellen*).' Thereupon the crew elected a Vogt and four *scabini*,
to act as their judges. At the end of the voyage the Vogt
and the *scabini* abdicated their functions and addressed the
crew as follows :—'What has happened on board ship, we
must pardon to each other and consider as dead (*todt und ab
sein lassen*). What we have judged right, was for the sake of
justice. This is why we beg you all, in the name of honest
justice, to forget all the animosity one may nourish against
another, and to swear on bread and salt that he will not think
of it in a bad spirit. If any one, however, considers himself
wronged, he must appeal to the land Vogt and ask justice
from him before sunset.' On landing, the Stock with the *fred-
fines* was handed over to the Vogt of the sea-port for distribution among the poor."[1]

This simple narrative, perhaps better than anything

---

Lambert, Hull, 1891. On the Georgian *amkari*, see S. Eghiazarov,
*Gorodskiye Tsekhi* ("Organization of Transcaucasian Amkari"), in
*Memoirs* of the Caucasian Geographical Society, xiv. 2, 1891.

[1] J. D. Wunderer's "Reisebericht" in Fichard's *Frankfurter Archiv*,
ii. 245 ; quoted by Janssen, *Geschichte des deutschen Volkes*, i. 355.

else, depicts the spirit of the mediæval guilds. Like organizations came into existence wherever a group of men—fishermen, hunters, travelling merchants, builders, or settled craftsmen—came together for a common pursuit. Thus, there was on board ship the naval authority of the captain ; but, for the very success of the common enterprise, all men on board, rich and poor, masters and crew, captain and sailors, agreed to be equals in their mutual relations, to be simply men, bound to aid each other and to settle their possible disputes before judges elected by all of them. So also when a number of craftsmen—masons, carpenters, stone-cutters, etc.—came together for building, say, a cathedral, they all belonged to a city which had its political organization, and each of them belonged moreover to his own craft ; but they were united besides by their common enterprise, which they knew better than any one else, and they joined into a body united by closer, although temporary, bonds ; they founded the guild for the building of the cathedral.[1] We may see the same till now in the Kabylian *çof* :[2] the Kabyles have their village community ; but this union is not sufficient for all political, commercial, and personal needs of union, and the closer brotherhood of the *çof* is constituted.

As to the social characters of the mediæval guild, any guild-statute may illustrate them. Taking, for instance, the *skraa* of some early Danish guild, we read in it, first, a statement of the general brotherly feelings which must reign in the guild ; next come the regulations relative to self-jurisdiction in cases of quarrels

---

[1] Dr. Leonard Ennen, *Der Dom zu Köln, Historische Einleitung*, Köln, 1871, pp. 46, 50.
[2] See previous chapter.

arising between two brothers, or a brother and a stranger; and then, the social duties of the brethren are enumerated. If a brother's house is burned, or he has lost his ship, or has suffered on a pilgrim's voyage, all the brethren must come to his aid. If a brother falls dangerously ill, two brethren must keep watch by his bed till he is out of danger, and if he dies, the brethren must bury him—a great affair in those times of pestilences—and follow him to the church and the grave. After his death they must provide for his children, if necessary; very often the widow becomes a sister to the guild.[1]

These two leading features appeared in every brotherhood formed for any possible purpose. In each case the members treated each other as, and named each other, brother and sister;[2] all were equals before the guild. They owned some "chattel" (cattle, land, buildings, places of worship, or "stock") in common. All brothers took the oath of abandoning all feuds of old; and, without imposing upon each other the obligation of never quarrelling again, they agreed that no quarrel should degenerate into a feud, or into a law-suit before another court than the tribunal of the brothers themselves. And if a brother was involved in a quarrel with a stranger to the guild, they agreed to support him for bad and for good; that is, whether he was unjustly accused of aggression, or really was the aggressor, they had to support him, and to bring

---

[1] Kofod Ancher, *Om gamle Danske Gilder og deres Undergång*, Copenhagen, 1785. Statutes of a Knu guild.

[2] Upon the position of women in guilds, see Miss Toulmin Smith's introductory remarks to the *English Guilds* of her father. One of the Cambridge statutes (p. 281) of the year 1503 is quite positive in the following sentence: "Thys statute is made by the comyne assent of all the bretherne and sisterne of alhallowe yelde."

things to a peaceful end. So long as his was not a secret aggression—in which case he would have been treated as an outlaw—the brotherhood stood by him.[1] If the relatives of the wronged man wanted to revenge the offence at once by a new aggression, the brotherhood supplied him with a horse to run away, or with a boat, a pair of oars, a knife and a steel for striking light; if he remained in town, twelve brothers accompanied him to protect him; and in the meantime they arranged the composition. They went to court to support by oath the truthfulness of his statements, and if he was found guilty they did not let him go to full ruin and become a slave through not paying the due compensation: they all paid it, just as the *gens* did in olden times. Only when a brother had broken the faith towards his guild-brethren, or other people, he was excluded from the brotherhood " with a Nothing's name " (*tha scal han maeles af brödrescap met nidings nafn*).[2]

Such were the leading ideas of those brotherhoods which gradually covered the whole of mediæval life. In fact, we know of guilds among all possible professions: guilds of serfs,[3] guilds of freemen, and guilds of both serfs and freemen; guilds called into life for the special purpose of hunting, fishing, or a trading

---

[1] In mediæval times, only secret aggression was treated as a murder. Blood-revenge in broad daylight was justice; and slaying in a quarrel was not murder, once the aggressor showed his willingness to repent and to repair the wrong he had done. Deep traces of this distinction still exist in modern criminal law, especially in Russia.

[2] Kofod Ancher, *l. c.* This old booklet contains much that has been lost sight of by later explorers.

[3] They played an important part in the revolts of the serfs, and were therefore prohibited several times in succession in the second half of the ninth century. Of course, the king's prohibitions remained a dead letter.

expedition, and dissolved when the special purpose had been achieved ; and guilds lasting for centuries in a given craft or trade. And, in proportion as life took an always greater variety of pursuits, the variety in the guilds grew in proportion. So we see not only merchants, craftsmen, hunters, and peasants united in guilds ; we also see guilds of priests, painters, teachers of primary schools and universities, guilds for performing the passion play, for building a church, for developing the "mystery" of a given school of art or craft, or for a special recreation—even guilds among beggars, executioners, and lost women, all organized on the same double principle of self-jurisdiction and mutual support.[1]  For Russia we have positive evidence showing that the very "making of Russia" was as much the work of its hunters', fishermen's, and traders' *artels* as of the budding village communities, and up to the present day the country is covered with *artels*.[2]

These few remarks show how incorrect was the view taken by some early explorers of the guilds when they wanted to see the essence of the institution in its

---

[1] The mediæval Italian painters were also organized in guilds, which became at a later epoch Academies of art. If the Italian art of those times is impressed with so much individuality that we distinguish, even now, between the different schools of Padua, Bassano, Treviso, Verona, and so on, although all these cities were under the sway of Venice, this was due—J. Paul Richter remarks—to the fact that the painters of each city belonged to a separate guild, friendly with the guilds of other towns, but leading a separate existence. The oldest guild-statute known is that of Verona, dating from 1303, but evidently copied from some much older statute. "Fraternal assistance in necessity of whatever kind," "hospitality towards strangers, when passing through the town, as thus information may be obtained about matters which one may like to learn," and "obligation of offering comfort in case of debility" are among the obligations of the members (*Nineteenth Century*, Nov. 1890, and Aug. 1892).

[2] The chief works on the *artels* are named in the article "Russia" of the *Encyclopædia Britannica*, 9th edition, p. 84.

yearly festival. In reality, the day of the common meal was always the day, or the morrow of the day, of election of aldermen, of discussion of alterations in the statutes, and very often the day of judgment of quarrels that had risen among the brethren,[1] or of renewed allegiance to the guild. The common meal, like the festival at the old tribal folkmote—the *mahl* or *malum* —or the Buryate *aba*, or the parish feast and the harvest supper, was simply an affirmation of brotherhood. It symbolized the times when everything was kept in common by the clan. This day, at least, all belonged to all; all sate at the same table and partook of the same meal. Even at a much later time the inmate of the almshouse of a London guild sat this day by the side of the rich alderman. As to the distinction which several explorers have tried to establish between the old Saxon " frith guild " and the so-called " social " or " religious " guilds—all were frith guilds in the sense above mentioned,[2] and all were

---

[1] See, for instance, the texts of the Cambridge guilds given by Toulmin Smith (*English Guilds*, London, 1870, pp. 274–276), from which it appears that the "generall and principall day" was the " eleccioun day ; " or, Ch. M. Clode's *The Early History of the Guild of the Merchant Taylors*, London, 1888, i. 45 ; and so on. For the renewal of allegiance, see the Jómsviking saga, mentioned in Pappenheim's *Altdänische Schutzgilden*, Breslau, 1885, p. 67. It appears very probable that when the guilds began to be prosecuted, many of them inscribed in their statutes the meal day only, or their pious duties, and only alluded to the judicial function of the guild in vague words ; but this function did not disappear till a very much later time. The question, "Who will be my judge?" has no meaning now, since the State has appropriated for its bureaucracy the organization of justice ; but it was of primordial importance in mediæval times, the more so as self-jurisdiction meant self-administration. It must also be remarked that the translation of the Saxon and Danish "guild-bretheren," or " brodræ," by the Latin *convivii* must also have contributed to the above confusion.

[2] See the excellent remarks upon the frith guild by J. R. Green and Mrs Green in *The Conquest of England*, London, 1883, pp. 229–230.

religious in the sense in which a village community or
a city placed under the protection of a special saint
is social and religious.    If the institution of the guild
has taken such an immense extension in Asia, Africa,
and Europe, if it has lived thousands of years, reappear-
ing again and again when similar conditions called it
into existence, it is because it was much more than
an eating association, or an association for going to
church on a certain day, or a burial club.    It answered
to a deeply inrooted want of human nature ; and it
embodied all the attributes which the State appropriated later on for its bureaucracy and police, and much
more than that.    It was an association for mutual
support in all circumstances and in all accidents of life,
"by deed and advise," and it was an organization for
maintaining justice—with this difference from the
State, that on all these occasions a humane, a brotherly
element was introduced instead of the formal element
which is the essential characteristic of State interference.
Even when appearing before the guild tribunal, the
guild-brother answered before men who knew him well
and had stood by him before in their daily work, at the
common meal, in the performance of their brotherly
duties : men who were his equals and brethren indeed,
not theorists of law nor defenders of some one
else's interests.[1]

It is evident that an institution so well suited to
serve the need of union, without depriving the
individual of his initiative, could but spread, grow, and
fortify.    The difficulty was only to find such form as
would permit to federate the unions of the guilds
without interfering with the unions of the village com-

[1] See Appendix X.

munities, and to federate all these into one harmonious whole. And when this form of combination had been found, and a series of favourable circumstances permitted the cities to affirm their independence, they did so with a unity of thought which can but excite our admiration, even in our century of railways, telegraphs, and printing. Hundreds of charters in which the cities inscribed their liberation have reached us, and through all of them—notwithstanding the infinite variety of details, which depended upon the more or less greater fulness of emancipation—the same leading ideas run. The city organized itself as a federation of both small village communities and guilds.

"All those who belong to the friendship of the town"—so runs a charter given in 1188 to the burghesses of Aire by Philip, Count of Flanders—"have promised and confirmed by faith and oath that they will aid each other as brethren, in whatever is useful and honest. That if one commits against another an offence in words or in deeds, the one who has suffered therefrom will not take revenge, either himself or his people . . . he will lodge a complaint and the offender will make good for his offence, according to what will be pronounced by twelve elected judges acting as arbiters. And if the offender or the offended, after having been warned thrice, does not submit to the decision of the arbiters, he will be excluded from the friendship as a wicked man and a perjuror.[1]

"Each one of the men of the commune will be faithful to his con-juror, and will give him aid and advice, according to what justice will dictate him"—the Amiens and Abbeville charters say. "All will aid each other, according to their powers, within the boundaries of the Commune, and will not suffer that any one takes anything from any one of them, or makes one pay contributions"—do we read in the charters of Soissons, Compiègne, Senlis, and many others of the same type.[2] And so on with countless variations on the same theme.

---

[1] *Recueil des ordonnances des rois de France*, t. xii. 562 ; quoted by Aug. Thierry in *Considérations sur l'histoire de France*, p. 196, ed. 12mo.

[2] A. Luchaire, *Les Communes françaises*, pp. 45–46.

"The Commune," Guilbert de Nogent wrote, "is an oath of mutual aid (*mutui adjutorii conjuratio*) . . . A new and detestable word.  Through it the serfs (*capite sensi*) are freed from all serfdom; through it, they can only be condemned to a legally determined fine for breaches of the law; through it, they cease to be liable to payments which the serfs always used to pay."[1]

The same wave of emancipation ran, in the twelfth century, through all parts of the continent, involving both rich cities and the poorest towns.  And if we may say that, as a rule, the Italian cities were the first to free themselves, we can assign no centre from which the movement would have spread.  Very often a small burg in central Europe took the lead for its region, and big agglomerations accepted the little town's charter as a model for their own.  Thus, the charter of a small town, Lorris, was adopted by eighty-three towns in south-west France, and that of Beaumont became the model for over five hundred towns and cities in Belgium and France.  Special deputies were dispatched by the cities to their neighbours to obtain a copy from their charter, and the constitution was framed upon that model.  However, they did not simply copy each other: they framed their own charters in accordance with the concessions they had obtained from their lords; and the result was that, as remarked by an historian, the charters of the mediæval communes offer the same variety as the Gothic architecture of their churches and cathedrals.  The same leading ideas in all of them—the cathedral symbolizing the union of parish and guild in the city,—and the same infinitely rich variety of detail.

Self-jurisdiction was the essential point, and self-jurisdiction meant self-administration.  But the com-

[1] Guilbert de Nogent, *De vita sua*, quoted by Luchaire, *l.c.*, p. 14.

mune was not simply an "autonomous" part of the State—such ambiguous words had not yet been invented by that time—it was a State in itself. It had the right of war and peace, of federation and alliance with its neighbours. It was sovereign in its own affairs, and mixed with no others. The supreme political power could be vested entirely in a democratic forum, as was the case in Pskov, whose *vyeche* sent and received ambassadors, concluded treaties, accepted and sent away princes, or went on without them for dozens of years ; or it was vested in, or usurped by, an aristocracy of merchants or even nobles, as was the case in hundreds of Italian and middle European cities. The principle, nevertheless, remained the same: the city was a State and—what was perhaps still more remarkable—when the power in the city was usurped by an aristocracy of merchants or even nobles, the inner life of the city and the democratism of its daily life did not disappear : they depended but little upon what may be called the political form of the State.

The secret of this seeming anomaly lies in the fact that a mediæval city was not a centralized State. During the first centuries of its existence, the city hardly could be named a State as regards its interior organization, because the middle ages knew no more of the present centralization of functions than of the present territorial centralization. Each group had its share of sovereignty. The city was usually divided into four quarters, or into five to seven sections radiating from a centre, each quarter or section roughly corresponding to a certain trade or profession which prevailed in it, but nevertheless containing inhabitants of different social positions and occupations—nobles, merchants, artisans, or even half-serfs ; and each section or quarter constituted a quite independent agglomeration. In Venice, each island was an independent political community. It had its own organized trades, its own commerce in salt, its own jurisdiction and administration, its own forum; and the nomination of a doge by the city changed nothing in the inner independence of the units.[1] In Cologne, we see the inhabitants divided into *Geburschaften* and *Heimschaften* (*viciniæ*), *i. e.* neighbour guilds, which dated from the Franconian period. Each of them had its judge (*Burrichter*) and the usual twelve elected sentence-finders (*Schöffen*), its Vogt, and its *greve* or commander of the local militia.[2] The story of early London before the Conquest—Mr. Green says—is that "of a number of little groups scattered here and there over the area within the walls, each growing up with its own life and institutions, guilds, sokes, religious houses and the like, and only slowly drawing together into a municipal union."[3] And if we refer to the annals of the Russian cities, Novgorod and Pskov, both of which are relatively rich in local details, we find the section (*konets*) consisting of independent streets (*ulitsa*), each of which, though chiefly peopled with artisans of a certain craft, had also merchants and landowners among its inhabitants, and was a separate community. It had the communal responsibility of all members in case of crime, its own jurisdiction and administration by street aldermen (*ulichanskiye starosty*), its own seal and, in case of need, its own

---

[1] Lebret, *Histoire de Venise*, i. 393; also Marin, quoted by Leo and Botta in *Histoire de l'Italie*, French edition, 1844, t. i. 500.

[2] Dr. W. Arnold, *Verfassungsgeschichte der deutschen Freistädte*, 1854, Bd. ii. 227 *seq.*; Ennen, *Geschichte der Stadt Koeln*, Bd. i. 228-229; also the documents published by Ennen and Eckert.

[3] *Conquest of England*, 1883, p. 453.

forum ; its own militia, as also its self-elected priests and its own collective life and collective enterprise.[1]

The mediæval city thus appears as a double federation : of all householders united into small territorial unions—the street, the parish, the section—and of individuals united by oath into guilds according to their professions ; the former being a produce of the village-community origin of the city, while the second is a subsequent growth called to life by new conditions.

To guarantee liberty, self-administration, and peace was the chief aim of the mediæval city ; and labour, as we shall presently see when speaking of the craft guilds, was its chief foundation. But "production" did not absorb the whole attention of the mediæval economist. With his practical mind, he understood that "consumption" must be guaranteed in order to obtain production ; and therefore, to provide for "the common first food and lodging of poor and rich alike" (*gemeine notdurft vnd gemach armer vnd richer*[2]) was the fundamental principle in each city. The purchase of food supplies and other first necessaries (coal, wood, etc.) before they had reached the market, or altogether in especially favourable conditions from which others would be excluded—the *preempcio*, in a word—was entirely prohibited. Everything had to go to the market and be offered there for every one's purchase, till the ringing of the bell had closed the market. Then only could the retailer buy the remainder, and even then his profit should be an "honest

---

[1] Byelaeff, *Russian History*, vols. ii. and iii.

[2] W. Gramich, *Verfassungs- und Verwaltungsgeschichte der Stadt Würzburg im 13. bis zum 15. Jahrhundert*, Würzburg, 1882, p. 34.

profit" only.[1]   Moreover, when corn was bought by a
baker wholesale after the close of the market, every
citizen had the right to claim part of the corn (about
half-a-quarter) for his own use, at wholesale price, if
he did so before the final conclusion of the bargain ;
and reciprocally, every baker could claim the same if
the citizen purchased corn for re-selling it.   In the
first case, the corn had only to be brought to the town
mill to be ground in its proper turn for a settled price,
and the bread could be baked in the *four banal,* or
communal oven.[2]   In short, if a scarcity visited the
city, all had to suffer from it more or less ; but apart
from the calamities, so long as the free cities existed
no one could die in their midst from starvation, as is
unhappily too often the case in our own times.

However, all such regulations belong to later periods
of the cities' life, while at an earlier period it was the
city itself which used to buy all food supplies for the
use of the citizens.   The documents recently published
by Mr. Gross are quite positive on this point and fully
support his conclusion to the effect that the cargoes of
subsistences "were purchased by certain civic officials
in the name of the town, and then distributed in shares

---

[1] When a boat brought a cargo of coal to Würzburg, coal could
only be sold in retail during the first eight days, each family being
entitled to no more than fifty basketfuls.  The remaining cargo
could be sold wholesale, but the retailer was allowed to raise a
*zittlicher* profit only, the *unzittlicher*, or dishonest profit, being strictly
forbidden (Gramich, *l. c.*).   Same in London (*Liber albus,* quoted by
Ochenkowski, p. 161), and, in fact, everywhere.

[2] See Fagniez, *Études sur l'industrie et la classe industrielle à Paris
au XIIIme et XIVme siècle,* Paris, 1877, pp. 155 *seq.*  It hardly need
be added that the tax on bread, and on beer as well, was settled after
careful experiments as to the quantity of bread and beer which could
be obtained from a given amount of corn.   The Amiens archives
contain the minutes of such experiences (A. de Calonne, *l. c.* pp. 77,
93).   Also those of London (Ochenkowski, *England's wirthschaftliche
Entwickelung, etc.,* Jena, 1879, p. 165).

among the merchant burgesses, no one being allowed
to buy wares landed in the port unless the municipal
authorities refused to purchase them. This seems—
he adds—to have been quite a common practice in
England, Ireland, Wales and Scotland."[1]  Even in
the sixteenth century we find that common purchases
of corn were made for the " comoditie and profitt in
all things of this . . . . Citie and Chamber of London,
and of all the Citizens and Inhabitants of the same as
moche as in us licth "—as the Mayor wrote in 1565.[2]
In Venice, the whole of the trade in corn is well known
to have been in the hands of the city ; the "quarters,"
on receiving the cereals from the board which adminis-
trated the imports, being bound to send to every
citizen's house the quantity allotted to him.[3]  In
France, the city of Amiens used to purchase salt and
to distribute it to all citizens at cost price ;[4] and even
now one sees in many French towns the *halles* which

---

[1] Ch. Gross, *The Guild Merchant*, Oxford, 1890, i. 135.  His
documents prove that this practice existed in Liverpool (ii. 148-150),
Waterford in Ireland, Neath in Wales, and Linlithgow and Thurso in
Scotland.  Mr. Gross's texts also show that the purchases were made
for distribution, not only among the merchant burgesses, but "upon
all citsains and commynalte" (p. 136, *note*), or, as the Thurso
ordinance of the seventeenth century runs, to "make offer to the
merchants, craftsmen, *and inhabitants* of the said burgh, that they
may have their proportion of the same, according to their necessitys
and ability."

[2] *The Early History of the Guild of Merchant Taylors*, by Charles
M. Clode, London, 1888, i. 361, appendix 10 ; also the following appendix which shows that the same purchases were made
in 1546.

[3] Cibrario, *Les conditions économiques de l'Italie au temps de Dante*,
Paris, 1865, p. 44.

[4] A. de Calonne, *La vie municipale au XVme siècle dans le Nord de
la France*, Paris, 1880, pp. 12-16.  In 1485 the city permitted the
export to Antwerp of a certain quantity of corn, "the inhabitants of
Antwerp being always ready to be agreeable to the merchants and
burgesses of Amiens " (*ibid.*, pp. 75-77 and texts).

formerly were municipal *dépôts* for corn and salt.[1]
In Russia it was a regular custom in Novgorod and
Pskov.

The whole matter relative to the communal purchases
for the use of the citizens, and the manner in which
they used to be made, seems not to have yet received
proper attention from the historians of the period ; but
there are here and there some very interesting facts
which throw a new light upon it. Thus there is,
among Mr. Gross's documents, a Kilkenny ordinance
of the year 1367, from which we learn how the prices
of the goods were established. "'The merchants and
the sailors," Mr. Gross writes, "were to state on oath
the first cost of the goods and the expenses of trans-
portation. Then the mayor of the town and two
discreet men were to name the price at which the
wares were to be sold." The same rule held good in
Thurso for merchandise coming " by sea or land."
This way of "naming the price" so well answers to
the very conceptions of trade which were current in
mediæval times that it must have been all but universal.
To have the price established by a third person was a
very old custom ; and for all interchange within the
city it certainly was a widely-spread habit to leave the
establishment of prices to "discreet men "—to a third
party—and not to the vendor or the buyer. But this
order of things takes us still further back in the
history of trade—namely, to a time when trade in
staple produce was carried on by the whole city,
and the merchants were only the commissioners, the
trustees, of the city for selling the goods which it
exported. A Waterford ordinance, published also by
Mr. Gross, says " that all manere of marchandis *what*

---

[1] A. Babeau, *La ville sous l'ancien régime,* Paris, 1880.

*so ever kynde thei be of* . . . shal be bought by the Maire and balives which bene commene biers [common buyers, for the town] for the time being, and to distribute the same on freemen of the citie (the propre goods of free citisains and inhabitants only excepted)." This ordinance can hardly be explained otherwise than by admitting that all the exterior trade of the town was carried on by its agents. Moreover, we have direct evidence of such having been the case for Novgorod and Pskov. It was the Sovereign Novgorod and the Sovereign Pskov who sent their caravans of merchants to distant lands.

We know also that in nearly all mediæval cities of Middle and Western Europe, the craft guilds used to buy, as a body, all necessary raw produce, and to sell the produce of their work through their officials, and it is hardly possible that the same should not have been done for exterior trade—the more so as it is well known that up to the thirteenth century, not only all merchants of a given city were considered abroad as responsible in a body for debts contracted by any one of them, but the whole city as well was responsible for the debts of each one of its merchants. Only in the twelfth and thirteenth century the towns on the Rhine entered into special treaties abolishing this responsibility.[1] And finally we have the remarkable Ipswich document published by Mr. Gross, from which document we learn that the merchant guild of this town was constituted by all who had the freedom of the city, and who wished to pay their contribution ("their hanse") to the guild, the whole community discussing all together how better to maintain the merchant guild, and giving it certain privileges. The merchant guild of Ipswich

[1] Ennen, *Geschichte der Stadt Köln*, i. 491, 492, also texts.

thus appears rather as a body of trustees of the town than as a common private guild.

In short, the more we begin to know the mediæval city the more we see that it was not simply a political organization for the protection of certain political liberties. It was an attempt at organizing, on a much grander scale than in a village community, a close union for mutual aid and support, for consumption and production, and for social life altogether, without imposing upon men the fetters of the State, but giving full liberty of expression to the creative genius of each separate group of individuals in art, crafts, science, commerce, and political organization. How far this attempt has been successful will be best seen when we have analyzed in the next chapter the organization of labour in the mediæval city and the relations of the cities with the surrounding peasant population.

# CHAPTER VI

## MUTUAL AID IN THE MEDIÆVAL CITY (*continued*)

Likeness and diversity among the mediæval cities.—The craft-guilds : State-attributes in each of them.—Attitude of the city towards the peasants ; attempts to free them.—The lords.—Results achieved by the mediæval city : in arts, in learning.—Causes of decay.

THE mediæval cities were not organized upon some preconceived plan in obedience to the will of an outside legislator. Each of them was a natural growth in the full sense of the word—an always varying result of struggle between various forces which adjusted and re-adjusted themselves in conformity with their relative energies, the chances of their conflicts, and the support they found in their surroundings. Therefore, there are not two cities whose inner organization and destinies would have been identical. Each one, taken separately, varies from century to century. And yet, when we cast a broad glance upon all the cities of Europe, the local and national unlikenesses disappear, and we are struck to find among all of them a wonderful resemblance, although each has developed for itself, independently from the others, and in different conditions. A small town in the north of Scotland, with its population of coarse labourers and fishermen ; a rich city of Flanders, with its world-wide commerce, luxury, love of amusement and animated life; an Italian city enriched by its intercourse with the East,

and breeding within its walls a refined artistic taste
and civilization ; and a poor, chiefly agricultural, city in
the marsh and lake district of Russia, seem to have
little in common.   And nevertheless, the leading lines
of their organization, and the spirit which animates
them, are imbued with a strong family likeness.
Everywhere we see the same federations of small
communities and guilds, the same " sub-towns " round
the mother city, the same folkmote, and the same
insigns of its independence.   The *defensor* of the city,
under different names and in different accoutrements,
represents the same authority and interests ; food
supplies, labour and commerce, are organized on
closely similar lines ; inner and outer struggles are
fought with like ambitions ; nay, the very formulæ
used in the struggles, as also in the annals, the ordinances, and the rolls, are identical ; and the architectural
monuments, whether Gothic, Roman, or Byzantine in
style, express the same aspirations and the same
ideals ; they are conceived and built in the same way.
Many dissemblances are mere differences of age, and
those disparities between sister cities which are real
are repeated in different parts of Europe.   The unity
of the leading idea and the identity of origin make
up for differences of climate, geographical situation,
wealth, language and religion.   This is why we can
speak of *the* mediæval city as of a well-defined phase
of civilization ; and while every research insisting upon
local and individual differences is most welcome, we
may still indicate the chief lines of development which
are common to all cities.[1]

---

[1] The literature of the subject is immense ; but there is no work
yet which treats of the mediæval city as of a whole.  For the French
Communes, Augustin Thierry's *Lettres* and *Considérations sur*

There is no doubt that the protection which used to be accorded to the market-place from the earliest barbarian times has played an important, though not an exclusive, part in the emancipation of the mediæval city. The early barbarians knew no trade within their village communities ; they traded with strangers only, at certain definite spots, on certain determined days. And, in order that the stranger might come to the barter-place without risk of being slain for some feud which might be running between two kins, the market was always placed under the special protection of all kins. It was inviolable, like the place of worship under the shadow of which it was held. With the Kabyles it is still *annaya*, like the footpath along

---

*l'histoire de France* still remain classical, and Luchaire's *Communes françaises* is an excellent addition on the same lines. For the cities of Italy, the great work of Sismondi (*Histoire des républiques italiennes du moyen âge*, Paris, 1826, 16 vols.), Leo and Botta's *History of Italy*, Ferrari's *Révolutions d'Italie*, and Hegel's *Geschichte der Städteverfassung in Italien*, are the chief sources of general information. For Germany we have Maurer's *Städteverfassung*, Barthold's *Geschichte der deutschen Städte*, and, of recent works, Hegel's *Städte und Gilden der germanischen Völker* (2 vols. Leipzig, 1891), and Dr. Otto Kallsen's *Die deutschen Städte im Mittelalter* (2 vols. Halle, 1891), as also Janssen's *Geschichte des deutschen Volkes* (5 vols. 1886), which, let us hope, will soon be translated into English (French translation in 1892). For Belgium, A. Wauters, *Les Libertés communales* (Bruxelles, 1869-78, 3 vols.). For Russia, Byelaeff's, Kostomaroff's and Sergievich's works. And finally, for England, we posses one of the best works on cities of a wider region in Mrs. J. R. Green's *Town Life in the Fifteenth Century* (2 vols. London, 1894). We have, moreover, a wealth of well-known local histories, and several excellent works of general or economical history which I have so often mentioned in this and the preceding chapter. The richness of literature consists, however, chiefly in separate, sometimes admirable, researches into the history of separate cities, especially Italian and German ; the guilds ; the land question ; the economical principles of the time ; the economical importance of guilds and crafts ; the leagues between cities (the Hansa) ; and communal art. An incredible wealth of information is contained in works of this second category, of which only some of the more important are named in these pages.

which women carry water from the well ; neither must
be trodden upon in arms, even during inter-tribal wars.
In mediæval times the market universally enjoyed the
same protection.[1]  No feud could be prosecuted on the
place whereto people came to trade, nor within a
certain radius from it ; and if a quarrel arose in the
motley crowd of buyers and sellers, it had to be
brought before those under whose protection the
market stood—the community's tribunal, or the
bishop's, the lord's, or the king's judge.  A stranger
who came to trade was a guest, and he went on under
this very name.  Even the lord who had no scruples
about robbing a merchant on the high road, respected
the *Weichbild*, that is, the pole which stood in the
market-place and bore either the king's arms, or a
glove, or the image of the local saint, or simply a
cross, according to whether the market was under the
protection of the king, the lord, the local church, or
the folkmote—the *vyeche*.[2]

It is easy to understand how the self-jurisdiction of
the city could develop out of the special jurisdiction in
the market-place, when this last right was conceded,
willingly or not, to the city itself.  And such an origin

[1] Kulischer, in an excellent essay on primitive trade (*Zeitschrift für
Völkerpsychologie*, Bd. x. 380), also points out that, according to
Herodotus, the Argippæans were considered inviolable, because the
trade between the Scythians and the northern tribes took place on
their territory.  A fugitive was sacred on their territory, and they
were often asked to act as arbiters for their neighbours.  See
Appendix XI.

[2] Some discussion has lately taken place upon the *Weichbild* and
the *Weichbild*-law, which still remain obscure (see Zöpfl, *Alterthümer
des deutschen Reichs und Rechts*, iii. 29 ; Kallsen, i. 316).  The above
explanation seems to be the more probable, but, of course, it must
be tested by further research.  It is also evident that, to use a Scotch
expression, the " mercet cross " could be considered as an emblem of
Church jurisdiction, but we find it both in bishop cities and in those
in which the folkmote was sovereign.

of the city's liberties, which can be traced in very many cases, necessarily laid a special stamp upon their subsequent development. It gave a predominance to the trading part of the community. The burghers who possessed a house in the city at the time being, and were co-owners in the town-lands, constituted very often a merchant guild which held in its hands the city's trade ; and although at the outset every burgher, rich and poor, could make part of the merchant guild, and the trade itself seems to have been carried on for the entire city by its trustees, the guild gradually became a sort of privileged body. It jealously prevented the outsiders who soon began to flock into the free cities from entering the guild, and kept the advantages resulting from trade for the few " families " which had been burghers at the time of the emancipation. There evidently was a danger of a merchant oligarchy being thus constituted. But already in the tenth, and still more during the two next centuries, the chief crafts, also organized in guilds, were powerful enough to check the oligarchic tendencies of the merchants.

The craft guild was then a common seller of its produce and a common buyer of the raw materials, and its members were merchants and manual workers at the same time. Therefore, the predominance taken by the old craft guilds from the very beginnings of the free city life guaranteed to manual labour the high position which it afterwards occupied in the city.[1] In fact, in a

---

[1] For all concerning the merchant guild see Mr. Gross's exhaustive work, *The Guild Merchant* (Oxford, 1890, 2 vols.); also Mrs. Green's remarks in *Town Life in the Fifteenth Century*, vol. ii. chaps. v. viii. x. ; and A. Doren's review of the subject in Schmoller's *Forschungen*, vol. xii. If the considerations indicated in the previous chapter (according to which trade was communal at its beginnings) prove to be correct, it will be permissible to suggest as a probable hypothesis that the guild merchant was a body entrusted

mediæval city manual labour was no token of inferiority ; it bore, on the contrary, traces of the high
respect it had been kept in in the village community.
Manual labour in a "mystery" was considered as a
pious duty towards the citizens : a public function
(*Amt*), as honourable as any other. An idea of
"justice" to the community, of "right" towards both
producer and consumer, which would seem so extravagant now, penetrated production and exchange.
The tanner's, the cooper's, or the shoemaker's work
must be "just," fair, they wrote in those times.
Wood, leather or thread which are used by the artisan
must be "right" ; bread must be baked "in justice,"
and so on. Transport this language into our present
life, and it would seem affected and unnatural ; but it
was natural and unaffected then, because the mediæval
artisan did not produce for an unknown buyer, or to
throw his goods into an unknown market. He produced for his guild first ; for a brotherhood of men
who knew each other, knew the technics of the craft,
and, in naming the price of each product, could
appreciate the skill displayed in its fabrication or the
labour bestowed upon it. Then the guild, not the
separate producer, offered the goods for sale in the
community, and this last, in its turn, offered to the
brotherhood of allied communities those goods which
were exported, and assumed responsibility for their

---

with commerce in the interest of the whole city, and only gradually
became a guild of merchants trading for themselves ; while the
merchant adventurers of this country, the Novgorod *povolniki* (free
colonizers and merchants) and the *mercati personati*, would be those
to whom it was left to open new markets and new branches of commerce for themselves. Altogether, it must be remarked that the
origin of the mediæval city can be ascribed to no separate agency.
It was a result of *many* agencies in different degrees.

quality. With such an organization, it was the ambition of each craft not to offer goods of inferior quality, and technical defects or adulterations became a matter concerning the whole community, because, an ordinance says, "they would destroy public confidence."[1] Production being thus a social duty, placed under the control of the whole *amitas*, manual labour could not fall into the degraded condition which it occupies now, so long as the free city was living.

A difference between master and apprentice, or between master and worker (*compayne, Geselle*), existed in the mediæval cities from their very beginnings ; but this was at the outset a mere difference of age and skill, not of wealth and power. After a seven years' apprenticeship, and after having proved his knowledge and capacities by a work of art, the apprentice became a master himself. And only much later, in the sixteenth century, after the royal power had destroyed the city and the craft organization, was it possible to become master in virtue of simple inheritance or wealth. But this was also the time of a general decay in mediæval industries and art.

There was not much room for hired work in the early flourishing periods of the mediæval cities, still less for individual hirelings. The work of the weavers, the archers, the smiths, the bakers, and so on, was performed for the craft and the city ; and when craftsmen were hired in the building trades, they worked as temporary corporations (as they still do in the Russian *artéls*), whose work was paid *en bloc*. Work for a master began to multiply only later on ; but even in this case the worker was paid better than he is paid now,

---

[1] Janssen's *Geschichte des deutschen Volkes*, i. 315; Gramich's *Würzburg ;* and, in fact, any collection of ordinances.

even in this country, and very much better than he used to be paid all over Europe in the first half of this century. Thorold Rogers has familiarized English readers with this idea; but the same is true for the Continent as well, as is shown by the researches of Falke and Schönberg, and by many occasional indications. Even in the fifteenth century a mason, a carpenter, or a smith worker would be paid at Amiens four *sols* a day, which corresponded to forty-eight pounds of bread, or to the eighth part of a small ox (*bouvard*). In Saxony, the salary of the *Geselle* in the building trade was such that, to put it in Falke's words, he could buy with his six days' wages three sheep and one pair of shoes.[1] The donations of workers (*Geselle*) to cathedrals also bear testimony of their relative well-being, to say nothing of the glorious donations of certain craft guilds nor of what they used to spend in festivities and pageants.[2] In fact, the more we learn about the mediæval city, the more we are convinced that at no time has labour enjoyed such conditions of prosperity and such respect as when city life stood at its highest.

More than that; not only many aspirations of our

[1] Falke, *Geschichtliche Statistik*, i. 373–393, and ii. 66; quoted in Janssen's *Geschichte*, i. 339; J. D. Blavignac, in *Comptes et dépenses de la construction du clocher de Saint-Nicolas à Fribourg en Suisse*, comes to a similar conclusion. For Amiens, De Calonne's *Vie Municipale*, p. 99 and Appendix. For a thorough appreciation and graphical representation of the mediæval wages in England and their value in bread and meat, see G. Steffen's excellent article and curves in *The Nineteenth Century* for 1891, and *Studier öfver lönsystemets historia i England*, Stockholm, 1895.

[2] To quote but one example out of many which may be found in Schönberg's and Falke's works, the sixteen shoemaker workers (*Schusterknechte*) of the town Xanten, on the Rhine, gave, for erecting a screen and an altar in the church, 75 guldens of subscriptions, and 12 guldens out of their box, which money was worth, according to the best valuations, ten times its present value.

modern radicals were already realized in the middle ages, but much of what is described now as Utopian was accepted then as a matter of fact. We are laughed at when we say that work must be pleasant, but— "every one must be pleased with his work," a mediæval Kuttenberg ordinance says, "and no one shall, while doing nothing (*mit nichts thun*), appropriate for himself what others have produced by application and work, because laws must be a shield for application and work." [1] And amidst all present talk about an eight hours' day, it may be well to remember an ordinance of Ferdinand the First relative to the Imperial coal mines, which settled the miner's day at eight hours, "as it used to be of old" (*wie vor Alters herkommen*), and work on Saturday afternoon was prohibited. Longer hours were very rare, we are told by Janssen, while shorter hours were of common occurrence. In this country, in the fifteenth century, Rogers says, "the workmen worked only forty-eight hours a week." [2] The Saturday half-holiday, too, which we consider as a modern conquest, was in reality an old mediæval institution ; it was bathing-time for a great part of the community, while Wednesday afternoon was bathing-time for the *Geselle*. [3] And although school meals did

---

[1] Quoted by Janssen, *l. c.* i. 343.

[2] *The Economical Interpretation of History*, London, 1891, p. 303.

[3] Janssen, *l. c.* See also Dr. Alwin Schultz, *Deutsches Leben im XIV. und XV. Jahrhundert*, grosse Ausgabe, Wien, 1892, pp. 67 *seq.* At Paris, the day of labour varied from seven to eight hours in the winter to fourteen hours in summer in certain trades, while in others it was from eight to nine hours in winter, to from ten to twelve in summer. All work was stopped on Saturdays and on about twenty-five other days (*jours de commun de vile foire*) at four o'clock, while on Sundays and thirty other holidays there was no work at all. The general conclusion is, that the mediæval worker worked *less* hours, all taken, than the present-day worker (Dr. E. Martin Saint-Léon, *Histoire des corporations*, p. 121).

not exist—probably because no children went hungry
to school—a distribution of bath-money to the children
whose parents found difficulty in providing it was
habitual in several places. As to Labour Congresses,
they also were a regular feature of the middles ages.
In some parts of Germany craftsmen of the same trade,
belonging to different communes, used to come together
every year to discuss questions relative to their trade,
the years of apprenticeship, the wandering years, the
wages, and so on; and in 1572, the Hanseatic towns
formally recognized the right of the crafts to come
together at periodical congresses, and to take any
resolutions, so long as they were not contrary to the
cities' rolls, relative to the quality of goods. Such
Labour Congresses, partly international like the Hansa
itself, are known to have been held by bakers, founders,
smiths, tanners, sword-makers and cask-makers.[1]

The craft organization required, of course, a close
supervision of the craftsmen by the guild, and special
jurates were always nominated for that purpose. But
it is most remarkable that, so long as the cities lived
their free life, no complaints were heard about the
supervision; while, after the State had stepped in,
confiscating the property of the guilds and destroying
their independence in favour of its own bureaucracy,
the complaints became simply countless.[2] On the
other hand, the immensity of progress realized in all

---

[1] W. Stieda, "Hansische Vereinbarungen über städtisches Gewerbe
im XIV. und XV. Jahrhundert," in *Hansische Geschichtsblätter*,
Jahrgang 1886, p. 121. Schönberg's *Wirthschaftliche Bedeutung der
Zünfte;* also, partly, Roscher.

[2] See Toulmin Smith's deeply-felt remarks about the royal spoliation of the guilds, in Miss Smith's Introduction to *English Guilds*.
In France the same royal spoliation and abolition of the guilds' jurisdiction was begun from 1306, and the final blow was struck in 1382
(Fagniez, *l. c.* pp. 52–54).

arts under the mediæval guild system is the best proof that the system was no hindrance to individual initiative.[1] The fact is, that the mediæval guild, like the mediæval parish, "street," or "quarter," was not a body of citizens, placed under the control of State function- aries; it was a union of all men connected with a given trade: jurate buyers of raw produce, sellers of manufactured goods, and artisans—masters, "com- paynes," and apprentices. For the inner organization of the trade its assembly was sovereign, so long as it did not hamper the other guilds, in which case the matter was brought before the guild of the guilds—the city. But there was in it something more than that. It had its own self-jurisdiction, its own military force, its own general assemblies, its own traditions of strug- gles, glory, and independence, its own relations with other guilds of the same trade in other cities : it had, in a word, a full organic life which could only result from the integrality of the vital functions. When the town was called to arms, the guild appeared as a separate company (*Schaar*), armed with its own arms (or its own guns, lovingly decorated by the guild, at a subsequent epoch), under its own self-elected com- manders. It was, in a word, as independent a unit of the federation as the republic of Uri or Geneva was fifty years ago in the Swiss Confederation. So that,

[1] Adam Smith and his contemporaries knew well what they were condemning when they wrote against the *State* interference in trade and the trade monopolies of *State* creation. Unhappily, their fol- lowers, with their hopeless superficiality, flung mediæval guilds and State interference into the same sack, making no distinction between a Versailles edict and a guild ordinance. It hardly need be said that the economists who have seriously studied the subject, like Schön- berg (the editor of the well-known course of *Political Economy*), never fell into such an error. But, till lately, diffuse discussions of the above type went on for economical "science."

to compare it with a modern trade union, divested of all attributes of State sovereignty, and reduced to a couple of functions of secondary importance, is as unreasonable as to compare Florence or Brügge with a French commune vegetating under the Code Napoléon, or with a Russian town placed under Catherine the Second's municipal law. Both have elected mayors, and the latter has also its craft corporations ; but the difference is—all the difference that exists between Florence and Fontenay-les-Oies or Tsarevokokshaisk, or between a Venetian doge and a modern mayor who lifts his hat before the *sous-préfet's* clerk.

The mediæval guilds were capable of maintaining their independence ; and, later on, especially in the fourteenth century, when, in consequence of several causes which shall presently be indicated, the old municipal life underwent a deep modification, the younger crafts proved strong enough to conquer their due share in the management of the city affairs. The masses, organized in " minor " arts, rose to wrest the power out of the hands of a growing oligarchy, and mostly succeeded in this task, opening again a new era of prosperity. True, that in some cities the uprising was crushed in blood, and mass decapitations of workers followed, as was the case in Paris in 1306, and in Cologne in 1371. In such cases the city's liberties rapidly fell into decay, and the city was gradually subdued by the central authority. But the majority of the towns had preserved enough of vitality to come out of the turmoil with a new life and vigour.[1] A new

[1] In Florence the seven minor arts made their revolution in 1270–82, and its results are fully described by Perrens (*Histoire de Florence*, Paris, 1877, 3 vols.), and especially by Gino Capponi (*Storia della*

period of rejuvenescence was their reward. New life was infused, and it found its expression in splendid architectural monuments, in a new period of prosperity, in a sudden progress of technics and invention, and in a new intellectual movement leading to the Renaissance and to the Reformation.

The life of a mediæval city was a succession of hard battles to conquer liberty and to maintain it. True, that a strong and tenacious race of burghers had developed during those fierce contests ; true, that love and worship of the mother city had been bred by these struggles, and that the grand things achieved by the mediæval communes were a direct outcome of that love. But the sacrifices which the communes had to sustain in the battle for freedom were, nevertheless, cruel, and left deep traces of division on their inner life as well. Very few cities had succeeded, under a concurrence of favourable circumstances, in obtaining

---

*repubblica di Firenze*, 2da edizione, 1876, i. 58–80 ; translated into German). In Lyons, on the contrary, where the movement of the minor crafts took place in 1402, the latter were defeated and lost the right of themselves nominating their own judges. The two parties came apparently to a compromise. In Rostock the same movement took place in 1313 ; in Zürich in 1336; in Bern in 1363 ; in Braunschweig in 1374, and next year in Hamburg ; in Lübeck in 1376–84; and so on. See Schmoller's *Strassburg zur Zeit der Zunftkämpfe* and *Strassburg's Blüthe ;* Brentano's *Arbeitergilden der Gegenwart,* 2 vols., Leipzig, 1871–72 ; Eb. Bain's *Merchant and Craft Guilds,* Aberdeen, 1887, pp. 26–47, 75, etc. As to Mr. Gross's opinion relative to the same struggles in England, see Mrs. Green's remarks in her *Town Life in the Fifteenth Century,* ii. 190–217 ; also the chapter on the Labour Question, and, in fact, the whole of this extremely interesting volume. Brentano's views on the crafts' struggles, expressed especially in §§ iii. and iv. of his essay "On the History and Development of Guilds," in Toulmin Smith's *English Guilds* remain classical for the subject, and may be said to have been again and again confirmed by subsequent research.

liberty at one stroke, and these few mostly lost it
equally easily ; while the great number had to fight
fifty or a hundred years. in succession, often more,
before their rights to free life had been recognized,
and another hundred years to found their liberty on a
firm basis—the twelfth century charters thus being but
one of the stepping-stones to freedom.[1]  In reality, the
mediæval city was a fortified oasis amidst a country
plunged into feudal submission, and it had to make
room for itself by the force of its arms.   In consequence of the causes briefly alluded to in the preceding chapter, each village community had gradually
fallen under the yoke of some lay or clerical lord.  His
house had grown to be a castle, and his brothers-in-
arms were now the scum of adventurers, always ready
to plunder the peasants.   In addition to three days a
week which the peasants had to work for the lord,
they had also to bear all sorts of exactions for the
right to sow and to crop, to be gay or sad, to live,
to marry, or to die.   And, worst of all, they were
continually plundered by the armed robbers of some
neighbouring lord, who chose to consider them as their
master's kin, and to take upon them, and upon their
cattle and crops, the revenge for a feud he was fighting
against their owner.   Every meadow, every field, every
river, and road around the city, and every man upon
the land was under some lord.

The hatred of the burghers towards the feudal
barons has found a most characteristic expression in

---

[1] To give but one example—Cambrai made its first revolution in
907, and, after three or four more revolts, it obtained its charter in
1076.   This charter was repealed twice (1107 and 1138), and twice
obtained again (in 1127 and 1180).  Total, 223 years of struggles
before conquering the right to independence.  Lyons—from 1195 to
1320.

the wording of the different charters which they compelled them to sign. Heinrich V. is made to sign in the charter granted to Speier in 1111, that he frees the burghers from "the horrible and execrable law of mortmain, through which the town has been sunk into deepest poverty" (*von dem scheusslichen und nichtswürdigen Gesetze, welches gemein Budel genannt wird*, Kallsen, i. 307). The *coutume* of Bayonne, written about 1273, contains such passages as these: "The people is anterior to the lords. It is the people, more numerous than all others, who, desirous of peace, has made the lords for bridling and knocking down the powerful ones," and so on (Giry, *Établissements de Rouen*, i. 117, quoted by Luchaire, p. 24). A charter submitted for King Robert's signature is equally characteristic. He is made to say in it: "I shall rob no oxen nor other animals. I shall seize no merchants, nor take their moneys, nor impose ransom. From Lady Day to the All Saints' Day I shall seize no horse, nor mare, nor foals, in the meadows. I shall not burn the mills, nor rob the flour. . . . I shall offer no protection to thieves," etc. (Pfister has published that document, reproduced by Luchaire). The charter "granted" by the Besançon Archbishop Hugues, in which he has been compelled to enumerate all the mischiefs due to his mortmain rights, is equally characteristic.[1] And so on.

Freedom could not be maintained in such surroundings, and the cities were compelled to carry on the war outside their walls. The burghers sent out emissaries to lead revolt in the villages; they received villages

[1] See Tuetey, "Étude sur le droit municipal . . . en Franche-Comté," in *Mémoires de la Société d'émulation de Montbéliard*, 2ᵉ série, ii. 129 *seq.*

into their corporations, and they waged direct war
against the nobles. It Italy, where the land was
thickly sprinkled with feudal castles, the war assumed
heroic proportions, and was fought with a stern
acrimony on both sides. Florence sustained for
seventy-seven years a succession of bloody wars, in
order to free its *contado* from the nobles; but when
the conquest had been accomplished (in 1181) all had
to begin anew. The nobles rallied; they constituted
their own leagues in opposition to the leagues of the
towns, and, receiving fresh support from either the
Emperor or the Pope, they made the war last for
another 130 years. The same took place in Rome, in
Lombardy, all over Italy.

Prodigies of valour, audacity, and tenaciousness
were displayed by the citizens in these wars. But the
bows and the hatchets of the arts and crafts had not
always the upper hand in their encounters with the
armour-clad knights, and many castles withstood the
ingenious siege-machinery and the perseverance of the
citizens. Some cities, like Florence, Bologna, and
many towns in France, Germany, and Bohemia, succeeded in emancipating the surrounding villages, and
they were rewarded for their efforts by an extraordinary
prosperity and tranquillity. But even here, and still
more in the less strong or less impulsive towns, the
merchants and artisans, exhausted by war, and mis-
understanding their own interests, bargained over the
peasants' heads. They compelled the lord to swear
allegiance to the city; his country castle was dis-
mantled, and he agreed to build a house and to
reside in the city, of which he became a co-burgher
(*com-bourgeois, con-cittadino*); but he maintained in
return most of his rights upon the peasants, who only

won a partial relief from their burdens. The burgher could not understand that equal rights of citizenship might be granted to the peasant upon whose food supplies he had to rely, and a deep rent was traced between town and village. In some cases the peasants simply changed owners, the city buying out the barons' rights and selling them in shares to her own citizens.[1] Serfdom was maintained, and only much later on, towards the end of the thirteenth century, it was the craft revolution which undertook to put an end to it, and abolished personal servitude, but dispossessed at the same time the serfs of the land.[2] It hardly need be added that the fatal results of such policy were soon felt by the cities themselves ; the country became the city's enemy.

The war against the castles had another bad effect. It involved the cities in a long succession of mutual wars, which have given origin to the theory, till lately in vogue, namely, that the towns lost their independence through their own jealousies and mutual fights. The imperialist historians have especially supported this theory, which, however, is very much undermined now by modern research. It is certain that in Italy cities fought each other with a stubborn animosity, but nowhere else did such contests attain the same proportions ; and in Italy itself the city wars,

---

[1] This seems to have been often the case in Italy. In Switzerland, Bern bought even the towns of Thun and Burgdorf.

[2] Such was, at least, the case in the cities of Tuscany (Florence, Lucca, Sienna, Bologna, etc.), for which the relations between city and peasants are best known. (Luchitzkiy, "Slavery and Russian Slaves in Florence," in Kieff University *Izvestia* for 1885, who has perused Rumohr's *Ursprung der Besitzlosigheit der Colonien in Toscana*, 1830.) The whole matter concerning the relations between the cities and the peasants requires much more study than has hitherto been done

especially those of the earlier period, had their special causes. They were (as was already shown by Sismondi and Ferrari) a mere continuation of the war against the castles—the free municipal and federative principle unavoidably entering into a fierce contest with feudalism, imperialism, and papacy. Many towns which had but partially shaken off the yoke of the bishop, the lord, or the Emperor, were simply driven against the free cities by the nobles, the Emperor, and Church, whose policy was to divide the cities and to arm them against each other. These special circumstances (partly reflected on to Germany also) explain why the Italian towns, some of which sought support with the Emperor to combat the Pope, while the others sought support from the Church to resist the Emperor, were soon divided into a Gibelin and a Guelf camp, and why the same division appeared in each separate city.[1]

The immense economical progress realized by most Italian cities just at the time when these wars were hottest,[2] and the alliances so easily concluded between towns, still better characterize those struggles and further undermine the above theory. Already in the years 1130–1150 powerful leagues came into existence ; and a few years later, when Frederick Barbarossa invaded Italy and, supported by the nobles and some retardatory cities, marched against Milan, popular enthusiasm was roused in many towns by popular

---

[1] Ferrari's generalizations are often too theoretical to be always correct ; but his views upon the part played by the nobles in the city wars are based upon a wide range of authenticated facts.

[2] Only such cities as stubbornly kept to the cause of the barons, like Pisa or Verona, lost through the wars. For many towns which fought on the barons' side, the defeat was also the beginning of liberation and progress.

preachers. Crema, Piacenza, Brescia, Tortona, etc., went to the rescue ; the banners of the guilds of Verona, Padua, Vicenza, and Trevisa floated side by side in the cities' camp against the banners of the Emperor and the nobles. Next year the Lombardian League came into existence, and sixty years later we see it reinforced by many other cities, and forming a lasting organization which had half of its federal war-chest in Genoa and the other half in Venice.[1] In Tuscany, Florence headed another powerful league, to which Lucca, Bologna, Pistoia, etc., belonged, and which played an important part in crushing down the nobles in middle Italy, while smaller leagues were of common occurrence. It is thus certain that although petty jealousies undoubtedly existed, and discord could be easily sown, they did not prevent the towns from uniting together for the common defence of liberty. Only later on, when separate cities became little States, wars broke out between them, as always must be the case when States struggle for supremacy or colonies.

Similar leagues were formed in Germany for the same purpose. When, under the successors of Conrad, the land was the prey of interminable feuds between the nobles, the Westphalian towns concluded a league against the knights, one of the clauses of which was never to lend money to a knight who would continue to conceal stolen goods.[2] When "the knights and the nobles lived on plunder, and murdered whom they chose to murder," as the *Wormser Zorn* complains, the cities on the Rhine (Mainz, Cologne, Speier, Strasburg, and Basel) took the initiative of a league which soon

---

[1] Ferrari, ii. 18, 104 *seq.* ; Leo and Botta, i. 432.
[2] Joh. Falke, *Die Hansa als Deutsche See- und Handelsmacht*, Berlin, 1863, pp. 31, 55.

numbered sixty allied towns, repressed the robbers,
and maintained peace. Later on, the league of the
towns of Suabia, divided into three "peace districts"
(Augsburg, Constance, and Ulm), had the same purpose. And even when such leagues were broken,[1]
they lived long enough to show that while the supposed
peacemakers—the kings, the emperors, and the Church
—fomented discord, and were themselves helpless
against the robber knights, it was from the cities that
the impulse came for re-establishing peace and union.
The cities—not the emperors—were the real makers
of the national unity.[2]

Similar federations were organized for the same
purpose among small villages, and now that attention
has been drawn to this subject by Luchaire we may
expect soon to learn much more about them. Villages
joined into small federations in the *contado* of Florence,
so also in the dependencies of Novgorod and Pskov.
As to France, there is positive evidence of a federation
of seventeen peasant villages which has existed in the
Laonnais for nearly a hundred years (till 1256), and
has fought hard for its independence. Three more
peasant republics, which had sworn charters similar to
those of Laon and Soissons, existed in the neighbourhood of Laon, and, their territories being contiguous,
they supported each other in their liberation wars.
Altogether, Luchaire is of the opinion that many such
federations must have come into existence in France
in the twelfth and thirteenth centuries, but that documents relative to them are mostly lost. Of course,

[1] For Aachen and Cologne we have direct testimony that the
bishops of these two cities—one of them bought by the enemy—
opened to him the gates.

[2] See the facts, though not always the conclusions, of Nitzsch, iii.
133 *seq.*; also Kallsen, i. 458, etc.

being unprotected by walls, they could easily be crushed
down by the kings and the lords; but in certain
favourable circumstances, when they found support in
a league of towns and protection in their mountains,
such peasant republics became independent units of the
Swiss Confederation.[1]

As to unions between cities for peaceful purposes,
they were of quite common occurrence. The intercourse which had been established during the period of
liberation was not interrupted afterwards. Sometimes,
when the *scabini* of a German town, having to pronounce
judgment in a new or complicated case, declared that
they knew not the sentence (*des Urtheiles nicht weise
zu sein*), they sent delegates to another city to get the
sentence. The same happened also in France;[2] while
Forli and Ravenna are known to have mutually
naturalized their citizens and granted them full rights
in both cities. To submit a contest arisen between
two towns, or within a city, to another commune which
was invited to act as arbiter, was also in the spirit of
the times.[3] As to commercial treaties between cities,
they were quite habitual.[4] Unions for regulating the
production and the sizes of casks which were used for

[1] On the Commune of the Laonnais, which, until Melleville's
researches (*Histoire de la Commune du Laonnais*, Paris, 1853), was
confounded with the Commune of Laon, see Luchaire, pp. 75 *seq.*
For the early peasants' guilds and subsequent unions see R. Wilman's
"Die ländlichen Schutzgilden Westphaliens," in *Zeitschrift für Kulturgeschichte*, neue Folge, Bd. iii., quoted in Henne-am-Rhyn's
*Kulturgeschichte*, iii. 249.

[2] Luchaire, p. 149.

[3] Two important cities, like Mainz and Worms, would settle a
political contest by means of arbitration. After a civil war broken
out in Abbeville, Amiens would act, in 1231, as arbiter (Luchaire,
149); and so on.

[4] See, for instance, W. Stieda, *Hansische Vereinbarungen, l. c.,*
p. 114.

the commerce in wine, " herring unions," and so on,
were mere precursors of the great commercial federations of the Flemish Hansa, and, later on, of the great
North German Hansa, the history of which alone
might contribute pages and pages to illustrate the
federation spirit which permeated men at that time.
It hardly need be added, that through the Hanseatic
unions the mediæval cities have contributed more to
the development of international intercourse, navigation, and maritime discovery than all the States of the
first seventeen centuries of our era.

In a word, federations between small territorial units,
as well as among men united by common pursuits
within their respective guilds, and federations between
cities and groups of cities constituted the very essence
of life and thought during that period.  The first five
of the second decade of centuries of our era may thus
be described as an immense attempt at securing mutual
aid and support on a grand scale, by means of the
principles of federation and association carried on
through all manifestations of human life and to all
possible degrees.  This attempt was attended with
success to a very great extent.  It united men formerly
divided ; it secured them a very great deal of freedom,
and it tenfolded their forces.  At a time when particularism was bred by so many agencies, and the causes
of discord and jealousy might have been so numerous,
it is gratifying to see that cities scattered over a wide
continent had so much in common, and were so ready
to confederate for the prosecution of so many common
aims.  They succumbed in the long run before powerful enemies ; not having understood the mutual-aid
principle widely enough, they themselves committed
fatal faults ; but they did not perish through their own

jealousies, and their errors were not a want of federation spirit among themselves.

The results of that new move which mankind made in the mediæval city were immense. At the beginning of the eleventh century the towns of Europe were small clusters of miserable huts, adorned but with low clumsy churches, the builders of which hardly knew how to make an arch ; the arts, mostly consisting of some weaving and forging, were in their infancy ; learning was found in but a few monasteries. Three hundred and fifty years later, the very face of Europe had been changed. The land was dotted with rich cities, surrounded by immense thick walls which were embellished by towers and gates, each of them a work of art in itself. The cathedrals, conceived in a grand style and profusely decorated, lifted their bell-towers to the skies, displaying a purity of form and a boldness of imagination which we now vainly strive to attain. The crafts and arts had risen to a degree of perfection which we can hardly boast of having superseded in many directions, if the inventive skill of the worker and the superior finish of his work be appreciated higher than rapidity of fabrication. The navies of the free cities furrowed in all directions the Northern and the Southern Mediterranean ; one effort more, and they would cross the oceans. Over large tracts of land well-being had taken the place of misery ; learning had grown and spread. The methods of science had been elaborated ; the basis of natural philosophy had been laid down ; and the way had been paved for all the mechanical inventions of which our own times are so proud. Such were the magic changes accomplished in Europe in less than four hundred years. And the losses which Europe sustained through the loss of its free cities can only be understood when we compare the seventeenth century with the fourteenth or the thirteenth. The prosperity which formerly characterized Scotland, Germany, the plains of Italy, was gone. The roads had fallen into an abject state, the cities were depopulated, labour was brought into slavery, art had vanished, commerce itself was decaying.[1]

If the mediæval cities had bequeathed to us no written documents to testify of their splendour, and left nothing behind but the monuments of building art which we see now all over Europe, from Scotland to Italy, and from Gerona in Spain to Breslau in Slavonian territory, we might yet conclude that the times of independent city life were times of the greatest development of human intellect during the Christian era down to the end of the eighteenth century. On looking, for instance, at a mediæval picture representing Nuremberg with its scores of towers and lofty spires, each of which bore the stamp of free creative art, we can hardly conceive that three hundred years before the town was but a collection of miserable hovels. And our admiration grows when we go into the details of the architecture and decorations of each of the countless churches, bell-towers, gates, and communal houses which are scattered all over Europe as far east as Bohemia and the now dead towns of Polish Galicia. Not only Italy, that mother of art, but all

[1] Cosmo Innes's *Early Scottish History* and *Scotland in Middle Ages*, quoted by Rev. Denton, *l.c.*, pp. 68, 69 ; Lamprecht's *Deutsches wirthschaftliche Leben im Mittelalter*, review by Schmoller in his *Jahrbuch*, Bd. xii. ; Sismondi's *Tableau de l'agriculture toscane*, pp. 226 *seq.* The dominions of Florence could be recognized at a glance through their prosperity.

Europe is full of such monuments. The very fact that of all arts architecture—a social art above all—had attained the highest development, is significant in itself. To be what it was, it must have originated from an eminently social life.

Mediæval architecture attained its grandeur—not only because it was a natural development of handicraft ; not only because each building, each architectural decoration, had been devised by men who knew through the experience of their own hands what artistic effects can be obtained from stone, iron, bronze, or even from simple logs and mortar ; not only because each monument was a result of collective experience, accumulated in each "mystery" or craft [1]—it was grand because it was born out of a grand idea. Like Greek art, it sprang out of a conception of brotherhood and unity fostered by the city. It had an audacity which could · only be won by audacious struggles and victories ; it had that expression of vigour, because vigour permeated all the life of the city. A cathedral or a communal house symbolized the grandeur of an organism of which every mason and stone-cutter was the builder, and a mediæval building ·appears—not as a solitary effort to which

[1] Mr. John J. Ennett (*Six Essays*, London, 1891) has excellent pages on this aspect of mediæval architecture. Mr. Willis, in his appendix to Whewell's *History of Inductive Sciences* (i. 261–262), has pointed out the beauty of the mechanical relations in mediæval buildings. "A new decorative construction was matured," he writes, "not thwarting and controlling, but assisting and harmonizing with the mechanical construction. Every member, every moulding, becomes a sustainer of weight ; and by the multiplicity of props assisting each other, and the consequent subdivision of weight, the eye was satisfied of the stability of the structure, notwithstanding curiously slender aspects of the separate parts." An art which sprang out of the *social* life of the city could not be better characterized.

thousands of slaves would have contributed the share assigned them by one man's imagination ; all the city contributed to it. The lofty bell-tower rose upon a structure, grand in itself, in which the life of the city was throbbing—not upon a meaningless scaffold like the Paris iron tower, not as a sham structure in stone intended to conceal the ugliness of an iron frame, as has been done in the Tower Bridge. Like the Acropolis of Athens, the cathedral of a mediæval city was intended to glorify the grandeur of the victorious city, to symbolize the union of its crafts, to express the glory of each citizen in a city of his own creation. After having achieved its craft revolution, the city often began a new cathedral in order to express the new, wider, and broader union which had been called into life.

The means at hand for these grand undertakings were disproportionately small. Cologne Cathedral was begun with a yearly outlay of but 500 marks ; a gift of 100 marks was inscribed as a grand donation ;[1] and even when the work approached completion, and gifts poured in in proportion, the yearly outlay in money stood at about 5,000 marks, and never exceeded 14,000. The cathedral of Basel was built with equally small means. But each corporation contributed its part of stone, work, and decorative genius to *their* common monument. Each guild expressed in it its political conceptions, telling in stone or in bronze the history of the city, glorifying the principles of " Liberty, equality, and fraternity,"[2] praising the city's allies, and sending to eternal fire its enemies.

[1] Dr. L. Ennen, *Der Dom zu Köln, seine Construction und Anstaltung*, Köln, 1871.

[2] The three statues are among the outer decorations of Nôtre Dame de Paris.

And each guild bestowed its *love* upon the communal monument by richly decorating it with stained windows, paintings, " gates, worthy to be the gates of Paradise," as Michel Angelo said, or stone decorations of each minutest corner of the building.[1] Small cities, even small parishes,[2] vied with the big agglomerations in this work, and the cathedrals of Laon and St. Ouen hardly stand behind that of Rheims, or the Communal House of Bremen, or the folkmote's bell-tower of Breslau. " No works must be begun by the commune but such as are conceived in response to the grand heart of the commune, composed of the hearts of all citizens, united in one common will "— such were the words of the Council of Florence ; and this spirit appears in all communal works of common utility, such as the canals, terraces, vineyards, and fruit gardens around Florence, or the irrigation canals which intersected the plains of Lombardy, or the port and aqueduct of Genoa, or, in fact, any works of the kind which were achieved by almost every city.[3]

All arts had progressed in the same way in the mediæval cities, those of our own days mostly being but a continuation of what had grown at that time. The prosperity of the Flemish cities was based upon the fine woollen cloth they fabricated. Florence, at the

[1] Mediæval art, like Greek art, did not know those curiosity-shops which we call a National Gallery or a Museum. A picture was painted, a statue was carved, a bronze decoration was cast to stand in its proper place in a monument of communal art. It lived there, it was part of a whole, and it contributed to give unity to the impression produced by the whole.

[2] Cf. J. T. Ennett's " Second Essay," p. 36.

[3] Sismondi, iv. 172 ; xvi. 356. The great canal, *Naviglio Grande*, which brings the water from the Tessino, was begun in 1179, *i.e.* after the conquest of independence, and it was ended in the thirteenth century. On the subsequent decay, see xvi. 355.

beginning of the fourteenth century, before the black death, fabricated from 70,000 to 100,000 *panni* of woollen stuffs, which were valued at 1,200,000 golden florins.[1]   The chiselling of precious metals, the art of casting, the fine forging of iron, were creations of the mediæval "mysteries" which had succeeded in attaining in their own domains all that could be made by the hand, without the use of a powerful prime motor.   By the hand and by invention, because, to use Whewell's words :

"Parchment and paper, printing and engraving, improved glass and steel, gunpowder, clocks, telescopes, the mariner's compass, the reformed calendar, the decimal notation ; algebra, trigonometry, chemistry, counterpoint (an invention equivalent to a new creation of music); these are all possessions which we inherit from that which has so disparagingly been termed the Stationary Period" (*History of Inductive Sciences*, i. 252).

True that no new principle was illustrated by any of these discoveries, as Whewell said ; but mediæval science had done something more than the actual discovery of new principles.   It had prepared the discovery of all the new principles which we know at the present time in mechanical sciences : it had accustomed the explorer to observe facts and to reason from them.   It was inductive science, even though it had not yet fully grasped the importance and the powers of induction ; and it laid the foundations of

---

[1] In 1336 it had 8,000 to 10,000 boys and girls in its primary schools, 1,000 to 1,200 boys in its seven middle schools, and from 550 to 600 students in its four universities.   The thirty communal hospitals contained over 1,000 beds for a population of 90,000 inhabitants (Capponi, ii. 249 *seq.*).   It has more than once been suggested by authoritative writers that education stood, as a rule, at a much higher level than is generally supposed.   Certainly so in democratic Nuremberg.

both mechanics and natural philosophy. Francis Bacon, Galileo, and Copernicus were the direct descendants of a Roger Bacon and a Michael Scot, as the steam engine was a direct product of the researches carried on in the Italian universities on the weight of the atmosphere, and of the mathematical and technical learning which characterized Nuremberg.

But why should one take trouble to insist upon the advance of science and art in the mediæval city? Is it not enough to point to the cathedrals in the domain of skill, and to the Italian language and the poem of Dante in the domain of thought, to give at once the measure of what the mediæval city *created* during the four centuries it lived?

The mediæval cities have undoubtedly rendered an immense service to European civilization. They have prevented it from being drifted into the theocracies and despotical states of old; they have endowed it with the variety, the self-reliance, the force of initiative, and the immense intellectual and material energies it now possesses, which are the best pledge for its being able to resist any new invasion of the East. But why did these centres of civilization, which attempted to answer to deeply-seated needs of human nature, and were so full of life, not live further on? Why were they seized with senile debility in the sixteenth century? and, after having repulsed so many assaults from without, and only borrowed new vigour from their interior struggles, why did they finally succumb to both?

Various causes contributed to this effect, some of them having their roots in the remote past, while others originated in the mistakes committed by the cities themselves. Towards the end of the fifteenth century, mighty States, reconstructed on the old Roman pattern, were already coming into existence. In each country and each region some feudal lord, more cunning, more given to hoarding, and often less scrupulous than his neighbours, had succeeded in appropriating to himself richer personal domains, more peasants on his lands, more knights in his following, more treasures in his chest. He had chosen for his seat a group of happily-situated villages, not yet trained into free municipal life—Paris, Madrid, or Moscow—and with the labour of his serfs he had made of them royal fortified cities, whereto he attracted war companions by a free distribution of villages, and merchants by the protection he offered to trade. The germ of a future State, which began gradually to absorb other similar centres, was thus laid. Lawyers, versed in the study of Roman law, flocked into such centres ; a tenacious and ambitious race of men issued from among the burgesses, who equally hated the naughtiness of the lords and what they called the lawlessness of the peasants. The very forms of the village community, unknown to their code, the very principles of federalism were repulsive to them as " barbarian" inheritances. Cæsarism, supported by the fiction of popular consent and by the force of arms, was their ideal, and they worked hard for those who promised to realize it.[1]

[1] Cf. L. Ranke's excellent considerations upon the essence of Roman law in his *Weltgeschichte*, Bd. iv. Abth. 2, pp. 20–31. Also Sismondi's remarks upon the part played by the *légistes* in the constitution of royal authority, *Histoire des Français*, Paris, 1826, viii. 85–99. The popular hatred against these "*weise Doktoren und Beutelschneider des Volks*" broke out with full force in the first years of the sixteenth century in the sermons of the early Reform movement.

The Christian Church, once a rebel against Roman law and now its ally, worked in the same direction. The attempt at constituting the theocratic Empire of Europe having proved a failure, the more intelligent and ambitious bishops now yielded support to those whom they reckoned upon for reconstituting the power of the Kings of Israel or of the Emperors of Constantinople. The Church bestowed upon the rising rulers her sanctity, she crowned them as God's representatives on earth, she brought to their service the learning and the statesmanship of her ministers, her blessings and maledictions, her riches, and the sympathies she had retained among the poor. The peasants, whom the cities had failed or refused to free, on seeing the burghers impotent to put an end to the interminable wars between the knights—which wars they had so dearly to pay for—now set their hopes upon the King, the Emperor, or the Great Prince ; and while aiding them to crush down the mighty feudal owners, they aided them to constitute the centralized State. And finally, the invasions of the Mongols and the Turks, the holy war against the Maures in Spain, as well as the terrible wars which soon broke out between the growing centres of sovereignty—Ile de France and Burgundy, Scotland and England, England and France, Lithuania and Poland, Moscow and Tver, and so on—contributed to the same end. Mighty States made their appearance ; and the cities had now to resist not only loose federations of lords, but strongly-organized centres, which had armies of serfs at their disposal.

The worst was, that the growing autocracies found support in the divisions which had grown within the cities themselves. The fundamental idea of the mediæval city was grand, but it was not wide enough. Mutual aid and support cannot be limited to a small association; they must spread to its surroundings, or else the surroundings will absorb the association. And in this respect the mediæval citizen had committed a formidable mistake at the outset. Instead of looking upon the peasants and artisans who gathered under the protection of his walls as upon so many aids who would contribute their part to the making of the city—as they really did—a sharp division was traced between the "families" of old burghers and the new-comers. For the former, all benefits from communal trade and communal lands were reserved, and nothing was left for the latter but the right of freely using the skill of their own hands. The city thus became divided into "the burghers" or "the commonalty," and "the inhabitants."[1] The trade, which was formerly communal, now became the privilege of the merchant and artisan "families," and the next step—that of becoming individual, or the privilege of oppressive trusts—was unavoidable.

The same division took place between the city proper and the surrounding villages. The commune had well tried to free the peasants, but her wars against the lords became, as already mentioned, wars for freeing the city itself from the lords, rather than for freeing the peasants. She left to the lord his rights over the villeins, on condition that he would molest the city no more and would become co-burgher. But the nobles "adopted" by the city, and now residing

---

[1] Brentano fully understood the fatal effects of the struggle between the "old burghers" and the new-comers. Miaskowski, in his work on the village communities of Switzerland, has indicated the same for village communities.

within its walls, simply carried on the old war within the very precincts of the city. They disliked to submit to a tribunal of simple artisans and merchants, and fought their old feuds in the streets. Each city had now its Colonnas and Orsinis, its Overstolzes and Wises. Drawing large incomes from the estates they had still retained, they surrounded themselves with numerous clients and feudalized the customs and habits of the city itself. And when discontent began to be felt in the artisan classes of the town, they offered their sword and their followers to settle the differences by a free fight, instead of letting the discontent find out the channels which it did not fail to secure itself in olden times.

The greatest and the most fatal error of most cities was to base their wealth upon commerce and industry, to the neglect of agriculture. They thus repeated the error which had once been committed by the cities of antique Greece, and they fell through it into the same crimes.[1] The estrangement of so many cities from the land necessarily drew them into a policy hostile to the land, which became more and more evident in the times of Edward the Third,[2] the French Jacqueries, the Hussite wars, and the Peasant War in Germany. On the other hand, a commercial policy involved them in distant enterprises. Colonies were founded by the Italians in the south-east, by German

---

[1] The trade in slaves kidnapped in the East was never discontinued in the Italian republics till the fifteenth century. Feeble traces of it are found also in Germany and elsewhere. See Cibrario. *Della schiavitù e del servaggio*, 2 vols. Milan, 1868; Professor Luchitzkiy, "Slavery and Russian Slaves in Florence in the Fourteenth and Fifteenth Centuries," in *Izvestia* of the Kieff University, 1885.

[2] J. R. Green's *History of the English People*, London, 1878, i. 455.

cities in the east, by Slavonian cities in the far north-east. Mercenary armies began to be kept for colonial wars, and soon for local defence as well. Loans were contracted to such an extent as to totally demoralize the citizens ; and internal contests grew worse and worse at each election, during which the colonial politics in the interest of a few families was at stake. The division into rich and poor grew deeper, and in the sixteenth century, in each city, the royal authority found ready allies and support among the poor.

And there is yet another cause of the decay of communal institutions, which stands higher and lies deeper than all the above. The history of the mediæval cities offers one of the most striking illustrations of the power of *ideas* and *principles* upon the destinies of mankind, and of the quite opposed results which are obtained when a deep modification of leading ideas has taken place. Self-reliance and federalism, the sovereignty of each group, and the construction of the political body from the simple to the composite, were the leading ideas in the eleventh century. But since that time the conceptions had entirely changed. The students of Roman law and the prelates of the Church, closely bound together since the time of Innocent the Third, had succeeded in paralyzing the idea—the antique Greek idea—which presided at the foundation of the cities. For two or three hundred years they taught from the pulpit, the University chair, and the judges' bench, that salvation must be sought for in a strongly-centralized State, placed under a semi-divine authority ;[1] that *one* man can and must be the saviour of society, and that in the name of public salvation he can com-

---

[1] See the theories expressed by the Bologna lawyers, already at the Congress of Roncaglia in 1158.

mit any violence : burn men and women at the stake,
make them perish under indescribable tortures, plunge
whole provinces into the most abject misery. Nor did
they fail to give object lessons to this effect on a
grand scale, and with an unheard-of cruelty, wherever
the king's sword and the Church's fire, or both at
once, could reach. By these teachings and examples,
continually repeated and enforced upon public attention, the very minds of the citizens had been shaped
into a new mould. They began to find no authority
too extensive, no killing by degrees too cruel, once it
was "for public safety." And, with this new direction
of mind and this new belief in one man's power, the
old federalist principle faded away, and the very
creative genius of the masses died out. The Roman
idea was victorious, and in such circumstances the
centralized State had in the cities a ready prey.

Florence in the fifteenth century is typical of this
change. Formerly a popular revolution was the signal
of a new departure. Now, when the people, brought
to despair, insurged, it had constructive ideas no
more ; no fresh idea came out of the movement. A
thousand representatives were put into the Communal
Council instead of 400 ; 100 men entered the *signoria*
instead of 80. But a revolution of figures could be
of no avail. The people's discontent was growing up,
and new revolts followed. A saviour—the "tyran"
—was appealed to ; he massacred the rebels, but the
disintegration of the communal body continued worse
than ever. And when, after a new revolt, the people
of Florence appealed to their most popular man,
Gieronimo Savonarola, for advice, the monk's answer
was :—" Oh, people mine, thou knowest that I cannot
go into State affairs . . . . purify thy soul, and if in

such a disposition of mind thou reformest thy city, then, people of Florence, thou shalt have inaugurated the reform in all Italy!" Carnival masks and vicious books were burned, a law of charity and another against usurers were passed—and the democracy of Florence remained where it was. The old spirit had gone. By too much trusting to government, they had ceased to trust to themselves; they were unable to open new issues. The State had only to step in and to crush down their last liberties.

And yet, the current of mutual aid and support did not die out in the masses, it continued to flow even after that defeat. It rose up again with a formidable force, in answer to the communist appeals of the first propagandists of the reform, and it continued to exist even after the masses, having failed to realize the life which they hoped to inaugurate under the inspiration of a reformed religion, fell under the dominions of an autocratic power. It flows still even now, and it seeks its way to find out a new expression which would not be the State, nor the mediæval city, nor the village community of the barbarians, nor the savage clan, but would proceed from all of them, and yet be superior to them in its wider and more deeply humane conceptions.

# CHAPTER VII

## MUTUAL AID AMONGST OURSELVES

Popular revolts at the beginning of the State-period.—Mutual Aid institutions of the present time.—The village community: its struggles for resisting its abolition by the State.—Habits derived from the village-community life, retained in our modern villages.—Switzerland, France, Germany, Russia.

THE mutual-aid tendency in man has so remote an origin, and is so deeply interwoven with all the past evolution of the human race, that it has been maintained by mankind up to the present time, notwithstanding all vicissitudes of history. It was chiefly evolved during periods of peace and prosperity; but when even the greatest calamities befell men—when whole countries were laid waste by wars, and whole populations were decimated by misery, or groaned under the yoke of tyranny—the same tendency continued to live in the villages and among the poorer classes in the towns; it still kept them together, and in the long run it reacted even upon those ruling, fighting, and devastating minorities which dismissed it as sentimental nonsense. And whenever mankind had to work out a new social organization, adapted to a new phasis of development, its constructive genius always drew the elements and the inspiration for the new departure from that same ever-living tendency. New economical and social institutions, in so far as they were a creation of the masses, new ethical systems, and new religions, all have originated from the same source, and the ethical progress of our race, viewed in its broad lines, appears as a gradual extension of the mutual-aid principles from the tribe to always larger and larger agglomerations, so as to finally embrace one day the whole of mankind, without respect to its divers creeds, languages, and races.

After having passed through the savage tribe, and next through the village community, the Europeans came to work out in mediæval times a new form of organization, which had the advantage of allowing great latitude for individual initiative, while it largely responded at the same time to man's need of mutual support. A federation of village communities, covered by a network of guilds and fraternities, was called into existence in the mediæval cities. The immense results achieved under this new form of union—in well-being for all, in industries, art, science, and commerce—were discussed at some length in two preceding chapters, and an attempt was also made to show why, towards the end of the fifteenth century, the mediæval republics —surrounded by domains of hostile feudal lords, unable to free the peasants from servitude, and gradually corrupted by ideas of Roman Cæsarism—were doomed to become a prey to the growing military States.

However, before submitting for three centuries to come, to the all-absorbing authority of the State, the masses of the people made a formidable attempt at reconstructing society on the old basis of mutual aid and support. It is well known by this time that the great movement of the reform was not a mere revolt against the abuses of the Catholic Church. It had its constructive ideal as well, and that ideal was life in

free, brotherly communities. Those of the early
writings and sermons of the period which found most
response with the masses were imbued with ideas of
the economical and social brotherhood of mankind.
The "Twelve Articles" and similar professions of
faith, which were circulated among the German and
Swiss peasants and artisans, maintained not only every
one's right to interpret the Bible according to his own
understanding, but also included the demand of communal lands being restored to the village communities
and feudal servitudes being abolished, and they always
alluded to the "true" faith—a faith of brotherhood.
At the same time scores of thousands of men and
women joined the communist fraternities of Moravia,
giving them all their fortune and living in numerous
and prosperous settlements constructed upon the principles of communism.[1]  Only wholesale massacres by
the thousand could put a stop to this widely-spread
popular movement, and it was by the sword, the fire,
and the rack that the young States secured their first
and decisive victory over the masses of the people.[2]

[1] A bulky literature, dealing with this formerly much-neglected
subject, is now growing in Germany. Keller's works, *Ein Apostel der
Wiedertäufer* and *Geschichte der Wiedertäufer*, Cornelius's *Geschichte
des münsterischen Aufruhrs*, and Janssen's *Geschichte des deutschen
Volkes* may be named as the leading sources.  The first attempt at
familiarizing English readers with the results of the wide researches
made in Germany in this direction has been made in an excellent little work by Richard Heath—"Anabaptism from its Rise
at Zwickau to its Fall at Münster, 1521–1536," London, 1895
(*Baptist Manuals*, vol. i.)—where the leading features of the movement are well indicated, and full bibliographical information is given.
Also K. Kautsky's "Communism in Central Europe in the Time of
the Reformation," London, 1897.

[2] Few of our contemporaries realize both the extent of this movement and the means by which it was suppressed.  But those who
wrote immediately after the great peasant war estimated at from
100,000 to 150,000 men the number of peasants slaughtered after

For the next three centuries the States, both on the
Continent and in these islands, systematically weeded
out all institutions in which the mutual-aid tendency
had formerly found its expression.   The village communities were bereft of their folkmotes, their courts
and independent administration ; their lands were
confiscated.   The guilds were spoliated of their possessions and liberties, and placed under the control, the
fancy, and the bribery of the State's official.   The
cities were divested of their sovereignty, and the very
springs of their inner life—the folkmote, the elected
justices and administration, the sovereign parish and
the sovereign guild—were annihilated ; the State's
functionary took possession of every link of what
formerly was an organic whole.   Under that fatal
policy and the wars it engendered, whole regions, once
populous and wealthy, were laid bare ; rich cities became insignificant boroughs ; the very roads which
connected them with other cities became impracticable.
Industry, art, and knowledge fell into decay.   Political
education, science, and law were rendered subservient
to the idea of State centralization.   It was taught in
the Universities and from the pulpit that the institutions in which men formerly used to embody their
needs of mutual support could not be tolerated in a
properly organized State ; that the State alone could
represent the bonds of union between its subjects ;
that federalism and " particularism " were the enemies
of progress, and the State was the only proper initiator
of further development.   By the end of the last century
the kings on the Continent, the Parliament in these

---

their defeat in Germany.   See Zimmermann's *Allgemeine Geschichte
des grossen Bauernkrieges*.   For the measures taken to suppress the
movement in the Netherlands see Richard Heath's *Anabaptism*.

isles, and the revolutionary Convention in France, although they were at war with each other, agreed in asserting that no separate unions between citizens must exist within the State ; that hard labour and death were the only suitable punishments to workers who dared to enter into " coalitions." " No state within the State!" The State alone, and the State's Church, must take care of matters of general interest, while the subjects must represent loose aggregations of individuals, connected by no particular bonds, bound to appeal to the Government each time that they feel a common need. Up to the middle of this century this was the theory and practice in Europe. Even commercial and industrial societies were looked at with suspicion. As to the workers, their unions were treated as unlawful almost within our own lifetime in this country and within the last twenty years on the Continent. The whole system of our State education was such that up to the present time, even in this country, a notable portion of society would treat as a revolutionary measure the concession of such rights as every one, freeman or serf, exercised five hundred years ago in the village folkmote, the guild, the parish, and the city.

The absorption of all social functions by the State necessarily favoured the development of an unbridled, narrow-minded individualism. In proportion as the obligations towards the State grew in numbers the citizens were evidently relieved from their obligations towards each other. In the guild—and in mediæval times every man belonged to some guild or fraternity—two "brothers" were bound to watch in turns a brother who had fallen ill; it would be sufficient now to give one's neighbour the address of the next paupers' hospital. In barbarian society, to assist at a fight between two men, arisen from a quarrel, and not to prevent it from taking a fatal issue, meant to be oneself treated as a murderer ; but under the theory of the all-protecting State the bystander need not intrude : it is the policeman's business to interfere, or not. And while in a savage land, among the Hottentots, it would be scandalous to eat without having loudly called out thrice whether there is not somebody wanting to share the food, all that a respectable citizen has to do now is to pay the poor tax and to let the starving starve. The result is, that the theory which maintains that men can, and must, seek their own happiness in a disregard of other people's wants is now triumphant all round— in law, in science, in religion. It is the religion of the day, and to doubt of its efficacy is to be a dangerous Utopian. Science loudly proclaims that the struggle of each against all is the leading principle of nature, and of human societies as well. To that struggle Biology ascribes the progressive evolution of the animal world. History takes the same line of argument ; and political economists, in their naïve ignorance, trace all progress of modern industry and machinery to the " wonderful " effects of the same principle. The very religion of the pulpit is a religion of individualism, slightly mitigated by more or less charitable relations to one's neighbours, chiefly on Sundays. " Practical " men and theorists, men of science and religious preachers, lawyers and politicians, all agree upon one thing—that individualism may be more or less softened in its harshest effects by charity, but that it is the only secure basis for the maintenance of society and its ulterior progress.

It seems, therefore, hopeless to look for mutual-aid institutions and practices in modern society. What

could remain of them?   And yet, as soon as we try to
ascertain how the millions of human beings live, and
begin to study their everyday relations, we are struck
with the immense part which the mutual-aid and mutual-
support principles play even now-a-days in human
life.   Although the destruction of mutual-aid institutions has been going on in practice and theory, for full
three or four hundred years, hundreds of millions of
men continue to live under such institutions; they
piously maintain them and endeavour to reconstitute
them where they have ceased to exist.   In our mutual
relations every one of us has his moments of revolt
against the fashionable individualistic creed of the day,
and actions in which men are guided by their mutual-
aid inclinations constitute so great a part of our daily
intercourse that if a stop to such actions could be put
all further ethical progress would be stopped at once.
Human society itself could not be maintained for even
so much as the lifetime of one single generation.
These facts, mostly neglected by sociologists and yet
of the first importance for the life and further elevation
of mankind, we are now going to analyze, beginning
with the standing institutions of mutual support, and
passing next to those acts of mutual aid which have
their origin in personal or social sympathies.

When we cast a broad glance on the present
constitution of European society we are struck at once
with the fact that, although so much has been done
to get rid of the village community, this form of union
continues to exist to the extent we shall presently
see, and that many attempts are now made either to
reconstitute it in some shape or another or to find
some substitute for it.   The current theory as regards

the village community is, that in Western Europe it has died out by a natural death, because the communal possession of the soil was found inconsistent with the modern requirements of agriculture. But the truth is that nowhere did the village community disappear of its own accord; everywhere, on the contrary, it took the ruling classes several centuries of persistent but not always successful efforts to abolish it and to confiscate the communal lands.

In France, the village communities began to be deprived of their independence, and their lands began to be plundered, as early as the sixteenth century. However, it was only in the next century, when the mass of the peasants was brought, by exactions and wars, to the state of subjection and misery which is vividly depicted by all historians, that the plundering of their lands became easy and attained scandalous proportions. "Every one has taken of them according to his powers . . . Imaginary debts have been claimed, in order to seize upon their lands;" so we read in an edict promulgated by Louis the Fourteenth in 1667.[1] Of course the State's remedy for such evils was to render the communes still more subservient to the State, and to plunder them itself. In fact, two years later all money revenue of the communes was confiscated by the King. As to the appropriation of communal lands, it grew worse and worse, and in the next century the nobles and the clergy had already taken possession of immense tracts of land—one-half of the cultivated area, according to certain estimates

---

[1] "Chacun s'en est accommodé selon sa bienséance . . . on les a partagés . . . pour dépouiller les communes, on s'est servi de dettes simulées" (Edict of Louis the Fourteenth, of 1667, quoted by several authors. Eight years before that date the communes had been taken under State management).

—mostly to let it go out of culture.[1] But the peasants still maintained their communal institutions, and until the year 1787 the village folkmotes, composed of all householders, used to come together in the shadow of the bell-tower or a tree, to allot and re-allot what they had retained of their fields, to assess the taxes, and to elect their executive, just as the Russian *mir* does at the present time. This is what Babeau's researches have proved to demonstration.[2]

The Government found, however, the folkmotes "too noisy," too disobedient, and in 1787, elected councils, composed of a mayor and three to six syndics, chosen from among the wealthier peasants, were introduced instead. Two years later the Revolutionary Assemblée Constituante, which was on this point at one with the old *régime*, fully confirmed this law (on the 14th of December, 1789), and the *bourgeois du village* had now their turn for the plunder of communal lands, which continued all through the Revolutionary period. Only on the 16th of August, 1792, the Convention, under the pressure of the peasants' insurrections, decided to return the enclosed lands to the communes;[3] but it ordered at the same time that they should be divided in equal parts among the wealthier peasants only—a measure which pro-

---

[1] "On a great landlord's estate, even if he has millions of revenue, you are sure to find the land uncultivated" (Arthur Young). "One-fourth part of the soil went out of culture;" "for the last hundred years the land has returned to a savage state;" "the formerly flourishing Sologne is now a big marsh;" and so on (Théron de Montaugé, quoted by Taine in *Origines de la France Contemporaine*, tome i. p. 441).

[2] A. Babeau, *Le Village sous l'Ancien Régime*, 3ᵉ édition. Paris, 1892.

[3] In Eastern France the law only confirmed what the peasants had already done themselves; in other parts of France it usually remained a dead letter.

voked new insurrections and was abrogated next year, in 1793, when the order came to divide the communal lands among all commoners, rich and poor alike, "active" and "inactive."

These two laws, however, ran so much against the conceptions of the peasants that they were not obeyed, and wherever the peasants had retaken possession of part of their lands they kept them undivided. But then came the long years of wars, and the communal lands were simply confiscated by the State (in 1794) as a mortgage for State loans, put up for sale, and plundered as such; then returned again to the communes and confiscated again (in 1813); and only in 1816 what remained of them, *i. e.* about 15,000,000 acres of the least productive land, was restored to the village communities.[1] Still this was not yet the end of the troubles of the communes. Every new *régime* saw in the communal lands a means for gratifying its supporters, and three laws (the first in 1837 and the last under Napoleon the Third) were passed to induce the village communities to divide their estates. Three

[1] After the triumph of the middle-class reaction the communal lands were declared (August 24, 1794) the States domains, and, together with the lands confiscated from the nobility, were put up for sale, and pilfered by the *bandes noires* of the small *bourgeoisie*. True that a stop to this pilfering was put next year (law of 2 Prairial, An V), and the preceding law was abrogated; but then the village communities were simply abolished, and cantonal councils were introduced instead. Only seven years later (9 Prairial, An XII), *i. e.* in 1801, the village communities were reintroduced, but not until after having been deprived of all their rights, the mayor and syndics being nominated by the Government in the 36,000 communes of France! This system was maintained till after the revolution of 1830, when elected communal councils were reintroduced under the law of 1787. As to the communal lands, they were again seized upon by the State in 1813, plundered as such, and only partly restored to the communes in 1816. See the classical collection of French laws, by Dalloz, *Répertoire de Jurisprudence;* also the works of Doniol, Dareste, Bonnemère, Babeau, and many others.

times these laws had to be repealed, in consequence of the opposition they met with in the villages; but something was snapped up each time, and Napoleon the Third, under the pretext of encouraging perfected methods of agriculture, granted large estates out of the communal lands to some of his favourites.

As to the autonomy of the village communities, what could be retained of it after so many blows? The mayor and the syndics were simply looked upon as unpaid functionaries of the State machinery. Even now, under the Third Republic, very little can be done in a village community without the huge State machinery, up to the *préfet* and the ministries, being set in motion. It is hardly credible, and yet it is true, that when, for instance, a peasant intends to pay in money his share in the repair of a communal road, instead of himself breaking the necessary amount of stones, no fewer than twelve different functionaries of the State must give their approval, and an aggregate of *fifty-two* different acts must be performed by them, and exchanged between them, before the peasant is permitted to pay that money to the communal council. All the remainder bears the same character.[1]

What took place in France took place everywhere in Western and Middle Europe. Even the chief dates of the great assaults upon the peasant lands are the same. For England the only difference is that the spoliation was accomplished by separate acts rather than by general sweeping measures—with less haste but more thoroughly than in France. The

---

[1] This procedure is so absurd that one would not believe it possible if the fifty-two different acts were not enumerated in full by a quite authoritative writer in the *Journal des Économistes* (1893, April, p. 94), and several similar examples were not given by the same author.

seizure of the communal lands by the lords also began in the fifteenth century, after the defeat of the peasant insurrection of 1380—as seen from Rossus's *Historia* and from a statute of Henry the Seventh, in which these seizures are spoken of under the heading of "enormitees and myschefes as be hurtfull . . . to the common wele."[1] Later on the Great Inquest, under Henry the Eighth, was begun, as is known, in order to put a stop to the enclosure of communal lands, but it ended in a sanction of what had been done.[2] The communal lands continued to be preyed upon, and the peasants were driven from the land. But it was especially since the middle of the eighteenth century that, in England as everywhere else, it became part of a systematic policy to simply weed out all traces of communal ownership; and the wonder is not that it has disappeared, but that it could be maintained, even in England, so as to be "generally prevalent so late as the grandfathers of this generation."[3] The very object of the Enclosure Acts, as shown by Mr. Seebohm, was to remove this system,[4] and it was so

---

[1] Dr. Ochenkowski, *Englands wirthschaftliche Entwickelung im Ausgange des Mittelalters* (Jena, 1879), pp. 35 *seq.*, where the whole question is discussed with full knowledge of the texts.

[2] Nasse, *Ueber die mittelalterliche Feldgemeinschaft und die Einhegungen des XVI. Jahrhunderts in England* (Bonn, 1869), pp. 4, 5; Vinogradov, *Villainage in England* (Oxford, 1892).

[3] F. Seebohm, *The English Village Community*, 3rd edition, 1884, pp. 13–15.

[4] "An examination into the details of an Enclosure Act will make clear the point that the system as above described [communal ownership] is the system which it was the object of the Enclosure Act to remove" (Seebohm, *l. c.* p. 13). And further on, "They were generally drawn in the same form, commencing with the recital that the open and common fields lie dispersed in small pieces, intermixed with each other and inconveniently situated; that divers persons own parts of them, and are entitled to rights of common on them . . . and that it is desired that they may be divided and

well removed by the nearly four thousand Acts passed between 1760 and 1844 that only faint traces of it remain now. The land of the village communities was taken by the lords, and the appropriation was sanctioned by Parliament in each separate case.

In Germany, in Austria, in Belgium the village community was also destroyed by the State. Instances of commoners themselves dividing their lands were rare,[1] while everywhere the States coerced them to enforce the division, or simply favoured the private appropriation of their lands. The last blow to communal ownership in Middle Europe also dates from the middle of the eighteenth century. In Austria sheer force was used by the Government, in 1768, to compel the communes to divide their lands—a special commission being nominated two years later for that purpose. In Prussia Frederick the Second, in several of his ordinances (in 1752, 1763, 1765, and 1769), recommended to the *Justizcollegien* to enforce the division. In Silesia a special resolution was issued to serve that aim in 1771. The same took place in Belgium, and, as the communes did not obey, a law was issued in 1847 empowering the Government to buy communal meadows in order to sell them in retail, and to make a forced sale of the communal land when there was a would-be buyer for it.[2]

---

enclosed, a specific share being let out and allowed to each owner" (p. 14). Porter's list contained 3867 such Acts, of which the greatest numbers fall upon the decades of 1770–1780 and 1800–1820, as in France.

[1] In Switzerland we see a number of communes, ruined by wars, which have sold part of their lands, and now endeavour to buy them back.

[2] A. Buchenberger, "Agrarwesen und Agrarpolitik," in A. Wagner's *Handbuch der politischen Oekonomie*, 1892, Band i. pp. 280 *seq.*

In short, to speak of the natural death of the village communities in virtue of economical laws is as grim a joke as to speak of the natural death of soldiers slaughtered on a battlefield. The fact was simply this: The village communities had lived for over a thousand years; and where and when the peasants were not ruined by wars and exactions they steadily improved their methods of culture. But as the value of land was increasing, in consequence of the growth of industries, and the nobility had acquired, under the State organization, a power which it never had had under the feudal system, it took possession of the best parts of the communal lands, and did its best to destroy the communal institutions.

However, the village-community institutions so well respond to the needs and conceptions of the tillers of the soil that, in spite of all, Europe is up to this date covered with *living* survivals of the village communities, and European country life is permeated with customs and habits dating from the community period. Even in England, notwithstanding all the drastic measures taken against the old order of things, it prevailed as late as the beginning of the nineteenth century. Mr. Gomme—one of the very few English scholars who have paid attention to the subject— shows in his work that many traces of the communal possession of the soil are found in Scotland, " runrig " tenancy having been maintained in Forfarshire up to 1813, while in certain villages of Inverness the custom was, up to 1801, to plough the land for the whole community, without leaving any boundaries, and to allot it after the ploughing was done. In Kilmorie the allotment and re-allotment of the fields was in full vigour "till the last twenty-five years," and the Crofters' Commission found it still in vigour in certain islands.[1]   In Ireland the system prevailed up to the great famine; and as to England, Marshall's works, which passed unnoticed until Nasse and Sir Henry Maine drew attention to them, leave no doubt as to the village-community system having been widely spread, in nearly all English counties, at the beginning of the nineteenth century.[2]   No more than twenty years ago Sir Henry Maine was "greatly surprised at the number of instances of abnormal property rights, necessarily implying the former existence of collective ownership and joint cultivation," which a comparatively brief inquiry brought under his notice.[3]   And, communal institutions having persisted so late as that, a great number of mutual-aid habits and customs would undoubtedly be discovered in English villages if the writers of this country only paid attention to village life.[4]

[1] G. L. Gomme, "The Village Community, with special reference to its Origin and Forms of Survival in Great Britain" (*Contemporary Science Series*), London, 1890, pp. 141-143; also his *Primitive Folkmoots* (London, 1880), pp. 98 *seq.*

[2] "In almost all parts of the country, in the Midland and Eastern counties particularly, but also in the west—in Wiltshire, for example —in the south, as in Surrey, in the north, as in Yorkshire,—there are extensive open and common fields.   Out of 316 parishes of Northamptonshire 89 are in this condition; more than 100 in Oxfordshire; about 50,000 acres in Warwickshire; in Berkshire half the county; more than half of Wiltshire; in Huntingdonshire out of a total area of 240,000 acres 130,000 were commonable meadows, commons, and fields" (Marshall, quoted in Sir Henry Maine's *Village Communities in the East and West*, New York edition, 1876, pp. 88, 89).

[3] *Ibid.* p. 88; also Fifth Lecture.   The wide extension of "commons" in Surrey, even now, is well known.

[4] In quite a number of books dealing with English country life which I have consulted I have found charming descriptions of country scenery and the like, but almost nothing about the daily life and customs of the labourers.

As to the Continent, we find the communal institutions fully alive in many parts of France, Switzerland,
Germany, Italy, the Scandinavian lands, and Spain, to
say nothing of Eastern Europe; the village life in
these countries is permeated with communal habits
and customs; and almost every year the Continental
literature is enriched by serious works dealing with
this and connected subjects. I must, therefore, limit
my illustrations to the most typical instances. Switzerland is undoubtedly one of them. Not only the five
republics of Uri, Schwytz, Appenzell, Glarus, and
Unterwalden hold their lands as undivided estates,
and are governed by their popular folkmotes, but in
all other cantons too the village communities remain
in possession of a wide self-government, and own
large parts of the Federal territory.[1] Two-thirds of
all the Alpine meadows and two-thirds of all the
forests of Switzerland are until now communal land;
and a considerable number of fields, orchards, vineyards, peat bogs, quarries, and so on, are owned in
common. In the Vaud, where all the householders
continue to take part in the deliberations of their
elected communal councils, the communal spirit is
especially alive. Towards the end of the winter all
the young men of each village go to stay a few days
in the woods, to fell timber and to bring it down the
steep slopes tobogganing way, the timber and the

---

[1] In Switzerland the peasants in the open land also fell under the
dominion of lords, and large parts of their estates were appropriated
by the lords in the sixteenth and seventeenth centuries. (See, for
instance, Dr. A. Miaskowski, in Schmoller's *Forschungen*, Bd. ii.
1879, pp. 12 *seq.*) But the peasant war in Switzerland did not end in
such a crushing defeat of the peasants as it did in other countries,
and a great deal of the communal rights and lands was retained.
The self-government of the communes is, in fact, the very foundation
of the Swiss liberties.

fuel wood being divided among all households or sold
for their benefit. These excursions are real *fêtes* of
manly labour. On the banks of Lake Leman part of
the work required to keep up the terraces of the vineyards is still done in common ; and in the spring,
when the thermometer threatens to fall below zero
before sunrise, the watchman wakes up all householders, who light fires of straw and dung and protect
their vine-trees from the frost by an artificial cloud.
In nearly all cantons the village communities possess
so-called *Bürgernutzen*—that is, they hold in common
a number of cows, in order to supply each family with
butter ; or they keep communal fields or vineyards, of
which the produce is divided between the burghers; or
they rent their land for the benefit of the community.[1]

It may be taken as a rule that where the communes
have retained a wide sphere of functions, so as to be
living parts of the national organism, and where they
have not been reduced to sheer misery, they never fail
to take good care of their lands. Accordingly the communal estates in Switzerland strikingly contrast with
the miserable state of " commons " in this country.
The communal forests in the Vaud and the Valais are
admirably managed, in conformity with the rules of
modern forestry. Elsewhere the "strips" of communal fields, which change owners under the system
of re-allotment, are very well manured, especially as
there is no lack of meadows and cattle. The high-
level meadows are well kept as a rule, and the rural
roads are excellent.[2] And when we admire the Swiss

---

[1] Miaskowski, in Schmoller's *Forschungen*, Bd. ii. 1879, p. 15.

[2] See on this subject a series of works, summed up in one of the
excellent and suggestive chapters (not yet translated into English)
which K. Bücher has added to the German translation of Laveleye's
*Primitive Ownership*. Also Meitzen, " Das Agrar- und Forst-Wesen,

*châlet,* the mountain road, the peasants' cattle, the terraces of vineyards, or the school-house in Switzerland, we must keep in mind that without the timber for the *châlet* being taken from the communal woods and the stone from the communal quarries, without the cows being kept on the communal meadows, and the roads being made and the school-houses built by communal work, there would be little to admire.

It hardly need be said that a great number of mutual-aid habits and customs continue to persist in the Swiss villages. The evening gatherings for shelling walnuts, which take place in turns in each household; the evening parties for sewing the dowry of the girl who is going to marry; the calling of "aids" for building the houses and taking in the crops, as well as for all sorts of work which may be required by one of the commoners; the custom of exchanging children from one canton to the other, in order to make them learn two languages, French and German; and so on—all these are quite habitual;[1] while, on the other side, divers modern requirements are met in the same spirit. Thus in Glarus most of the Alpine meadows have been sold during a time of calamity; but the communes still continue to buy field land, and after the newly-bought fields have been left in the possession of separate commoners for ten, twenty, or thirty years, as the case might be, they return to the common stock, which is re-allotted

---

die Allmenden und die Landgemeinden der Deutschen Schweiz," in *Jahrbuch für Staatswissenschaft,* 1880, iv. (analysis of Miaskowsky's works); O'Brien, "Notes in a Swiss village," in *Macmillan's Magazine,* October 1885.

[1] The wedding gifts, which often substantially contribute in this country to the comfort of the young households, are evidently a remainder of the communal habits.

according to the needs of all. A great number of small associations are formed to produce some of the necessaries for life—bread, cheese, and wine—by common work, be it only on a limited scale ; and agricultural co-operation altogether spreads in Switzerland with the greatest ease. Associations formed between ten to thirty peasants, who buy meadows and fields in common, and cultivate them as co-owners, are of common occurrence ; while dairy associations for the sale of milk, butter, and cheese are organized everywhere. In fact, Switzerland was the birthplace of that form of co-operation. It offers, moreover, an immense field for the study of all sorts of small and large societies, formed for the satisfaction of all sorts of modern wants. In certain parts of Switzerland one finds in almost every village a number of associations—for protection from fire, for boating, for maintaining the quays on the shores of a lake, for the supply of water, and so on ; and the country is covered with societies of archers, sharpshooters, topographers, footpath explorers, and the like, originated from modern militarism.

Switzerland is, however, by no means an exception in Europe, because the same institutions and habits are found in the villages of France, of Italy, of Germany, of Denmark, and so on. We have just seen what has been done by the rulers of France in order to destroy the village community and to get hold of its lands; but notwithstanding all that onetenth part of the whole territory available for culture, i. e. 13,500,000 acres, including one-half of all the natural meadows and nearly a fifth part of all the forests of the country, remain in communal possession. The woods supply the communers with fuel, and the timber wood is cut, mostly by communal work, with all desirable regularity; the grazing lands are free for the commoners' cattle; and what remains of communal fields is allotted and re-allotted in certain parts of France—namely, in the Ardennes—in the usual way.[1]

These additional sources of supply, which aid the poorer peasants to pass through a year of bad crops without parting with their small plots of land and without running into irredeemable debts, have certainly their importance for both the agricultural labourers and the nearly three millions of small peasant proprietors. It is even doubtful whether small peasant proprietorship could be maintained without these additional resources. But the ethical importance of the communal possessions, small as they are, is still greater than their economical value. They maintain in village life a nucleus of customs and habits of mutual aid which undoubtedly acts as a mighty check upon the development of reckless individualism and greediness, which small land-ownership is only too prone to develop. Mutual aid in all possible circumstances of village life is part of the routine life in all parts of the country. Everywhere we meet, under different names, with the *charroi*, i. e. the free aid of the neighbours for taking in a crop, for vintage, or for building a house; everywhere we find the same evening gatherings as have just been mentioned in Switzerland; and everywhere the commoners associate for all sorts of work. Such habits are mentioned by

---

[1] The communes own, 4,554,100 acres of woods out of 24,813,000 in the whole territory, and 6,936,300 acres of natural meadows out of 11,394,000 acres in France. The remaining 2,000,000 acres are fields, orchards, and so on.

nearly all those who have written upon French village
life. But it will perhaps be better to give in this place
some abstracts from letters which I have just received
from a friend of mine whom I have asked to communicate to me his observations on this subject.
They come from an aged man who for years has been
the mayor of his commune in South France (in
Ariège); the facts he mentions are known to him
from long years of personal observation, and they
have the advantage of coming from one neighbourhood instead of being skimmed from a large area.
Some of them may seem trifling, but as a whole they
depict quite a little world of village life.

" In several communes in our neighbourhood," my
friend writes, "the old custom of *l'emprount* is in
vigour. When many hands are required in a *métairie*
for rapidly making some work—dig out potatoes or
mow the grass—the youth of the neighbourhood is
convoked; young men and girls come in numbers,
make it gaily and for nothing ; and in the evening,
after a gay meal, they dance.

" In the same communes, when a girl is going to
marry, the girls of the neighbourhood come to aid in
sewing the dowry. In several communes the women
still continue to spin a good deal. When the winding
off has to be done in a family it is done in one evening—all friends being convoked for that work. In
many communes of the Ariège and other parts of the
south-west the shelling of the Indian corn-sheaves is
also done by all the neighbours. They are treated
with chestnuts and wine, and the young people dance
after the work has been done. The same custom is
practised for making nut oil and crushing hemp. In
the commune of L. the same is done for bringing in

the corn crops. These days of hard work become *fête* days, as the owner stakes his honour on serving a good meal. No remuneration is given; all do it for each other.[1]

"In the commune of S. the common grazing-land is every year increased, so that nearly the whole of the land of the commune is now kept in common. The shepherds are elected by all owners of the cattle, including women. The bulls are communal.

"In the commune of M. the forty to fifty small sheep flocks of the commoners are brought together and divided into three or four flocks before being sent to the higher meadows. Each owner goes for a week to serve as shepherd.

"In the hamlet of C. a threshing machine has been bought in common by several households; the fifteen to twenty persons required to serve the machine being supplied by all the families. Three other threshing machines have been bought and are rented out by their owners, but the work is performed by outside helpers, invited in the usual way.

"In our commune of R. we had to raise the wall of the cemetery. Half of the money which was required for buying lime and for the wages of the skilled workers was supplied by the county council, and the other half by subscription. As to the work of carrying sand and water, making mortar, and serving the masons, it was done entirely by volunteers [just as in the Kabyle *djemmâa*]. The rural roads were repaired in the same way, by volunteer days of work given by the commoners. Other communes have built in

---

[1] In Caucasia they even do better among the Georgians. As the meal costs, and a poor man cannot afford to give it, a sheep is bought by those same neighbours who come to aid in the work.

the same way their fountains. The wine-press and other smaller appliances are frequently kept by the commune."

Two residents of the same neighbourhood, questioned by my friend, add the following :—

"At O. a few years ago there was no mill. The commune has built one, levying a tax upon the commoners. As to the miller, they decided, in order to avoid frauds and partiality, that he should be paid two francs for each bread-eater, and the corn be ground free.

"At St. G. few peasants are insured against fire. When a conflagration has taken place—so it was lately—all give something to the family which has suffered from it—a chaldron, a bed-cloth, a chair, and so on—and a modest household is thus reconstituted. All the neighbours aid to build the house, and in the meantime the family is lodged free by the neighbours."

Such habits of mutual support—of which many more examples could be given—undoubtedly account for the easiness with which the French peasants associate for using, in turn, the plough with its team of horses, the wine-press, and the threshing machine, when they are kept in the village by one of them only, as well as for the performance of all sorts of rural work in common. Canals were maintained, forests were cleared, trees were planted, and marshes were drained by the village communities from time immemorial ; and the same continues still. Quite lately, in *La Borne* of Lozère barren hills were turned into rich gardens by communal work. "The soil was brought on men's backs ; terraces were made and planted with chestnut trees, peach trees, and orchards, and water was brought for irrigation in canals two or three miles long." Just now they have dug a new canal, eleven miles in length.[1]

To the same spirit is also due the remarkable success lately obtained by the *syndicats agricoles*, or peasants' and farmers' associations. It was not until 1884 that associations of more than nineteen persons were permitted in France, and I need not say that when this "dangerous experiment" was ventured upon—so it was styled in the Chambers—all due "precautions" which functionaries can invent were taken. Notwithstanding all that, France begins to be covered with syndicates. At the outset they were only formed for buying manures and seeds, falsification having attained colossal proportions in these two branches;[2] but gradually they extended their functions in various directions, including the sale of agricultural produce and permanent improvements of the land. In South France the ravages of the phylloxera have called into existence a great number of wine-growers' associations. Ten to thirty growers form a syndicate, buy a steam-engine for pumping water, and make the necessary arrangements for inundating their vineyards in turn.[3] New associations for protecting the land

---

[1] Alfred Baudrillart, in H. Baudrillart's *Les Populations Rurales de la France*, 3rd series (Paris, 1893), p. 479.

[2] The *Journal des Économistes* (August 1892, May and August 1893) has lately given some of the results of analyses made at the agricultural laboratories at Ghent and at Paris. The extent of falsification is simply incredible; so also the devices of the "honest traders." In certain seeds of grass there was 32 per cent. of grains of sand, coloured so as to deceive even an experienced eye; other samples contained from 52 to 22 per cent. only of pure seed, the remainder being weeds. Seeds of vetch contained 11 per cent. of a poisonous grass (*nielle*); a flour for cattle-fattening contained 36 per cent. of sulphates; and so on *ad infinitum*.

[3] A. Baudrillart, *l. c.* p. 309. Originally one grower would undertake to supply water, and several others would agree to make use of it. "What especially characterises such associations," A. Bau-

from inundations, for irrigation purposes, and for maintaining canals are continually formed, and the unanimity
of all peasants of a neighbourhood, which is required
by law, is no obstacle. Elsewhere we have the
*fruitières*, or dairy associations, in some of which all
butter and cheese is divided in equal parts, irrespective
of the yield of each cow. In the Ariège we find an
association of eight separate communes for the common culture of their lands, which they have put
together; syndicates for free medical aid have been
formed in 172 communes out of 337 in the same
department; associations of consumers arise in connection with the syndicates; and so on.[1]   "Quite a
revolution is going on in our villages," Alfred Baudrillart writes, "through these associations, which take
in each region their own special characters."

Very much the same must be said of Germany.
Wherever the peasants could resist the plunder of
their lands, they have retained them in communal
ownership, which largely prevails in Württemberg,
Baden, Hohenzollern, and in the Hessian province of
Starkenberg.[2]  The communal forests are kept, as a

---

drillart remarks, "is that no sort of written agreement is concluded.
All is arranged in words. There was, however, not one single case
of difficulties having arisen between the parties."

[1] A. Baudrillart, *l. c.* pp. 300, 341, etc. M. Terssac, president of
the St. Gironnais syndicate (Ariège), wrote to my friend in substance
as follows:—"For the exhibition of Toulouse our association has
grouped the owners of cattle which seemed to us worth exhibiting.
The society undertook to pay one-half of the travelling and exhibition
expenses; one-fourth was paid by each owner, and the remaining
fourth by those exhibitors who had got prizes. The result was that
many took part in the exhibition who never would have done it
otherwise. Those who got the highest awards (350 francs) have
contributed 10 per cent. of their prizes, while those who have got no
prize have only spent 6 to 7 francs each."

[2] In Württemberg 1,629 communes out of 1,910 have communal
property. They owned in 1863 over 1,000,000 acres of land. In

rule, in an excellent state, and in thousands of communes timber and fuel wood are divided every year among all inhabitants; even the old custom of the *Lesholztag* is widely spread: at the ringing of the village bell all go to the forest to take as much fuel wood as they can carry.[1] In Westphalia one finds communes in which all the land is cultivated as one common estate, in accordance with all requirements of modern agronomy. As to the old communal customs and habits, they are in vigour in most parts of Germany. The calling in of *aids*, which are real *fêtes* of labour, is known to be quite habitual in Westphalia, Hesse, and Nassau. In well-timbered regions the timber for a new house is usually taken from the communal forest, and all the neighbours join in building the house. Even in the suburbs of Frankfort it is a regular custom among the gardeners that in case of one of them being ill all come on Sunday to cultivate his garden.[2]

In Germany, as in France, as soon as the rulers of the people repealed their laws against the peasant associations—that was only in 1884–1888—these unions began to develop with a wonderful rapidity, notwithstanding all legal obstacles which were put in

---

Baden 1,256 communes out of 1,582 have communal land; in 1884–1888 they held 121,500 acres of fields in communal culture, and 675,000 acres of forests, *i.e.* 46 per cent. of the total area under woods. In Saxony 39 per cent. of the total area is in communal ownership (Schmoller's *Jahrbuch*, 1886, p. 359). In Hohenzollern nearly two-thirds of all meadow land, and in Hohenzollern-Hechingen 41 per cent. of all landed property, are owned by the village communities (Buchenberger, *Agrarwesen*, vol. i. p. 300).

[1] See K. Bücher, who, in a special chapter added to Laveleye's *Ureigenthum*, has collected all information relative to the village community in Germany.

[2] K. Bücher, *ibid.* pp. 89, 90.

their way.[1]  "It is a fact," Buchenberger says, "that in *thousands* of village communities, in which no sort of chemical manure or rational fodder was ever known, both have become of everyday use, to a quite unforeseen extent, owing to these associations" (vol. ii. p. 507).  All sorts of labour-saving implements and agricultural machinery, and better breeds of cattle, are bought through the associations, and various arrangements for improving the quality of the produce begin to be introduced.  Unions for the sale of agricultural produce are also formed, as well as for permanent improvements of the land.[2]

From the point of view of social economics all these efforts of the peasants certainly are of little importance. They cannot substantially, and still less permanently, alleviate the misery to which the tillers of the soil are doomed all over Europe.  But from the ethical point of view, which we are now considering, their importance cannot be overrated.  They prove that even under the system of reckless individualism which now prevails the agricultural masses piously maintain their mutual-support inheritance ; and as soon as the States relax the iron laws by means of which they have broken all bonds between men, these bonds are at once reconstituted, notwithstanding the difficulties, political, economical, and social, which are many, and

---

[1] For this legislation and the numerous obstacles which were put in the way, in the shape of red-tapeism and supervision, see Buchenberger's *Agrarwesen und Agrarpolitik*, Bd. ii. pp. 342–363, and p. 506, note.

[2] Buchenberger, *l. c.* Bd. ii. p. 510.  The General Union of Agricultural Co-operation comprises an aggregate of 1,679 societies. In Silesia an aggregate of 32,000 acres of land has been lately drained by 73 associations; 454,800 acres in Prussia by 516 associations; in Bavaria there are 1,715 drainage and irrigation unions.

in such forms as best answer to the modern requirements of production. They indicate in which direction and in which form further progress must be
expected.

I might easily multiply such illustrations, taking
them from Italy, Spain, Denmark, and so on, and
pointing out some interesting features which are
proper to each of these countries.[1] The Slavonian
populations of Austria and the Balkan peninsula,
among whom the "compound family," or "undivided
household," is found in existence, ought also to be
mentioned.[2] But I hasten to pass on to Russia, where
the same mutual-support tendency takes certain new
and unforeseen forms. Moreover, in dealing with the
village community in Russia we have the advantage
of possessing an immense mass of materials, collected
during the colossal house-to-house inquest which was
lately made by several *zemstvos* (county councils), and
which embraces a population of nearly 20,000,000
peasants in different parts of the country.[3]

Two important conclusions may be drawn from the
bulk of evidence collected by the Russian inquests.
In Middle Russia, where fully one-third of the peasants
have been brought to utter ruin (by heavy taxation,
small allotments of unproductive land, rack rents, and

[1] See Appendix XII.
[2] For the Balkan peninsula see Laveleye's *Propriété Primitive*.
[3] The facts concerning the village community, contained in nearly
a hundred volumes (out of 450) of these inquests, have been classified
and summed up in an excellent Russian work by "V. V.," *The
Peasant Community* (*Krestianskaya Obschina*), St. Petersburg, 1892,
which, apart from its theoretical value, is a rich compendium of data
relative to this subject. The above inquests have also given origin
to an immense literature, in which the modern village-community
question for the first time emerges from the domain of generalities
and is put on the solid basis of reliable and sufficiently detailed facts.

very severe tax-collecting after total failures of crops), there was, during the first five-and-twenty years after the emancipation of the serfs, a decided tendency towards the constitution of individual property in land within the village communities. Many impoverished "horseless" peasants abandoned their allotments, and this land often became the property of those richer peasants, who borrow additional incomes from trade, or of outside traders, who buy land chiefly for exacting rack rents from the peasants. It must also be added that a flaw in the land-redemption law of 1861 offered great facilities for buying peasants' lands at a very small expense,[1] and that the State officials mostly used their weighty influence in favour of individual as against communal ownership. However, for the last twenty years a strong wind of opposition to the individual appropriation of the land blows again through the Middle Russian villages, and strenuous efforts are being made by the bulk of those peasants who stand between the rich and the very poor to uphold the village community. As to the fertile steppes of the South, which are now the most populous and the richest part of European Russia, they were mostly colonized, during the present century, under the system of individual ownership or occupation, sanctioned in that form by the State. But since improved methods of agriculture with the aid of machinery have been introduced in the region, the peasant owners have

---

[1] The redemption had to be paid by annuities for forty-nine years. As years went, and the greatest part of it was paid, it became easier and easier to redeem the smaller remaining part of it, and, as each allotment could be redeemed individually, advantage was taken of this disposition by traders, who bought land for half its value from the ruined peasants. A law was consequently passed to put a stop to such sales.

gradually begun themselves to transform their individual ownership into communal possession, and one
finds now, in that granary of Russia, a very great
number of spontaneously formed village communities
of recent origin.[1]

The Crimea and the part of the mainland which lies
to the north of it (the province of Taurida), for which
we have detailed data, offer an excellent illustration of
that movement. This territory began to be colonized,
after its annexation in 1783, by Great, Little, and
White Russians—Cossacks, freemen, and runaway
serfs—who came individually or in small groups from
all corners of Russia. They took first to cattle-breed-
ing, and when they began later on to till the soil, each
one tilled as much as he could afford to. But when—
immigration continuing, and perfected ploughs being
introduced—land stood in great demand, bitter disputes
arose among the settlers. They lasted for years, until
these men, previously tied by no mutual bonds,
gradually came to the idea that an end must be put to
disputes by introducing village-community ownership.
They passed decisions to the effect that the land which
they owned individually should henceforward be their
common property, and they began to allot and to
re-allot it in accordance with the usual village-community rules. The movement gradually took a great
extension, and on a small territory, the Taurida
statisticians found 161 villages in which communal
ownership had been introduced by the peasant proprietors themselves, chiefly in the years 1855–1885, *in*

---

[1] Mr. V. V., in his *Peasant Community*, has grouped together all
facts relative to this movement. About the rapid agricultural development of South Russia and the spread of machinery English readers
will find information in the Consular Reports (Odessa, Taganrog).

*lieu* of individual ownership. Quite a variety of village-community types has been freely worked out in this way by the settlers.[1] What adds to the interest of this transformation is that it took place, not only among the Great Russians, who are used to village-community life, but also among Little Russians, who have long since forgotten it under Polish rule, among Greeks and Bulgarians, and even among Germans, who have long since worked out in their prosperous and half-industrial Volga colonies their own type of village community.[2] It is evident that the Mussulman Tartars of Taurida hold their land under the Mussulman customary law, which is limited personal occupation; but even with them the European village community has been introduced in a few cases. As to other nationalities in Taurida, individual ownership has been abolished in six Esthonian, two Greek, two Bulgarian, one Czech, and one German village.

This movement is characteristic for the whole of the fertile steppe region of the south. But separate instances of it are also found in Little Russia. Thus in a number of villages of the province of Chernigov the peasants were formerly individual owners of their plots; they had separate legal documents for their plots and used to rent and to sell their land at will. But in the fifties of the nineteenth century a movement began among them in favour of communal possession, the chief

---

[1] In some instances they proceeded with great caution. In one village they began by putting together all meadow land, but only a small portion of the fields (about five acres per soul) was rendered communal; the remainder continued to be owned individually. Later on, in 1862–1864, the system was extended, but only in 1884 was communal possession introduced in full.—V. V.'s *Peasant Community*, pp. 1–14.

[2] On the Mennonite village community see A. Klaus, *Our Colonies* (*Nashi Kolonii*), St. Petersburg, 1869.

argument being the growing number of pauper families. The initiative of the reform was taken in one village, and the others followed suit, the last case on record dating from 1882. Of course there were struggles between the poor, who usually claim for communal possession, and the rich, who usually prefer individual ownership; and the struggles often lasted for years. In certain places the unanimity required then by the law being impossible to obtain, the village divided into two villages, one under individual ownership and the other under communal possession; and so they remained until the two coalesced into one community, or else they remained divided still. As to Middle Russia, it is a fact that in many villages which were drifting towards individual ownership there began since 1880 a mass movement in favour of re-establishing the village community. Even peasant proprietors who had lived for years under the individualist system returned *en masse* to the communal institutions. Thus, there is a considerable number of ex-serfs who have received one-fourth part only of the regulation allotments, but they have received them free of redemption and in individual ownership. There was in 1890 a wide-spread movement among them (in Kursk, Ryazan, Tambov, Orel, etc.) towards putting their allotments together and introducing the village community. The "free agriculturists" (*volnyie khlebopashtsy*), who were liberated from serfdom under the law of 1803, and had *bought* their allotments—each family separately—are now nearly all under the village-community system, which they have introduced themselves. All these movements are of recent origin, and non-Russians too join them. Thus the Bulgares in the district of Tiraspol, after having remained for sixty years under the personal-property system, introduced the village community in the years 1876–1882. The German Mennonites of Berdyansk fought in 1890 for introducing the village community, and the small peasant proprietors (*Kleinwirthschaftliche*) among the German Baptists were agitating in their villages in the same direction. One instance more: In the province of Samara the Russian government created in the forties, by way of experiment, 103 villages on the system of individual ownership. Each household received a splendid property of 105 acres. In 1890, out of the 103 villages the peasants in 72 had already notified the desire of introducing the village community. I take all these facts from the excellent work of V. V., who simply gives, in a classified form, the facts recorded in the above-mentioned house-to-house inquest.

This movement in favour of communal possession runs badly against the current economical theories, according to which intensive culture is incompatible with the village community. But the most charitable thing that can be said of these theories is that they have never been submitted to the test of experiment: they belong to the domain of political metaphysics. The facts which we have before us show, on the contrary, that wherever the Russian peasants, owing to a concurrence of favourable circumstances, are less miserable than they are on the average, and wherever they find men of knowledge and initiative among their neighbours, the village community becomes the very means for introducing various improvements in agriculture and village life altogether. Here, as elsewhere, mutual aid is a better leader to progress than the war of each against all, as may be seen from the following facts.

Under Nicholas the First's rule many Crown officials and serf-owners used to compel the peasants to introduce the communal culture of small plots of the village lands, in order to refill the communal storehouses after loans of grain had been granted to the poorest commoners. Such cultures, connected in the peasants' minds with the worst reminiscences of serfdom, were abandoned as soon as serfdom was abolished; but now the peasants begin to reintroduce them on their own account. In one district (Ostrogozhsk, in Kursk) the initiative of one person was sufficient to call them to life in four-fifths of all the villages. The same is met with in several other localities. On a given day the commoners come out, the richer ones with a plough or a cart and the poorer ones single-handed, and no attempt is made to discriminate one's share in the work. The crop is afterwards used for loans to the poorer commoners, mostly free grants, or for the orphans and widows, or for the village church, or for the school, or for repaying a communal debt.[1]

That all sorts of work which enters, so to say, in the routine of village life (repair of roads and bridges, dams, drainage, supply of water for irrigation, cutting of wood, planting of trees, etc.) are made by whole communes, and that land is rented and meadows are mown by whole communes—the work being accomplished by old and young, men and women, in the way described by Tolstoi—is only what one may expect

---

[1] Such communal cultures are known to exist in 159 villages out of 195 in the Ostrogozhsk district; in 150 out of 187 in Slavyano-serbsk; in 107 village communities in Alexandrovsk, 93 in Niko-layevsk, 35 in Elisabethgrad. In a German colony the communal culture is made for repaying a communal debt. All join in the work, although the debt was contracted by 94 householders out of 155.

from people living under the village-community system.[1]
They are of everyday occurrence all over the country.
But the village community is also by no means averse
to modern agricultural improvements, when it can
stand the expense, and when knowledge, hitherto kept
for the rich only, finds its way into the peasant's house.

It has just been said that perfected ploughs rapidly
spread in South Russia, and in many cases the village
communities were instrumental in spreading their use.
A plough was bought by the community, experimented
upon on a portion of the communal land, and the
necessary improvements were indicated to the makers,
whom the communes often aided in starting the manu-
facture of cheap ploughs as a village industry. In the
district of Moscow, where 1,560 ploughs were lately
bought by the peasants during five years, the impulse
came from those communes which rented lands as a
body for the special purpose of improved culture.

In the north-east (Vyatka) small associations of
peasants, who travel with their winnowing machines
(manufactured as a village industry in one of the iron
districts), have spread the use of such machines in the
neighbouring governments. The very wide spread of
threshing machines in Samara, Saratov, and Kherson
is due to the peasant associations, which can afford to
buy a costly engine, while the individual peasant
cannot. And while we read in nearly all economical
treatises that the village community was doomed to
disappear when the three-fields system had to be
substituted by the rotation of crops system, we see
in Russia many village communities taking the initiative
of introducing the rotation of crops. Before accepting

[1] Lists of such works which came under the notice of the *zemstvo*
statisticians will be found in V. V.'s *Peasant Community*, pp. 459-600.

S

it the peasants usually set apart a portion of the communal fields for an experiment in artificial meadows, and the commune buys the seeds.[1]   If the experiment proves successful they find no difficulty whatever in re-dividing their fields, so as to suit the five or six fields system.

This system is now in use in *hundreds* of villages of Moscow, Tver, Smolensk, Vyatka, and Pskov.[2] And where land can be spared the communities give also a portion of their domain to allotments for fruit-growing.   Finally, the sudden extension lately taken in Russia by the little model farms, orchards, kitchen gardens, and silkworm-culture grounds—which are started at the village school-houses, under the conduct of the school-master, or of a village volunteer—is also due to the support they found with the village communities.

Moreover, such permanent improvements as drainage and irrigation are of frequent occurrence.   For instance, in three districts of the province of Moscow —industrial to a great extent—drainage works have been accomplished within the last ten years on a large

---

[1] In the government of Moscow the experiment was usually made on the field which was reserved for the above-mentioned communal culture.

[2] Several instances of such and similar improvements were given in the *Official Messenger*, 1894, Nos. 256–258.   Associations between "horseless" peasants begin to appear also in South Russia.   Another extremely interesting fact is the sudden development in Southern West Siberia of very numerous co-operative creameries for making butter.   Hundreds of them spread in Tobolsk and Tomsk, without any one knowing wherefrom the initiative of the movement came. It came from the Danish co-operators, who used to export their own butter of higher quality, and to buy butter of a lower quality for their own use in Siberia.   After a several years' intercourse, they introduced creameries there.   Now, a great export trade has grown out of their endeavours.

scale in no less than 180 to 200 different villages—the
commoners working themselves with the spade. At
another extremity of Russia, in the dry Steppes of
Novouzen, over a thousand dams for ponds were built
and several hundreds of deep wells were sunk by the
communes; while in a wealthy German colony of the
south-east the commoners worked, men and women
alike, for five weeks in succession, to erect a dam, two
miles long, for irrigation purposes. What could
isolated men do in that struggle against the dry
climate? What could they obtain through individual
effort when South Russia was struck with the marmot
plague, and all people living on the land, rich and
poor, commoners and individualists, had to work with
their hands in order to conjure the plague? To call
in the policeman would have been of no use; to
associate was the only possible remedy.

And now, after having said so much about mutual
aid and support which are practised by the tillers of
the soil in "civilized" countries, I see that I might
fill an octavo volume with illustrations taken from the
life of the hundreds of millions of men who also live
under the tutorship of more or less centralized States,
but are out of touch with modern civilization and
modern ideas. I might describe the inner life of a
Turkish village and its network of admirable mutual-
aid customs and habits. On turning over my leaflets
covered with illustrations from peasant life in Caucasia,
I come across touching facts of mutual support. I
trace the same customs in the Arab *djemmâa* and the
Afghan *purra*, in the villages of Persia, India, and
Java, in the undivided family of the Chinese, in the
encampments of the semi-nomads of Central Asia and

the nomads of the far North. On consulting notes taken at random in the literature of Africa, I find them replete with similar facts—of aids convoked to take in the crops, of houses built by all inhabitants of the village—sometimes to repair the havoc done by civilized filibusters—of people aiding each other in case of accident, protecting the traveller, and so on. And when I peruse such works as Post's compendium of African customary law I understand why, notwithstanding all tyranny, oppression, robberies and raids, tribal wars, glutton kings, deceiving witches and priests, slave-hunters, and the like, these populations have not gone astray in the woods; why they have maintained a certain civilization, and have remained men, instead of dropping to the level of straggling families of decaying orang-outans. The fact is, that the slave-hunters, the ivory robbers, the fighting kings, the Matabele and the Madagascar "heroes" pass away, leaving their traces marked with blood and fire; but the nucleus of mutual-aid institutions, habits, and customs, grown up in the tribe and the village community, remains; and it keeps men united in societies, open to the progress of civilization, and ready to receive it when the day comes that they shall receive civilization instead of bullets.

The same applies to our civilized world. The natural and social calamities pass away. Whole populations are periodically reduced to misery or starvation; the very springs of life are crushed out of millions of men, reduced to city pauperism; the understanding and the feelings of the millions are vitiated by teachings worked out in the interest of the few. All this is certainly a part of our existence. But the nucleus of mutual-support institutions, habits, and customs remains alive with the millions ; it keeps them together ; and they prefer to cling to their customs, beliefs, and traditions rather than to accept the teachings of a war of each against all, which are offered to them under the title of science, but are no science at all.

# CHAPTER VIII

## MUTUAL AID AMONGST OURSELVES (*continued*)

Labour-unions grown after the destruction of the guilds by the State.—Their struggles.—Mutual Aid in strikes.—Co-operation.— Free associations for various purposes.—Self-sacrifice.—Countless societies for combined action under all possible aspects.—Mutual Aid in slum-life.—Personal aid.

WHEN we examine the every-day life of the rural populations of Europe, we find that, notwithstanding all that has been done in modern States for the destruction of the village community, the life of the peasants remains honeycombed with habits and customs of mutual aid and support ; that important vestiges of the communal possession of the soil are still retained ; and that, as soon as the legal obstacles to rural association were lately removed, a network of free unions for all sorts of economical purposes rapidly spread among the peasants—the tendency of this young movement being to reconstitute some sort of union similar to the village community of old. Such being the conclusions arrived at in the preceding chapter, we have now to consider, what institutions for mutual support can be found at the present time amongst the industrial populations.

For the last three hundred years, the conditions for the growth of such institutions have been as unfavourable in the towns as they have been in the villages.

It is well known, indeed, that when the mediæval cities were subdued in the sixteenth century by growing military States, all institutions which kept the artisans, the masters, and the merchants together in the guilds and the cities were violently destroyed. The self-government and the self-jurisdiction of both the guild and the city were abolished ; the oath of allegiance between guild-brothers became an act of felony towards the State ; the properties of the guilds were confiscated in the same way as the lands of the village communities ; and the inner and technical organization of each trade was taken in hand by the State. Laws, gradually growing in severity, were passed to prevent artisans from combining in any way. For a time, some shadows of the old guilds were tolerated : merchants' guilds were allowed to exist under the condition of freely granting subsidies to the kings, and some artisan guilds were kept in existence as organs of administration. Some of them still drag on their meaningless existence. But what formerly was the vital force of mediæval life and industry has long since disappeared under the crushing weight of the centralized State.

In Great Britain, which may be taken as the best illustration of the industrial policy of the modern States, we see the Parliament beginning the destruction of the guilds as early as the fifteenth century ; but it was especially in the next century that decisive measures were taken. Henry the Eighth not only ruined the organization of the guilds, but also confiscated their properties, with even less excuse and manners, as Toulmin Smith wrote, than he had produced for confiscating the estates of the monasteries.[1]

---

[1] Toulmin Smith, *English Guilds*, London, 1870, Introd. p. xliii.

Edward the Sixth completed his work,[1] and already in the second part of the sixteenth century we find the Parliament settling all the disputes between craftsmen and merchants, which formerly were settled in each city separately. The Parliament and the king not only legislated in all such contests, but, keeping in view the interests of the Crown in the exports, they soon began to determine the number of apprentices in each trade and minutely to regulate the very technics of each fabrication—the weights of the stuffs, the number of threads in the yard of cloth, and the like. With little success, it must be said; because contests and technical difficulties which were arranged for centuries in succession by agreement between closely-interdependent guilds and federated cities lay entirely beyond the powers of the centralized State. The continual interference of its officials paralyzed the trades, bringing most of them to a complete decay; and the last century economists, when they rose against the State regulation of industries, only ventilated a widely-felt discontent. The abolition of that interference by the French Revolution was greeted as an act of liberation, and the example of France was soon followed elsewhere.

With the regulation of wages the State had no better success. In the mediæval cities, when the distinction between masters and apprentices or journeymen became more and more apparent in the

---

[1] The Act of Edward the Sixth—the first of his reign—ordered to hand over to the Crown "all fraternities, brotherhoods, and guilds being within the realm of England and Wales and other of the king's dominions; and all manors, lands, tenements, and other hereditaments belonging to them or any of them" (*English Guilds*, Introd. p. xliii). See also Ockenkowski's *Englands wirtschaftliche Entwickelung im Ausgange des Mittelalters*, Jena, 1879, chaps. ii.-v.

fifteenth century, unions of apprentices (*Gesellenver-bände*), occasionally assuming an international character, were opposed to the unions of masters and merchants. Now it was the State which undertook to settle their griefs, and under the Elizabethan Statute of 1563 the Justices of Peace had to settle the wages, so as to guarantee a "convenient" livelihood to journeymen and apprentices. The Justices, however, proved helpless to conciliate the conflicting interests, and still less to compel the masters to obey their decisions. The law gradually became a dead letter, and was repealed by the end of the eighteenth century. But while the State thus abandoned the function of regulating wages, it continued severely to prohibit all combinations which were entered upon by journeymen and workers in order to raise their wages, or to keep them at a certain level. All through the eighteenth century it legislated against the workers' unions, and in 1799 it finally prohibited all sorts of combinations, under the menace of severe punishments. In fact, the British Parliament only followed in this case the example of the French Revolutionary Convention, which had issued a draconic law against coalitions of workers—coalitions between a number of citizens being considered as attempts against the sovereignty of the State, which was supposed equally to protect all its subjects. The work of destruction of the mediæval unions was thus completed. Both in the town and in the village the State reigned over loose aggregations of individuals, and was ready to prevent by the most stringent measures the reconstitution of any sort of separate unions among them. These were, then, the conditions under which the mutual-aid tendency had to make its way in the nineteenth century.

Need it be said that no such measures could destroy that tendency? Throughout the eighteenth century, the workers' unions were continually reconstituted.[1] Nor were they stopped by the cruel prosecutions which took place under the laws of 1797 and 1799. Every flaw in supervision, every delay of the masters in denouncing the unions was taken advantage of. Under the cover of friendly societies, burial clubs, or secret brotherhoods, the unions spread in the textile industries, among the Sheffield cutlers, the miners, and vigorous federal organizations were formed to support the branches during strikes and prosecutions.[2]

The repeal of the Combination Laws in 1825 gave a new impulse to the movement. Unions and national federations were formed in all trades;[3] and when Robert Owen started his Grand National Consolidated Trades' Union, it mustered half a million members in a few months. True that this period of relative liberty did not last long. Prosecution began anew in the thirties, and the well-known ferocious condemnations of 1832–1844 followed. The Grand National Union was disbanded, and all over the country, both the private employers and the Government in its own workshops began to compel the workers to resign all

[1] See Sidney and Beatrice Webb, *History of Trade-Unionism*, London, 1894. pp. 21–38.

[2] See in Sidney Webb's work the associations which existed at that time. The London artisans are supposed to have never been better organized than in 1810-20.

[3] The National Association for the Protection of Labour included about 150 separate unions, which paid high levies, and had a membership of about 100,000. The Builders' Union and the Miners' Unions also were big organizations (Webb, *l.c.* p. 107).

connection with unions, and to sign "the Document" to that effect. Unionists were prosecuted wholesale under the Master and Servant Act—workers being summarily arrested and condemned upon a mere complaint of misbehaviour lodged by the master.[1] Strikes were suppressed in an autocratic way, and the most astounding condemnations took place for merely having announced a strike or acted as a delegate in it —to say nothing of the military suppression of strike riots, nor of the condemnations which followed the frequent outbursts of acts of violence. To practise mutual support under such circumstances was anything but an easy task. And yet, notwithstanding all obstacles, of which our own generation hardly can have an idea, the revival of the unions began again in 1841, and the amalgamation of the workers has been steadily continued since. After a long fight, which lasted for over a hundred years, the right of combining together was conquered, and at the present time nearly one-fourth part of the regularly-employed workers, *i. e.* about 1,500,000, belong to trade unions.[2]

As to the other European States, sufficient to say that up to a very recent date, all sorts of unions were

[1] I follow in this Mr. Webb's work, which is replete with documents to confirm his statements.

[2] Great changes have taken place since the forties in the attitude of the richer classes towards the unions. However, even in the sixties, the employers made a formidable concerted attempt to crush them by locking out whole populations. Up to 1869 the simple agreement to strike, and the announcement of a strike by placards, to say nothing of picketing, were often punished as intimidation. Only in 1875 the Master and Servant Act was repealed, peaceful picketing was permitted, and "violence and intimidation" during strikes fell into the domain of common law. Yet, even during the dock-labourers' strike in 1887, relief money had to be spent for fighting before the Courts for the right of picketing, while the prosecutions of the last few years menace once more to render the conquered rights illusory.

prosecuted as conspiracies ; and that nevertheless they exist everywhere, even though they must often take the form of secret societies ; while the extension and the force of labour organizations, and especially of the Knights of Labour, in the United States and in Belgium, have been sufficiently illustrated by strikes in the nineties. It must, however, be borne in mind that, prosecution apart, the mere fact of belonging to a labour union implies considerable sacrifices in money, in time, and in unpaid work, and continually implies the risk of losing employment for the mere fact of being a unionist.[1] There is, moreover, the strike, which a unionist has continually to face ; and the grim reality of a strike is, that the limited credit of a worker's family at the baker's and the pawnbroker's is soon exhausted, the strike-pay goes not far even for food, and hunger is soon written on the children's faces. For one who lives in close contact with workers, a protracted strike is the most heartrending sight ; while what a strike meant forty years ago in this country, and still means in all but the wealthiest parts of the continent, can easily be conceived. Continually, even now, strikes will end with the total ruin and the forced emigration of whole populations, while the shooting down of strikers on the slightest provocation, or even without any provocation,[2] is quite habitual still on the continent.

[1] A weekly contribution of 6d. out of an 18s. wage, or of 1s. out of 25s., means much more than 9l. out of a 300l. income : it is mostly taken upon food ; and the levy is soon doubled when a strike is declared in a brother union. The graphic description of trade-union life, by a skilled craftsman, published by Mr. and Mrs. Webb (pp. 431 seq.), gives an excellent idea of the amount of work required from a unionist.

[2] See the debates upon the strikes of Falkenau in Austria before the Austrian Reichstag on the 10th of May, 1894, in which debates

And yet, every year there are thousands of strikes and lock-outs in Europe and America—the most severe and protracted contests being, as a rule, the so-called "sympathy strikes," which are entered upon to support locked-out comrades or to maintain the rights of the unions. And while a portion of the Press is prone to explain strikes by "intimidation," those who have lived among strikers speak with admiration of the mutual aid and support which are constantly practised by them. Every one has heard of the colossal amount of work which was done by volunteer workers for organizing relief during the London dock-labourers' strike; of the miners who, after having themselves been idle for many weeks, paid a levy of four shillings a week to the strike fund when they resumed work; of the miner widow who, during the Yorkshire labour war of 1894, brought her husband's life-savings to the strike-fund; of the last loaf of bread being always shared with neighbours; of the Radstock miners, favoured with larger kitchen-gardens, who invited four hundred Bristol miners to take their share of cabbage and potatoes, and so on. All newspaper correspondents, during the great strike of miners in Yorkshire in 1894, knew heaps of such facts, although not all of them could report such "irrelevant" matters to their respective papers.[1]

Unionism is not, however, the only form in which the worker's need of mutual support finds its expression. There are, besides, the political associations, whose activity many workers consider as more con-

the fact is fully recognized by the Ministry and the owner of the colliery. Also the English Press of that time.

[1] Many such facts will be found in the *Daily Chronicle* and partly the *Daily News* for October and November 1894.

ducive to general welfare than the trade-unions,
limited as they are now in their purposes. Of course
the mere fact of belonging to a political body cannot
be taken as a manifestation of the mutual-aid tendency.
We all know that politics are the field in which the
purely egotistic elements of society enter into the most
entangled combinations with altruistic aspirations.
But every experienced politician knows that all great
political movements were fought upon large and often
distant issues, and that those of them were the strongest which provoked most disinterested enthusiasm.
All great historical movements have had this character,
and for our own generation Socialism stands in that
case. "Paid agitators" is, no doubt, the favourite
refrain of those who know nothing about it. The
truth, however, is that—to speak only of what I know
personally—if I had kept a diary for the last twenty-
four years and inscribed in it all the devotion and
self-sacrifice which I came across in the Socialist
movement, the reader of such a diary would have had
the word "heroism" constantly on his lips. But the
men I would have spoken of were not heroes; they
were average men, inspired by a grand idea. Every
Socialist newspaper—and there are hundreds of them
in Europe alone—has the same history of years of
sacrifice without any hope of reward, and, in the overwhelming majority of cases, even without any personal
ambition. I have seen families living without knowing
what would be their food to-morrow, the husband boy-
cotted all round in his little town for his part in the
paper, and the wife supporting the family by sewing,
and such a situation lasting for years, until the family
would retire, without a word of reproach, simply
saying: "Continue; we can hold on no more!" I

have seen men, dying from consumption, and knowing it, and yet knocking about in snow and fog to prepare meetings, speaking at meetings within a few weeks from death, and only then retiring to the hospital with the words: "Now, friends, I am done; the doctors say I have but a few weeks to live. Tell the comrades that I shall be happy if they come to see me." I have seen facts which would be described as "idealization" if I told them in this place; and the very names of these men, hardly known outside a narrow circle of friends, will soon be forgotten when the friends, too, have passed away. In fact, I don't know myself which most to admire, the unbounded devotion of these few, or the sum total of petty acts of devotion of the great number. Every quire of a penny paper sold, every meeting, every hundred votes which are won at a Socialist election, represent an amount of energy and sacrifices of which no outsider has the faintest idea. And what is now done by Socialists has been done in every popular and advanced party, political and religious, in the past. All past progress has been promoted by like men and by a like devotion.

Co-operation, especially in Britain, is often described as "joint-stock individualism"; and such as it is now, it undoubtedly tends to breed a co-operative egotism, not only towards the community at large, but also among the co-operators themselves. It is, nevertheless, certain that at its origin the movement had an essentially mutual-aid character. Even now, its most ardent promoters are persuaded that co-operation leads mankind to a higher harmonic stage of economical relations, and it is not possible to stay in some of the strongholds of co-operation in the North without realizing that the great number of the rank and file hold the same opinion. Most of them would lose interest in the movement if that faith were gone ; and it must be owned that within the last few years broader ideals of general welfare and of the producers' solidarity have begun to be current among the co-operators. There is undoubtedly now a tendency towards establishing better relations between the owners of the co-operative workshops and the workers.

The importance of co-operation in this country, in Holland and in Denmark is well known ; while in Germany, and especially on the Rhine, the co-operative societies are already an important factor of industrial life.[1] It is, however, Russia which offers perhaps the best field for the study of co-operation under an infinite variety of aspects. In Russia, it is a natural growth, an inheritance from the middle ages ; and while a formally established co-operative society would have to cope with many legal difficulties and official sus-picion, the informal co-operation—the *artél*—makes the very substance of Russian peasant life. The history of "the making of Russia," and of the coloniz-ation of Siberia, is a history of the hunting and trading *artéls* or guilds, followed by village com-munities, and at the present time we find the *artél* everywhere ; among each group of ten to fifty peasants who come from the same village to work at a factory, in all the building trades, among fishermen and hunters, among convicts on their way to and in Siberia, among railway porters, Exchange messengers, Customs House labourers, everywhere in the village

---

[1] The 31,473 productive and consumers' associations on the Middle Rhine showed, about 1890, a yearly expenditure of 18,437,500*l*. ; 3,675,000*l*. were granted during the year in loans.

industries, which give occupation to 7,000,000 men—from top to bottom of the working world, permanent and temporary, for production and consumption under all possible aspects. Until now, many of the fishing-grounds on the tributaries of the Caspian Sea are held by immense *artéls*, the Ural river belonging to the whole of the Ural Cossacks, who allot and re-allot the fishing-grounds—perhaps the richest in the world—among the villages, without any interference of the authorities. Fishing is always made by *artéls* in the Ural, the Volga, and all the lakes of Northern Russia. Besides these permanent organizations, there are the simply countless temporary *artéls*, constituted for each special purpose. When ten or twenty peasants come from some locality to a big town, to work as weavers, carpenters, masons, boat-builders, and so on, they always constitute an *artél*. They hire rooms, hire a cook (very often the wife of one of them acts in this capacity), elect an elder, and take their meals in common, each one paying his share for food and lodging to the *artél*. A party of convicts on its way to Siberia always does the same, and its elected elder is the officially-recognized intermediary between the convicts and the military chief of the party. In the hard-labour prisons they have the same organization. The railway porters, the messengers at the Exchange, the workers at the Custom House, the town messengers in the capitals, who are collectively responsible for each member, enjoy such a reputation that any amount of money or bank-notes is trusted to the *artél*-member by the merchants. In the building trades, *artéls* of from 10 to 200 members are formed; and the serious builders and railway contractors always prefer to deal with an *artél* than with separately-hired workers. The

last attempts of the Ministry of War to deal directly
with productive *artéls*, formed *ad hoc* in the domestic
trades, and to give them orders for boots and all sorts
of brass and iron goods, are described as most satis-
factory; while the renting of a Crown iron work
(*Votkinsk*) to an *artél* of workers, which took place
seven or eight years ago, has been a decided success.

We can thus see in Russia how the old mediæval
institution, having not been interfered with by the
State (in its informal manifestations), has fully survived until now, and takes the greatest variety of
forms in accordance with the requirements of modern
industry and commerce. As to the Balkan peninsula,
the Turkish Empire and Caucasia, the old guilds are
maintained there in full. The *esnafs* of Servia have
fully preserved their mediæval character; they include
both masters and journeymen, regulate the trades, and
are institutions for mutual support in labour and sickness;[1] while the *amkari* of Caucasia, and especially at
Tiflis, add to these functions a considerable influence
in municipal life.[2]

In connection with co-operation, I ought perhaps to
mention also the friendly societies, the unities of odd-
fellows, the village and town clubs organized for meeting the doctors' bills, the dress and burial clubs, the
small clubs very common among factory girls, to which
they contribute a few pence every week, and afterwards
draw by lot the sum of one pound, which can at least
be used for some substantial purchase, and many others.

[1] British Consular Report, April 1889.
[2] A capital research on this subject has been published in Russian
in the *Zapiski* (*Memoirs*) of the Caucasian Geographical Society,
vol. vi. 2, Tiflis, 1891, by C. Egiazaroff.

A not inconsiderable amount of sociable or jovial spirit is alive in all such societies and clubs, even though the "credit and debit" of each member are closely watched over. But there are so many associations based on the readiness to sacrifice time, health, and life if required, that we can produce numbers of illustrations of the best forms of mutual support.

The Lifeboat Association in this country, and similar institutions on the Continent, must be mentioned in the first place. The former has now over three hundred boats along the coasts of these isles, and it would have twice as many were it not for the poverty of the fishermen, who cannot afford to buy lifeboats. The crews consist, however, of volunteers, whose readiness to sacrifice their lives for the rescue of absolute strangers to them is put every year to a severe test; every winter the loss of several of the bravest among them stands on record. And if we ask these men what moves them to risk their lives, even when there is no reasonable chance of success, their answer is something on the following lines. A fearful snowstorm, blowing across the Channel, raged on the flat, sandy coast of a tiny village in Kent, and a small smack, laden with oranges, stranded on the sands near by. In these shallow waters only a flat-bottomed lifeboat of a simplified type can be kept, and to launch it during such a storm was to face an almost certain disaster. And yet the men went out, fought for hours against the wind, and the boat capsized twice. One man was drowned, the others were cast ashore. One of these last, a refined coastguard, was found next morning, badly bruised and half frozen in the snow. I asked him, how they came to make that desperate attempt? " I don't know myself," was his reply. "*There* was the wreck; all the people from the village stood on the beach, and all said it would be foolish to go out; we never should work through the surf. We saw five or six men clinging to the mast, making desperate signals. We all felt that something must be done, but what could we do? One hour passed, two hours, and we all stood there. We all felt most uncomfortable. Then, all of a sudden, through the storm, it seemed to us as if we heard their cries—they had a boy with them. We could not stand that any longer. All at once we said, "We must go!" The women said so too; they would have treated us as cowards if we had not gone, although next day they said we had been fools to go. As one man, we rushed to the boat, and went. The boat capsized, but we took hold of it. The worst was to see poor —— drowning by the side of the boat, and we could do nothing to save him. Then came a fearful wave, the boat capsized again, and we were cast ashore. The men were still rescued by the D. boat, ours was caught miles away. I was found next morning in the snow."

The same feeling moved also the miners of the Rhonda Valley, when they worked for the rescue of their comrades from the inundated mine. They had pierced through thirty-two yards of coal in order to reach their entombed comrades; but when only three yards more remained to be pierced, fire-damp enveloped them. The lamps went out, and the rescue-men retired. To work in such conditions was to risk being blown up at every moment. But the raps of the entombed miners were still heard, the men were still alive and appealed for help, and several miners volunteered to work at any risk ; and as they went down the mine, their wives had only silent tears to follow them—not one word to stop them.

There is the gist of human psychology. Unless men are maddened in the battlefield, they "cannot stand it" to hear appeals for help, and not to respond to them. The hero goes ; and what the hero does, *all* feel that they ought to have done as well. The sophisms of the brain cannot resist the mutual-aid feeling, because this feeling has been nurtured by thousands of years of human social life and hundreds of thousands of years of pre-human life in societies.

" But what about those men who were drowned in the Serpentine in the presence of a crowd, out of which no one moved for their rescue ? " it may be asked. "What about the child which fell into the Regent's Park Canal—also in the presence of a holiday crowd—and was only saved through the presence of mind of a maid who let out a Newfoundland dog to the rescue ? " The answer is plain enough. Man is a result of both his inherited instincts and his education. Among the miners and the seamen, their common occupations and their every-day contact with one another create a feeling of solidarity, while the surrounding dangers maintain courage and pluck. In the cities, on the contrary, the absence of common interest nurtures indifference, while courage and pluck, which seldom find their opportunities, disappear, or take another direction. Moreover, the tradition of the hero of the mine and the sea lives in the miners' and fishermen's villages, adorned with a poetical halo. But what are the traditions of a motley London crowd? The only tradition they might have in common ought to be created by literature, but a literature which would correspond to the village epics hardly exists. The clergy are so anxious to prove that all that comes from human nature is sin, and that all good in man has a supernatural origin, that they mostly ignore the facts which cannot be produced as an example of higher inspiration or grace, coming from above. And as to the lay-writers, their attention is chiefly directed towards one sort of heroism, the heroism which promotes the idea of the State. Therefore, they admire the Roman hero, or the soldier in the battle, while they pass by the fisherman's heroism, hardly paying attention to it. The poet and the painter might, of course, be taken by the beauty of the human heart in itself; but both seldom know the life of the poorer classes, and while they can sing or paint the Roman or the military hero in conventional surroundings, they can neither sing nor paint impressively the hero who acts in those modest surroundings which they ignore. If they venture to do so, they produce a mere piece of rhetoric.[1]

[1] Escape from a French prison is extremely difficult; nevertheless a prisoner escaped from one of the French prisons in 1884 or 1885. He even managed to conceal himself during the whole day, although the alarm was given and the peasants in the neighbourhood were on the look-out for him. Next morning found him concealed in a ditch, close by a small village. Perhaps he intended to steal some food, or some clothes in order to take off his prison uniform. As he was lying in the ditch a fire broke out in the village. He saw a woman running out of one of the burning houses, and heard her desperate appeals to rescue a child in the upper storey of the burning house. No one moved to do so. Then the escaped prisoner dashed out of his retreat, made his way through the fire, and, with a scalded face and burning clothes, brought the child safe out of the fire, and handed it to its mother. Of course he was arrested on the spot by the village *gendarme*, who now made his appearance. He was taken back to the prison. The fact was reported in all French papers, but none of them bestirred itself to obtain his release. If he had shielded a warder from a comrade's blow, he would have been made a hero of. But his act was simply humane, it did not promote the

The countless societies, clubs, and alliances, for the enjoyment of life, for study and research, for education, and so on, which have lately grown up in such numbers that it would require many years to simply tabulate them, are another manifestation of the same ever-working tendency for association and mutual support. Some of them, like the broods of young birds of different species which come together in the autumn, are entirely given to share in common the joys of life. Every village in this country, in Switzerland, Germany, and so on, has its cricket, football, tennis, nine-pins, pigeon, musical or singing clubs. Other societies are much more numerous, and some of them, like the Cyclists' Alliance, have suddenly taken a formidable development. Although the members of this alliance have nothing in common but the love of cycling, there is already among them a sort of freemasonry for mutual help, especially in the remote nooks and corners which are not flooded by cyclists; they look upon the "C.A.C."—the Cyclists' Alliance Club—in a village as a sort of home ; and at the yearly Cyclists' Camp many a standing friendship has been established. The *Kegelbrüder*, the Brothers of the Nine Pins, in Germany, are a similar association ; so also the Gymnasts'. Societies (300,000 members in Germany), the informal brotherhood of paddlers in France, the yacht clubs, and so on. Such associations certainly do not alter the economical stratification of society, but, especially in the small towns, they contribute to smooth social distinctions, and as they all tend to

---

State's ideal ; he himself did not attribute it to a sudden inspiration of divine grace ; and that was enough to let the man fall into oblivion. Perhaps, six or twelve months were added to his sentence for having stolen—"the State's property "—the prison's dress.

join in large national and international federations, they certainly aid the growth of personal friendly intercourse between all sorts of men scattered in different parts of the globe.

The Alpine Clubs, the *Jagdschutzverein* in Germany, which has over 100,000 members—hunters, educated foresters, zoologists, and simple lovers of Nature—and the International Ornithological Society, which includes zoologists, breeders, and simple peasants in Germany, have the same character. Not only have they done in a few years a large amount of very useful work, which large associations alone could do properly (maps, refuge huts, mountain roads; studies of animal life, of noxious insects, of migrations of birds, and so on), but they create new bonds between men. Two Alpinists of different nationalities who meet in a refuge hut in the Caucasus, or the professor and the peasant ornithologist who stay in the same house, are no more strangers to each other; while the Uncle Toby's Society at Newcastle, which has already induced over 260,000 boys and girls never to destroy birds' nests and to be kind to all animals, has certainly done more for the development of human feelings and of taste in natural science than lots of moralists and most of our schools.

We cannot omit, even in this rapid review, the thousands of scientific, literary, artistic, and educational societies. Up till now, the scientific bodies, closely controlled and often subsidized by the State, have generally moved in a very narrow circle, and they often came to be looked upon as mere openings for getting State appointments, while the very narrowness of their circles undoubtedly bred petty jealousies. Still it is a fact that the distinctions of birth, political parties and creeds are smoothed to some extent by such associations; while in the smaller and remote towns the scientific, geographical, or musical societies, especially those of them which appeal to a larger circle of amateurs, become small centres of intellectual life, a sort of link between the little spot and the wide world, and a place where men of very different conditions meet on a footing of equality. To fully appreciate the value of such centres, one ought to know them, say, in Siberia. As to the countless educational societies which only now begin to break down the State's and the Church's monopoly in education, they are sure to become before long the leading power in that branch. To the "Froebel Unions" we already owe the Kindergarten system; and to a number of formal and informal educational associations we owe the high standard of women's education in Russia, although all the time these societies and groups had to act in strong opposition to a powerful government.[1] As to the various pedagogical societies in Germany, it is well known that they have done the best part in the working out of the modern methods of teaching science in popular schools. In such associations the teacher finds also his best support. How miserable the overworked and underpaid village teacher would have been without their aid![2]

[1] The Medical Academy for Women (which has given to Russia a large portion of her 700 graduated lady doctors), the four Ladies' Universities (about 1,000 pupils in 1887 ; closed that year, and reopened in 1895), and the High Commercial School for Women are *entirely* the work of such private societies. To the same societies we owe the high standard which the girls' gymnasia attained since they were opened in the sixties. The 100 gymnasia now scattered over the Empire (over 70,000 pupils), correspond to the High Schools for Girls in this country ; all teachers are, however, graduates of the universities.

[2] The *Verein für Verbreitung gemeinnützlicher Kenntnisse*, although

All these associations, societies, brotherhoods, alliances, institutes, and so on, which must now be counted by the ten thousand in Europe alone, and each of which represents an immense amount of voluntary, unambitious, and unpaid or underpaid work—what are they but so many manifestations, under an infinite variety of aspects, of the same ever-living tendency of man towards mutual aid and support? For nearly three centuries men were prevented from joining hands even for literary, artistic, and educational purposes. Societies could only be formed under the protection of the State, or the Church, or as secret brotherhoods, like free-masonry. But now that the resistance has been broken, they swarm in all directions, they extend over all multifarious branches of human activity, they become international, and they undoubtedly contribute, to an extent which cannot yet be fully appreciated, to break down the screens erected by States between different nationalities. Notwithstanding the jealousies which are bred by commercial competition, and the provocations to hatred which are sounded by the ghosts of a decaying past, there is a conscience of international solidarity which is growing both among the leading spirits of the world and the masses of the workers, since they also have conquered the right of international intercourse; and in the preventing of a European war during the last quarter of a century, this spirit has undoubtedly had its share.

The religious charitable associations, which again represent a whole world, certainly must be mentioned in this place. There is not the slightest doubt that

it has only 5,500 members, has already opened more than 1,000 public and school libraries, organized thousands of lectures, and published most valuable books.

the great bulk of their members are moved by the same mutual-aid feelings which are common to all mankind. Unhappily the religious teachers of men prefer to ascribe to such feelings a supernatural origin. Many of them pretend that man does not consciously obey the mutual-aid inspiration so long as he has not been enlightened by the teachings of the special religion which they represent, and, with St. Augustin, most of them do not recognize such feelings in the "pagan savage." Moreover, while early Christianity, like all other religions, was an appeal to the broadly human feelings of mutual aid and sympathy, the Christian Church has aided the State in wrecking all standing institutions of mutual aid and support which were anterior to it, or developed outside of it; and, instead of the *mutual aid* which every savage considers as due to his kinsman, it has preached *charity* which bears a character of inspiration from above, and, accordingly, implies a certain superiority of the giver upon the receiver. With this limitation, and without any intention to give offence to those who consider themselves as a body elect when they accomplish acts simply humane, we certainly may consider the immense numbers of religious charitable associations as an outcome of the same mutual-aid tendency.

All these facts show that a reckless prosecution of personal interests, with no regard to other people's needs, is not the only characteristic of modern life. By the side of this current which so proudly claims leadership in human affairs, we perceive a hard struggle sustained by both the rural and industrial populations in order to reintroduce standing institutions of mutual aid and support; and we discover, in all classes of

society, a widely-spread movement towards the establishment of an infinite variety of more or less permanent institutions for the same purpose. But when
we pass from public life to the private life of the modern
individual, we discover another extremely wide world of
mutual aid and support, which only passes unnoticed
by most sociologists because it is limited to the narrow
circle of the family and personal friendship.[1]

Under the present social system, all bonds of union
among the inhabitants of the same street or neighbourhood have been dissolved. In the richer parts of the
large towns, people live without knowing who are their
next-door neighbours. But in the crowded lanes people
know each other perfectly, and are continually brought
into mutual contact. Of course, petty quarrels go their
course, in the lanes as elsewhere; but groupings in
accordance with personal affinities grow up, and within
their circle mutual aid is practised to an extent of which
the richer classes have no idea. If we take, for instance,
the children of a poor neighbourhood who play in a
street or a churchyard, or on a green, we notice at once

[1] Very few writers in sociology have paid attention to it. Dr.
Ihering is one of them, and his case is very instructive. When the
great German writer on law began his philosophical work, *Der Zweck
im Rechte* ("Purpose in Law"), he intended to analyze "the active
forces which call forth the advance of society and maintain it," and
to thus give "the theory of the sociable man." He analyzed, first,
the egotistic forces at work, including the present wage-system and
coercion in its variety of political and social laws; and in a carefully-
worked-out scheme of his work he intended to give the last paragraph
to the ethical forces—the sense of duty and mutual love—which contribute to the same aim. When he came, however, to discuss the
social functions of these two factors, he had to write a second volume,
twice as big as the first; and yet he treated only of the *personal*
factors which will take in the following pages only a few lines. L. Dargun
took up the same idea in *Egoismus und Altruismus in der Nationalö-
konomie*, Leipzig, 1885, adding some new facts. Büchner's *Love*, and
the several paraphrases of it published here and in Germany, deal
with the same subject.

that a close union exists among them, notwithstanding the temporary fights, and that that union protects them from all sorts of misfortunes.   As soon as a mite bends inquisitively over the opening of a drain—" Don't stop there," another mite shouts out, "fever sits in the hole!"   " Don't climb over that wall, the train will kill you if you tumble down!  Don't come near to the ditch! Don't eat those berries—poison! you will die!"  Such are the first teachings imparted to the urchin when he joins his mates out-doors.   How many of the children whose play-grounds are the pavements around " model workers' dwellings," or the quays and bridges of the canals, would be crushed to death by the carts or drowned in the muddy waters, were it not for that sort of mutual support!   And when a fair Jack has made a slip into the unprotected ditch at the back of the milkman's yard, or a cherry-cheeked Lizzie has, after all, tumbled down into the canal, the young brood raises such cries that all the neighbourhood is on the alert and rushes to the rescue.

Then comes in the alliance of the mothers.   " You could not imagine " (a lady-doctor who lives in a poor neighbourhood told me lately) " how much they help each other.   If a woman has prepared nothing, or could prepare nothing, for the baby which she expected— and how often that happens!—all the neighbours bring something for the new-comer.   One of the neighbours always takes care of the children, and some other always drops in to take care of the household, so long as the mother is in bed."   This habit is general.   It is mentioned by all those who have lived among the poor. In a thousand small ways the mothers support each other and bestow their care upon children that are not their own.   Some training—good or bad, let them decide it for themselves—is required in a lady of the richer classes to render her able to pass by a shivering and hungry child in the street without noticing it. But the mothers of the poorer classes have not that training. They cannot stand the sight of a hungry child ; they *must* feed it, and so they do. " When the school children beg bread, they seldom or rather never meet with a refusal "—a lady-friend, who has worked several years in Whitechapel in connection with a workers' club, writes to me. But I may, perhaps, as well transcribe a few more passages from her letter :—

" Nursing neighbours, in cases of illness, without any shade of remuneration, is quite general among the workers. Also, when a woman has little children, and goes out for work, another mother always takes care of them.

" If, in the working classes, they would not help each other, they could not exist. I know families which continually help each other—with money, with food, with fuel, for bringing up the little children, in cases of illness, in cases of death.

" The 'mine' and 'thine' is much less sharply observed among the poor than among the rich. Shoes, dress, hats, and so on,—what may be wanted on the spot—are continually borrowed from each other, also all sorts of household things.

" Last winter the members of the United Radical Club had brought together some little money, and began after Christmas to distribute free soup and bread to the children going to school. Gradually they had 1,800 children to attend to. The money came from outsiders, but all the work was done by the members of the club. Some of them, who were out of work, came at four in the morning to wash and to peel the vegetables ; five women came at nine or ten (after having done their own household work) for cooking, and stayed till six or seven to wash the dishes. And at meal time, between twelve and half-past one, twenty to thirty workers came in to aid in serving the soup, each one staying what he could spare of his meal time. This lasted for two months. No one was paid."

My friend also mentions various individual cases, of which the following are typical :—

> "Annie W. was given by her mother to be boarded by an old person in Wilmot Street. When her mother died, the old woman, who herself was very poor, kept the child without being paid a penny for that. When the old lady died too, the child, who was five years old, was of course neglected during her illness, and was ragged ; but she was taken at once by Mrs. S., the wife of a shoemaker, who herself has six children. Lately, when the husband was ill, they had not much to eat, all of them.
>
> "The other day, Mrs. M., mother of six children, attended Mrs. M—g throughout her illness, and took to her own rooms the elder child. . . . But do you need such facts? They are quite general. . . . I know also Mrs. D. (Oval, Hackney Road), who has a sewing machine and continually sews for others, without ever accepting any remuneration, although she has herself five children and her husband to look after. . . . And so on."

For every one who has any idea of the life of the labouring classes it is evident that without mutual aid being practised among them on a large scale they never could pull through all their difficulties. It is only by chance that a worker's family can live its life-time without having to face such circumstances as the crisis described by the ribbon weaver, Joseph Gutteridge, in his autobiography.[1] And if all do not go to the ground in such cases, they owe it to mutual help. In Gutteridge's case it was an old nurse, miserably poor herself, who turned up at the moment when the family was slipping towards a final catastrophe, and brought in some bread, coal, and bedding, which she had obtained on credit. In other cases, it will be some one else, or the neighbours will take steps to save the family. But without some aid from other poor,

---

[1] *Light and Shadows in the Life of an Artisan.* Coventry, 1893.

how many more would be brought every year to irreparable ruin ![1]

Mr. Plimsoll, after he had lived for some time among the poor, on 7s. 6d. a week, was compelled to recognize that the kindly feelings he took with him when he began this life "changed into hearty respect and admiration" when he saw how the relations between the poor are permeated with mutual aid and support, and learned the simple ways in which that support is given. After a many years' experience, his conclusion was that "when you come to think of it, such as these men were, so were the vast majority of the working classes."[2] As to bringing up orphans, even by the poorest families, it is so widely-spread a habit, that it may be described as a general rule; thus among the miners it was found, after the two explosions at Warren Vale and at Lund Hill, that " nearly one-third of the men killed, as the respective committees

---

[1] Many rich people cannot understand how the very poor *can* help each other, because they do not realize upon what infinitesimal amounts of food or money often hangs the life of one of the poorest classes. Lord Shaftesbury had understood this terrible truth when he started his Flowers and Watercress Girls' Fund, out of which loans of one pound, and only occasionally two pounds, were granted, to enable the girls to buy a basket and flowers when the winter sets in and they are in dire distress. The loans were given to girls who had "not a sixpence," but never failed to find some other poor to go bail for them. "Of all the movements I have ever been connected with," Lord Shaftesbury wrote, " I look upon this Watercress Girls' movement as the most successful. . . . It was begun in 1872, and we have had out 800 to 1,000 loans, and have not lost 50l. during the whole period. . . . What has been lost—and it has been very little, under the circumstances—has been by reason of death or sickness, not by fraud" (*The Life and Work of the Seventh Earl of Shaftesbury*, by Edwin Hodder, vol. iii. p. 322. London, 1885-86). Several more facts in point in Ch. Booth's *Life and Labour in London*, vol. i.; in Miss Beatrice Potter's "Pages from a Work Girl's Diary" (*Nineteenth Century*, September 1888, p. 310); and so on.

[2] Samuel Plimsoll, *Our Seamen*, cheap edition, London, 1870, p. 110.

can testify, were thus supporting relations other than
wife and child." " Have you reflected," Mr. Plimsoll
added, "what this is ? Rich men, even comfortably-
to-do men do this, I don't doubt. But consider the
difference." Consider what a sum of one shilling, sub-
scribed by each worker to help a comrade's widow, or
6*d*. to help a fellow-worker to defray the extra expense
of a funeral, means for one who earns 16*s*. a week and
has a wife, and in some cases five or six children to
support.[1] But such subscriptions are a general practice
among the workers all over the world, even in much
more ordinary cases than a death in the family, while
aid in work is the commonest thing in their lives.

Nor do the same practices of mutual aid and support
fail among the richer classes. Of course, when one
thinks of the harshness which is often shown by the
richer employers towards their employees, one feels
inclined to take the most pessimist view of human
nature. Many must remember the indignation which
was aroused during the great Yorkshire strike of
1894, when old miners who had picked coal from an
abandoned pit were prosecuted by the colliery owners.
And, even if we leave aside the horrors of the periods

---

[1] *Our Seamen*, u.s., p. 110. Mr. Plimsoll added : " I don't wish to
disparage the rich, but I think it may be reasonably doubted whether
these qualities are so fully developed in them ; for, notwithstanding
that not a few of them are not unacquainted with the claims, reasonable or unreasonable, of poor relatives, these qualities are not in such
constant exercise. Riches seem in so many cases to smother the
manliness of their possessors, and their sympathies become, not so
much narrowed as—so to speak—stratified : they are reserved for
the sufferings of their own class, and also the woes of those above
them. They seldom tend downwards much, and they are far more
likely to admire an act of courage . . . than to admire the constantly
exercised fortitude and the tenderness which are the daily characteristics
of a British workman's life "—and of the workmen all over the world
as well.

of struggle and social war, such as the extermination
of thousands of workers' prisoners after the fall of the
Paris Commune—who can read, for instance, revela-
tions of the labour inquest which was made here in
the forties, or what Lord Shaftesbury wrote about
"the frightful waste of human life in the factories, to
which the children taken from the workhouses, or
simply purchased all over this country to be sold as
factory slaves, were consigned"[1]—who can read that
without being vividly impressed by the baseness which
is possible in man when his greediness is at stake?
But it must also be said that all fault for such treatment must not be thrown entirely upon the criminality
of human nature.    Were not the teachings of men of
science, and even of a notable portion of the clergy,
up to a quite recent time, teachings of distrust, despite
and almost hatred towards the poorer classes?    Did
not science teach that since serfdom has been abolished,
no one need be poor unless for his own vices?    And
how few in the Church had the courage to blame the
children-killers, while the great numbers taught that
the sufferings of the poor, and even the slavery of the
negroes, were part of the Divine Plan!    Was not
Nonconformism itself largely a popular protest against
the harsh treatment of the poor at the hand of the
Established Church?

With such spiritual leaders, the feelings of the
richer classes necessarily became, as Mr. Pimsoll
remarked, not so much blunted as "stratified."    They
seldom went downwards towards the poor, from whom
the well-to-do-people are separated by their manner of
life, and whom they do not know under their best

---

[1] *Life of the Seventh Earl of Shaftesbury*, by Edwin Hodder, vol.
i. pp. 137-138.

aspects, in their every-day life. But among themselves
—allowance being made for the effects of the wealth-
accumulating passions and the futile expenses imposed
by wealth itself—among themselves, in the circle of
family and friends, the rich practise the same mutual
aid and support as the poor. Dr. Ihering and L.
Dargun are perfectly right in saying that if a statistical
record could be taken of all the money which passes
from hand to hand in the shape of friendly loans and
aid, the sum total would be enormous, even in comparison with the commercial transactions of the world's
trade. And if we could add to it, as we certainly
ought to, what is spent in hospitality, petty mutual
services, the management of other people's affairs,
gifts and charities, we certainly should be struck by
the importance of such transfers in national economy.
Even in the world which is ruled by commercial
egotism, the current expression, "We have been
harshly treated by that firm," shows that there is also
the friendly treatment, as opposed to the harsh, i.e. the
legal treatment; while every commercial man knows
how many firms are saved every year from failure by
the friendly support of other firms.

As to the charities and the amounts of work for
general well-being which are voluntarily done by so
many well-to-do persons, as well as by workers, and
especially by professional men, every one knows the
part which is played by these two categories of
benevolence in modern life. If the desire of acquiring
notoriety, political power, or social distinction often
spoils the true character of that sort of benevolence,
there is no doubt possible as to the impulse coming in
the majority of cases from the same mutual-aid feelings. Men who have acquired wealth very often do

not find in it the expected satisfaction. Others begin to feel that, whatever economists may say about wealth being the reward of capacity, their own reward is exaggerated. The conscience of human solidarity begins to tell; and, although society life is so arranged as to stifle that feeling by thousands of artful means, it often gets the upper hand; and then they try to find an outcome for that deeply human need by giving their fortune, or their forces, to something which, in their opinion, will promote general welfare.

In short, neither the crushing powers of the centralized State nor the teachings of mutual hatred and pitiless struggle which came, adorned with the attributes of science, from obliging philosophers and sociologists, could weed out the feeling of human solidarity, deeply lodged in men's understanding and heart, because it has been nurtured by all our preceding evolution. What was the outcome of evolution since its earliest stages cannot be overpowered by one of the aspects of that same evolution. And the need of mutual aid and support which had lately taken refuge in the narrow circle of the family, or the slum neighbours, in the village, or the secret union of workers, re-asserts itself again, even in our modern society, and claims its rights to be, as it always has been, the chief leader towards further progress. Such are the conclusions which we are necessarily brought to when we carefully ponder over each of the groups of facts briefly enumerated in the last two chapters.

# CONCLUSION

IF we take now the teachings which can be borrowed from the analysis of modern society, in connection with the body of evidence relative to the importance of mutual aid in the evolution of the animal world and of mankind, we may sum up our inquiry as follows.

In the animal world we have seen that the vast majority of species live in societies, and that they find in association the best arms for the struggle for life: understood, of course, in its wide Darwinian sense—not as a struggle for the sheer means of existence, but as a struggle against all natural conditions unfavourable to the species. The animal species, in which individual struggle has been reduced to its narrowest limits, and the practice of mutual aid has attained the greatest development, are invariably the most numerous, the most prosperous, and the most open to further progress. The mutual protection which is obtained in this case, the possibility of attaining old age and of accumulating experience, the higher intellectual development, and the further growth of sociable habits, secure the maintenance of the species, its extension, and its further progressive evolution. The unsociable species, on the contrary, are doomed to decay.

Going next over to man, we found him living in clans and tribes at the very dawn of the stone age;

we saw a wide series of social institutions developed
already in the lower savage stage, in the clan and the
tribe ; and we found that the earliest tribal customs
and habits gave to mankind the embryo of all the
institutions which made later on the leading aspects of
further progress.   Out of the savage tribe grew up
the barbarian village community ; and a new, still
wider, circle of social customs, habits, and institutions,
numbers of which are still alive among ourselves, was
developed under the principles of common possession
of a given territory and common defence of it, under
the jurisdiction of the village folkmote, and in the
federation of villages belonging, or supposed to belong,
to one stem.   And when new requirements induced
men to make a new start, they made it in the city,
which represented a double network of territorial
units (village communities), connected with guilds—
these latter arising out of the common prosecution of
a given art or craft, or for mutual support and defence.

And finally, in the last two chapters facts were
produced to show that although the growth of the
State on the pattern of Imperial Rome had put a
violent end to all mediæval institutions for mutual
support, this new aspect of civilization could not last.
The State, based upon loose aggregations of individuals
and undertaking to be their only bond of union, did
not answer its purpose.   The mutual-aid tendency
finally broke down its iron rules ; it reappeared and
reasserted itself in an infinity of associations which
now tend to embrace all aspects of life and to take
possession of all that is required by man for life and
for reproducing the waste occasioned by life.

It will probably be remarked that mutual aid, even
though it may represent one of the factors of evolution,

covers nevertheless one aspect only of human relations ;
that by the side of this current, powerful though it
may be, there is, and always has been, the other
current—the self-assertion of the individual, not only
in its efforts to attain personal or caste superiority,
economical, political, and spiritual, but also in its much
more important although less evident function of
breaking through the bonds, always prone to become
crystallized, which the tribe, the village community,
the city, and the State impose upon the individual.
In other words, there is the self-assertion of the
individual taken as a progressive element.

It is evident that no review of evolution can be
complete, unless these two dominant currents are
analyzed. However, the self-assertion of the individual
or of groups of individuals, their struggles for superiority, and the conflicts which resulted therefrom, have
already been analyzed, described, and glorified from
time immemorial. In fact, up to the present time,
this current alone has received attention from the epical
poet, the annalist, the historian, and the sociologist.
History, such as it has hitherto been written, is almost
entirely a description of the ways and means by which
theocracy, military power, autocracy, and, later on, the
richer classes' rule have been promoted, established,
and maintained. The struggles between these forces
make, in fact, the substance of history. We may thus
take the knowledge of the individual factor in human
history as granted—even though there is full room for
a new study of the subject on the lines just alluded to ;
while, on the other side, the mutual-aid factor has
been hitherto totally lost sight of; it was simply
denied, or even scoffed at, by the writers of the present
and past generation. It was therefore necessary to

show, first of all, the immense part which this factor plays in the evolution of both the animal world and human societies. Only after this has been fully recognized will it be possible to proceed to a comparison between the two factors.

To make even a rough estimate of their relative importance by any method more or less statistical, is evidently impossible. One single war—we all know —may be productive of more evil, immediate and subsequent, than hundreds of years of the unchecked action of the mutual-aid principle may be productive of good. But when we see that in the animal world, progressive development and mutual aid go hand in hand, while the inner struggle within the species is concomitant with retrogressive development; when we notice that with man, even success in struggle and war is proportionate to the development of mutual aid in each of the two conflicting nations, cities, parties, or tribes, and that in the process of evolution war itself (so far as it can go this way) has been made subservient to the ends of progress in mutual aid within the nation, the city or the clan—we already obtain a perception of the dominating influence of the mutual-aid factor as an element of progress. But we see also that the practice of mutual aid and its successive developments have created the very conditions of society life in which man was enabled to develop his arts, knowledge, and intelligence ; and that the periods when institutions based on the mutual-aid tendency took their greatest development were also the periods of the greatest progress in arts, industry, and science. In fact, the study of the inner life of the mediæval city and of the ancient Greek cities reveals the fact that the combination of mutual aid, as it was practised within the guild and the Greek clan, with a large initiative which was left to the individual and the group by means of the federative principle, gave to mankind the two greatest periods of its history—the ancient Greek city and the mediæval city periods; while the ruin of the above institutions during the State periods of history, which followed, corresponded in both cases to a rapid decay.

As to the sudden industrial progress which has been achieved during our own century, and which is usually ascribed to the triumph of individualism and competition, it certainly has a much deeper origin than that. Once the great discoveries of the fifteenth century were made, especially that of the pressure of the atmosphere, supported by a series of advances in natural philosophy—and they were made under the mediæval city organization,—once these discoveries were made, the invention of the steam-motor, and all the revolution which the conquest of a new power implied, had necessarily to follow. If the mediæval cities had lived to bring their discoveries to that point, the ethical consequences of the revolution effected by steam might have been different; but the same revolution in technics and science would have inevitably taken place. It remains, indeed, an open question whether the general decay of industries which followed the ruin of the free cities, and was especially noticeable in the first part of the eighteenth century, did not considerably retard the appearance of the steam-engine as well as the consequent revolution in arts. When we consider the astounding rapidity of industrial progress from the twelfth to the fifteenth centuries—in weaving, working of metals, architecture and navigation, and ponder over the scientific discoveries which that industrial progress led to at the end of the fifteenth century—

we must ask ourselves whether mankind was not delayed
in its taking full advantage of these conquests when a
general depression of arts and industries took place
in Europe after the decay of mediæval civilization.
Surely it was not the disappearance of the artist-
artisan, nor the ruin of large cities and the extinction
of intercourse between them, which could favour the
industrial revolution ; and we know indeed that James
Watt spent twenty or more years of his life in order to
render his invention serviceable, because he could not
find in the last century what he would have readily
found in mediæval Florence or Brügge, that is, the
artisans capable of realizing his devices in metal, and
of giving them the artistic finish and precision which
the steam-engine requires.

To attribute, therefore, the industrial progress of
our century to the war of each against all which it has
proclaimed, is to reason like the man who, knowing
not the causes of rain, attributes it to the victim he
has immolated before his clay idol. For industrial
progress, as for each other conquest over nature,
mutual aid and close intercourse certainly are, as they
have been, much more advantageous than mutual
struggle.

However, it is especially in the domain of ethics
that the dominating importance of the mutual-aid
principle appears in full. That mutual aid is the real
foundation of our ethical conceptions seems evident
enough. But whatever the opinions as to the first
origin of the mutual-aid feeling or instinct may be—
whether a biological or a supernatural cause is ascribed
to it—we must trace its existence as far back as to the
lowest stages of the animal world ; and from these
stages we can follow its uninterrupted evolution, in

opposition to a number of contrary agencies, through all degrees of human development, up to the present times. Even the new religions which were born from time to time—always at epochs when the mutual-aid principle was falling into decay in the theocracies and despotic States of the East, or at the decline of the Roman Empire—even the new religions have only reaffirmed that same principle. They found their first supporters among the humble, in the lowest, down-trodden layers of society, where the mutual-aid principle is the necessary foundation of every-day life ; and the new forms of union which were introduced in the earliest Buddhist and Christian communities, in the Moravian brotherhoods and so on, took the character of a return to the best aspects of mutual aid in early tribal life.

Each time, however, that an attempt to return to this old principle was made, its fundamental idea itself was widened. From the clan it was extended to the stem, to the federation of stems, to the nation, and finally—in ideal, at least—to the whole of mankind. It was also refined at the same time. In primitive Buddhism, in primitive Christianity, in the writings of some of the Mussulman teachers, in the early movements of the Reform, and especially in the ethical and philosophical movements of the last century and of our own times, the total abandonment of the idea of revenge, or of "due reward"—of good for good and evil for evil—is affirmed more and more vigorously. The higher conception of "no revenge for wrongs," and of freely giving more than one expects to receive from his neighbours, is proclaimed as being the real principle of morality—a principle superior to mere equivalence, equity, or justice, and more conducive to happiness. And man is appealed to to be guided in his acts, not merely by love, which is always personal, or at the best tribal, but by the perception of his oneness with each human being. In the practice of mutual aid, which we can retrace to the earliest beginnings of evolution, we thus find the positive and undoubted origin of our ethical conceptions; and we can affirm that in the ethical progress of man, mutual support— not mutual struggle—has had the leading part. In its wide extension, even at the present time, we also see the best guarantee of a still loftier evolution of our race.

# APPENDIX

I.—SWARMS OF BUTTERFLIES, DRAGON-FLIES, ETC.

(To p. 10.)

M. C. PIEPERS has published in *Natuurkunding Tijdschrift voor Neederlandsch Indië*, 1891, Deel L. p. 198 (analyzed in *Naturwissenschaftliche Rundschau*, 1891, vol. vi. p. 573), interesting researches into the mass-flights of butterflies which occur in Dutch East India, seemingly under the influence of great draughts occasioned by the west monsoon. Such mass-flights usually take place in the first months after the beginning of the monsoon, and it is usually individuals of both sexes of *Catopsilia (Callidryas) crocale*, Cr., which join in it, but occasionally the swarms consist of individuals belonging to three different species of the genus *Euphœa*. Copulation seems also to be the purpose of such flights. That these flights are not the result of concerted action but rather a consequence of imitation, or of a desire of following all others, is, of course, quite possible.

Bates saw, on the Amazon, the yellow and the orange *Callidryas* "assembling in densely packed masses, sometimes two or three yards in circumference, their wings all held in an upright position, so that the beach looked as though variegated with beds of crocuses." Their migrating columns, crossing the river from north to south, "were uninterrupted, from an early hour in the morning till sunset" (*Naturalist on the Amazon*, p. 131).

Dragon-flies, in their long migrations across the Pampas, come together in countless numbers, and their immense swarms contain individuals belonging to different species (Hudson, *Naturalist on the La Plata*, pp. 130 *seq.*).

The grasshoppers (*Zoniopoda tarsata*) are also eminently gregarious (Hudson, *l. c.* p. 125).

———

## II.—THE ANTS.

### (To p. 13.)

Pierre Huber's *Les fourmis indigènes* (Genève, 1810), of which a cheap edition was issued in 1861 by Cherbuliez, in the *Bibliothèque Genevoise*, and of which translations ought to be circulated in cheap editions in every language, is not only the best work on the subject, but also a model of really scientific research. Darwin was quite right in describing Pierre Huber as an even greater naturalist than his father. This book ought to be read by every young naturalist, not only for the facts it contains but as a lesson in the methods of research. The rearing of ants in artificial glass nests, and the test experiments made by subsequent explorers, including Lubbock, will all be found in Huber's admirable little work. Readers of the books of Forel and Lubbock are, of course, aware that both the Swiss professor and the British writer began their work in a critical mood, with the intention of disproving Huber's assertions concerning the admirable mutual-aid instincts of the ants; but that after a careful investigation they could only confirm them. However, it is unfortunately characteristic of human nature gladly to believe any affirmation concerning men being able to change at will the action of the forces of Nature, but to refuse to admit well-proved scientific facts tending to reduce the distance between man and his animal brothers.

Mr. Sutherland (*Origin and Growth of Moral Instinct*) evidently began his book with the intention of proving that all moral feelings have originated from parental care and familial love, which both appeared only in warm-blooded animals; consequently he tries to minimize the importance of sympathy and co-operation among ants. He quotes Büchner's book, *Mind in Animals*, and knows Lubbock's experiments. As to the works of Huber and Forel, he dismisses them in the following sentence; "but they [Büchner's instances of sympathy among ants] are all, or mostly all, marred by a certain air of sentimentalism . . . which renders them better suited for school books than for cautious works of science, and *the same is to be remarked* [italics are mine] of some of Huber's and Forel's best-known *anecdotes*" (vol. i. p. 298).

Mr. Sutherland does not specify which "anecdotes" he means, but it seems to me that he could never have had the opportunity of perusing the works of Huber and Forel. Naturalists who know these works find no "anecdotes" in them.

The recent work of Professor Gottfried Adlerz on the ants in Sweden, *Myrmecologiska Studier: Svenska Myror och des Lefnadsförhållanden* (*Bihang till Svenska Akademiens Handlingar*, Bd. xi. No. 18, 1886), may be mentioned in this place. It hardly need be said that all the observations of Huber and Forel concerning the mutual-aid life of ants, including the one concerning the sharing of food, felt to be so striking by those who previously had paid no attention to the subject, are fully confirmed by the Swedish professor (pp. 136-137).

Professor G. Adlerz gives also very interesting experiments to prove what Huber had already observed; namely, that ants from two different nests do not always attack each other. He has made one of his experiments with the ant, *Tapinoma erraticum.* Another was made with the common *Rufa* ant. Taking a whole nest in a sack, he emptied it at a distance of six feet from another nest. There was no battle, but the ants of the second nest began to carry the pupæ of the former. As a rule, when Professor Adlerz brought together workers with their pupæ, both taken from different nests, there was no battle; but if the workers were without their pupæ, a battle ensued (pp. 185–186).

He also completes Forel's and MacCook's observations about the "nations" of ants, composed of many nests, and, taking his own estimates, which brought him to take an average of 300,000 *Formica exsecta* ants in each nest, he concludes that such "nations" may reach scores and even hundreds of millions of inhabitants.

Maeterlinck's admirably written book on bees, although it contains no new observations, would be very useful, if it were less marred with metaphysical "words."

---

### III.—NESTING ASSOCIATIONS.

(To p. 35.)

Audubon's Journals (*Audubon and his Journals,* New York, 1898), especially those relating to his life on the coasts of Labrador and the St. Lawrence river in the thirties, contain excellent descriptions of the nesting associations of aquatic birds. Speaking of "The Rock," one of the Magdalene or Amherst Islands, he wrote :—" At eleven I could distinguish its top plainly from the deck, and thought it covered with snow to the depth of several feet; this appearance

existed on every portion of the flat, projecting shelves."
But it was not snow: it was gannets, all calmly seated
on their eggs or newly-hatched brood—their heads all
turned windwards, almost touching each other, and in
regular lines. The air above, for a hundred yards and
for some distance round the rock, "was filled with
gannets on the wing, as if a heavy fall of snow was
directly above us." Kittiwake gulls and foolish
guillemots bred on the same rock (*Journals*, vol. i. pp.
360–363).

In sight of Anticosti Island, the sea "was literally
covered with foolish guillemots and with razor-
billed auks (*Alca torva*)." Further on, the air was
filled with velvet ducks. On the rocks of the Gulf,
the herring gulls, the terns (great, Arctic, and
probably Foster's), the *Tringa pusilla*, the sea-gulls,
the auks, the Scoter ducks, the wild geese (*Anser
canadensis*), the red-breasted merganser, the cormor-
ants, etc., were all breeding. The sea-gulls were
extremely abundant there ; "they are for ever harass-
ing every other bird, sucking their eggs and devouring
their young ;" "they take here the place of eagles
and hawks."

On the Missouri, above Saint Louis, Audubon saw,
in 1843, vultures and eagles nesting in colonies. Thus
he mentioned "long lines of elevated shore, surmounted
by stupendous rocks of limestone, with many curious
holes in them, where we saw vultures and eagles
enter towards dusk"—that is, Turkey buzzards
(*Cathartes aura*) and bald eagles (*Haliaëtus leuco-
cephalus*), E. Coües remarks in a footnote (vol. i. p.
458).

One of the best breeding-grounds along the British
shores are the Farne Islands, and one will find in
Charles Dixon's work, *Among the Birds in Northern
Shires*, a lively description of these grounds, where
scores of thousands of gulls, terns, eider-ducks, cor-
morants, ringed plovers, oyster-catchers, guillemots,

and puffins come together every year. "On approaching some of the islands the first impression is that this gull (the lesser black-backed gull) monopolizes the whole of the ground, as it occurs in such vast abundance. The air seems full of them, the ground and bare rocks are crowded ; and as our boat finally grates against the rough beach and we eagerly jump ashore all becomes noisy excitement—a perfect babel of protesting cries that is persistently kept up until we leave the place" (p. 219).

## IV.—SOCIABILITY OF ANIMALS.

(To p. 42.)

That the sociability of animals was greater when they were less hunted by man, is confirmed by many facts showing that those animals who now live isolated in countries inhabited by man continue to live in herds in uninhabited regions. Thus on the waterless plateau deserts of Northern Thibet Prjevalsky found bears living in societies. He mentions numerous "herds of yaks, khulans, antelopes, and even bears." The latter, he says, feed upon the extremely numerous small rodents, and are so numerous that, "as the natives assured me, they have found a hundred or a hundred and fifty of them asleep in the same cave" (*Yearly Report* of the Russian Geographical Society for 1885, p. 11 ; Russian). Hares (*Lepus Lehmani*) live in large societies in the Transcaspian territory (N. Zarudnyi, *Recherches zoologiques dans la contrée Transcaspienne*, in *Bull. Soc. Natur. Moscou*, 1889, 4). The small Californian foxes, who, according to E. S. Holden, live round the Lick observatory "on a mixed diet of Manzanita berries and astronomers' chickens"

(*Nature*, Nov. 5, 1891), seem also to be very sociable.

Some very interesting instances of the love of society among animals have lately been given by Mr. C. J. Cornish (*Animals at Work and Play*, London, 1896). All animals, he truly remarks, hate solitude. He gives also an amusing instance of the habit of the prairie dogs of keeping sentries. It is so great that they always keep a sentinel on duty, even at the London Zoological Garden, and in the Paris Jardin d'Acclimatation (p. 46).

Professor Kessler was quite right in pointing out that the young broods of birds, keeping together in autumn, contribute to the development of feelings of sociability. Mr. Cornish (*Animals at Work and Play*) has given several examples of the plays of young mammals, such as, for instance, lambs playing at "follow my leader," or at "I'm the king of the castle," and their love of steeplechases; also the fawns playing a kind of "cross-touch," the touch being given by the nose. We have, moreover, the excellent work by Karl Gross, *The Play of Animals*.

---

## V.—CHECKS TO OVER-MULTIPLICATION.

### (To p. 72.)

Hudson, in his *Naturalist on the La Plata* (Chapter III.), has a very interesting account of a sudden increase of a species of mice and of the consequences of that sudden "wave of life."

"In the summer of 1872–73," he writes, "we had plenty of sunshine, with frequent showers, so that the hot months brought no dearth of wild flowers, as in most years." The season was very favourable for

mice, and "these prolific little creatures were soon
so abundant that the dogs and the cats subsisted
almost exclusively on them.   Foxes, weasels and
opossums fared sumptuously ; even the insectivorous
armadillo took to mice-hunting."   The fowls became
quite rapacious, "while the sulphur tyrant-birds
(*Pitangus*) and the *Guira* cuckoos preyed on nothing
but mice."   In the autumn, countless numbers of
storks and of short-eared owls made their appearance, coming also to assist at the general feast.
Next came a winter of continued drought ; the dry
grass was eaten, or turned to dust ; and the mice,
deprived of cover and food, began to die out.   The
cats sneaked back to the houses ; the short-eared
owls—a wandering species—left ; while the little
burrowing owls became so reduced as scarcely to
be able to fly, "and hung about the houses all day
long on the look-out for some stray morsel of food."
Incredible numbers of sheep and cattle perished the
same winter, during a month of cold that followed
the drought.   As to the mice, Hudson makes the
remark that "scarcely a hard-pressed remnant remains
after the great reaction, to continue the species."

This illustration has an additional interest in its
showing how, on flat plains and plateaus, the sudden
increase of a species immediately attracts enemies
from other parts of the plains, and how species unprotected by their social organization must necessarily
succumb before them.

Another excellent illustration in point is given
by the same author from the Argentine Republic.
The coypù (*Myopotamus coypù*) is there a very common rodent—a rat in shape, but as large as an otter.
It is aquatic in its habits and very sociable.  "Of an
evening," Hudson writes, "they are all out swimming
and playing in the water, conversing together in
strange tunes, which sound like the moans and cries
of wounded and suffering men.   The coypù, which

has a fine fur under the long coarse hair, was largely exported to Europe; but some sixty years ago the Dictator Rosas issued a decree prohibiting the hunting of this animal. The result was that the animals increased and multiplied exceedingly, and, abandoning their aquatic habits, they became terrestrial and migratory, and swarmed everywhere in search of food. Suddenly a mysterious malady fell on them, from which they quickly perished, and became almost extinct" (p. 12).

Extermination by man on the one side, and contagious diseases on the other side, are thus the main checks which keep the species down—not competition for the means of existence, which may not exist at all.

Facts, proving that regions enjoying a far more congenial climate than Siberia are equally under-populated, could be produced in numbers. But in Bates' well-known work we find the same remark concerning even the shores of the Amazon river.

"There is, in fact," Bates wrote, "a great variety of mammals, birds and reptiles, but they are widely scattered and all excessively shy of man. The region is so extensive and uniform in the forest-clothing of its surface, that it is only at long intervals that animals are seen in abundance, where some particular spot is found which is more attractive than the others" (*Naturalist on the Amazon*, 6th ed., p. 31).

This fact is the more striking as the Brazilian fauna, which is poor in mammals, is not poor at all in birds, and the Brazilian forests afford ample food for birds, as may be seen from a quotation, already given on a previous page, about birds' societies. And yet, the forests of Brazil, like those of Asia and Africa, are not over-populated, but rather under-populated. The same is true concerning the pampas of South America, about which W. H. Hudson remarks that it is really astonishing that only one small ruminant should be found on this immense grassy area, so admirably suited to herbivorous quadrupeds. Millions of sheep, cattle and horses, introduced by man, graze now, as is known, upon a portion of these prairies. Land-birds on the pampas are also few in species and in numbers.

---

## VI.—ADAPTATIONS TO AVOID COMPETITION.

### (To p. 75.)

Numerous examples of such adaptations can be found in the works of all field-naturalists. One of them, very interesting, may be given in the hairy armadillo, of which W. H. Hudson says, that "it has struck a line for itself, and consequently thrives, while its congeners are fast disappearing. Its food is most varied. It preys on all kinds of insects, discovering worms and larvæ several inches beneath the surface. It is fond of eggs and fledglings; it feeds on carrion as readily as a vulture; and, failing animal food, it subsists on vegetable diet—clover, and even grains of maize. Therefore, when other animals are starving, the hairy armadillo is always fat and vigorous" (*Naturalist on the La Plata*, p. 71).

The adaptivity of the lapwing makes it a species of which the range of extension is very wide. In England, it "makes itself at home on arable land as readily as in wilder areas." Ch. Dixon says in his *Birds of Northern Shires* (p. 67), "Variety of food is still more the rule with the birds of prey." Thus, for instance, we learn from the same author (pp. 60, 65), "that the hen harrier of the British moors feeds not only on small birds, but also on moles and mice, and on frogs, lizards and insects, while most of the smaller falcons subsist largely on insects."

The very suggestive chapter which W. H. Hudson gives to the family of the South American tree-creepers, or woodhewers, is another excellent illustration of the ways in which large portions of the animal population avoid competition, while at the same time they succeed in becoming very numerous in a given region, without being possessed of any of the weapons usually considered as essential in the struggle for existence. The above family covers an immense range, from South Mexico to Patagonia, and no fewer than 290 species, referable to about 46 genera, are already known from this family, the most striking feature of which is the great diversity of habits of its members. Not only the different genera and the different species possess habits peculiarly their own, but even the same species is often found to differ in its manner of life in different localities. "Some species of *Xenops* and *Magarornis*, like woodpeckers, climb vertically on tree-trunks in search of insect prey, but also, like tits, explore the smaller twigs and foliage at the extremity of the branches; so that the whole tree, from the root to its topmost foliage, is hunted over by them. The *Sclerurus*, although an inhabitant of the darkest forest, and provided with sharply-curved claws, never seeks its food on trees, but exclusively on the ground, among the decaying fallen leaves; but, strangely enough, when alarmed, it flies to the trunk of the nearest tree, to which it clings in a vertical position, and, remaining silent and motionless, escapes observation by means of its dark protective colour." And so on. In their nesting habits they also vary immensely. Thus, in one single genus, three species build an oven-shaped clay-nest, a fourth builds a nest of sticks in the trees, and a fifth burrows in the side of a bank, like a kingfisher.

Now, this extremely large family, of which Hudson says that "every portion of the South American continent is occupied by them; for there is really no climate, and no kind of soil or vegetation, which does not possess its appropriate species, belongs"—to use his own words—"to the most defenceless of birds." Like the ducks which were mentioned by Syevertsoff (see in the text), they display no powerful beak or claws; "they are timid, unresisting creatures, without strength or weapons; their movements are less quick and vigorous than those of other kinds, and their flight is exceedingly feeble." But they possess— both Hudson and Asara observe—"the social disposition in an eminent degree," although "the social habit is kept down in them by the conditions of a life which makes solitude necessary." They cannot make those large breeding associations which we see in the sea-birds, because they live on the tree-insects, and they, must carefully explore separately every tree—which they do in a most business-like way; but they continually call each other in the woods, "conversing with one another over long distances;" and they associate in those "wandering bands" which are well known from Bates' picturesque description, while Hudson was led to believe "that everywhere in South America the Dendrocolaptidæ are the first in combining to act in concert, and that the birds of other families follow their march and associate with them, knowing from experience that a rich harvest may be reaped." It hardly need be added that Hudson pays them also a high compliment concerning their intelligence. Sociability and intelligence always go hand in hand.

## VII.—THE ORIGIN OF THE FAMILY.

### (To p. 86.)

At the time when I wrote the chapter inserted in the text, a certain accord seemed to have been established amongst anthropologists concerning the relatively late appearance, in the institutions of men, of the patriarchal family, such as we know it among the Hebrews, or in Imperial Rome. However, works have been published since, in which the ideas promulgated by Bachofen and MacLennan, systematized especially by Morgan, and further developed and confirmed by Post, Maxim Kovalevsky, and Lubbock, were contested—the most important of such works being by the Danish Professor, C. N. Starcke (*Primitive Family*, 1889), and by the Helsingfors Professor, Edward Westermarck (*The History of Human Marriage*, 1891; 2nd ed. 1894). The same has happened with this question of primitive marriage institutions as it happened with the question of the primitive land-ownership institutions.. When the ideas of Maurer and Nasse on the village community, developed by quite a school of gifted explorers, and those of all modern anthropologists upon the primitively communistic constitution of the clan had nearly won general acceptance—they called forth the appearance of such works as those of Fustel de Coulanges in France, Frederic Seebohm in England, and several others, in which an attempt was made —with more brilliancy than real depth of investigation—to undermine these ideas and to cast a doubt upon the conclusions arrived at by modern research (see Prof. Vinogradov's Preface to his remarkable work, *Villainage in England*). Similarly, when the ideas about the non-existence of the family at the early tribal stage of mankind began to be accepted by most anthropologists and students of ancient law, they necessarily called forth such works as those of Starcke and Westermarck, in which man was represented, in accordance with the Hebrew tradition, as having started with the family, evidently patriarchal, and never having passed through the stages described by MacLennan, Bachofen, or Morgan. These works, of which the brilliantly-written *History of Human Marriage* has especially been widely read, have undoubtedly produced a certain effect: those who have not had the opportunity of reading the bulky volumes related to the controversy became hesitating; while some anthropologists, well acquainted with the matter, like the French Professor Durkheim, took a conciliatory, but somewhat undefined attitude.

For the special purpose of a work on Mutual Aid, this controversy may be irrelevant. The fact that men have lived in *tribes* from the earliest stages of mankind, is not contested, even by those who feel shocked at the idea that man may have passed through a stage when the family as we understand it did not exist. The subject, however, has its own interest and deserves to be mentioned, although it must be remarked that a volume would be required to do it full justice.

When we labour to lift the veil that conceals from us ancient institutions, and especially such institutions as have prevailed at the first appearance of beings of the human type, we are bound—in the necessary absence of direct testimony—to accomplish a most painstaking work of tracing backwards every institution, carefully noting even its faintest traces in habits, customs, traditions, songs, folk-lore, and so on; and then, combining the separate results of each of these separate studies, to mentally reconstitute the society which would answer to the co-existence of all these institutions. One can consequently understand what a formidable array of facts, and what a vast number of minute studies of particular points is required to come to any safe conclusion. This is exactly what one finds in the monumental work of Bachofen and his followers, but fails to find in the works of the other school. The mass of facts ransacked by Prof. Westermarck is undoubtedly great enough, and his work is certainly very valuable as a criticism; but it hardly will induce those who know the works of Bachofen, Morgan, MacLennan, Post, Kovalevsky, etc., in the originals, and are acquainted with the village-community school, to change their opinions and accept the patriarchal family theory.

Thus the arguments borrowed by Westermarck from the familiar habits of the primates have not, I dare say, the value which he attributes to them. Our knowledge about the family relations amongst the sociable species of monkeys of our own days is extremely uncertain, while the two unsociable species of orang-outan and gorilla must be ruled out of discussion, both being evidently, as I have indicated in the text, decaying species. Still less do we know about the relations which existed between males and females amongst the primates towards the end of the Tertiary period. The species which lived then are probably all extinct, and we have not the slightest idea as to which of them was the ancestral form which Man sprung from. All we can say with any approach to probability is, that various family and tribe relations must have existed in the different ape species, which were extremely numerous at that time; and that great changes must have taken place since in the habits of the primates, similarly to the changes that took place, even within the last two centuries, in the habits of many other mammal species.

The discussion must consequently be limited entirely to human institutions; and in the minute discussion of each separate trace of each early institution, *in connection with all that we know about every other institution of the same people or the same tribe*, lies the main force of the argument of the school which maintains that the patriarchal family is an institution of a relatively late origin.

There is, in fact, *quite a cycle of institutions* amongst primitive men, which become fully comprehensible if we accept the ideas of Bachofen and Morgan, but are utterly incomprehensible otherwise. Such are: the communistic life of the clan, so long as it was not split up into separate paternal families; the life in *long houses*, and in *classes* occupying separate long houses according to the age and stage of initiation of the youth (M. Maclay, H. Schurz); the restrictions to personal accumulation of property of which several illustrations are given above, in the text; the fact that women taken from another tribe belonged to the whole tribe before becoming private property; and many similar institutions analyzed by Lubbock. This wide cycle of institutions, which fell into decay and finally disappeared in the village-community phase of human development, stand in perfect accord with the "tribal marriage" theory; but they are mostly left unnoticed by the followers of the patriarchal family school. This is certainly not the proper way of discussing the problem. Primitive men have not several superposed or juxtaposed institutions as we have now. They have but *one* institution, the clan, which embodies *all* the mutual relations of the members of the clan. Marriage-relations and possession-relations are clan-relations. And the last that we might expect from the defenders of the patriarchal family theory would be to show us how the just mentioned cycle of institutions (which disappear later on) could have existed in an agglomeration of men living under a system contradictory of such institutions —the system of separate families governed by the *pater familias.*

Again, one cannot recognize scientific value in the

way in which certain serious difficulties are set aside
by the promoters of the patriarchal family theory.
Thus, Morgan has proved by a considerable amount
of evidence that a strictly-kept "classificatory group
system" exists with many primitive tribes, and that
all the individuals of the same category address each
other as if they were brothers and sisters, while the
individuals of a younger category will address their
mothers' sisters as mothers, and so on.   To say that
this must be a simple *façon de parler*—a way of
expressing respect to age—is certainly an easy method
of getting rid of the difficulty of explaining, why
this special mode of expressing respect, and not some
other, has prevailed among so many peoples of different
origin, so as to survive with many of them up to the
present day?   One may surely admit that *ma* and *pa*
are the syllables which are easiest to pronounce for a
baby, but the question is—Why this part of "baby
language" is used by full-grown people, and is
applied to a certain strictly-defined category of
persons?   Why, with so many tribes in which the
mother and her sisters are called *ma*, the father is
designated by *tiatia* (similar to *diadia*—uncle), *dad*,
*da* or *pa?*   Why the appellation of mother given to
maternal aunts is supplanted later on by a separate
name?   And so on.   But when we learn that with
many savages the mother's sister takes as responsible a part in bringing up a child as the mother
itself, and that, if death takes away a beloved child,
the other "mother" (the mother's sister) will sacrifice
herself to accompany the child in its journey into the
other world—we surely see in these names something
much more profound than a mere *façon de parler*, or
a way of testifying respect.   The more so when we
learn of the existence of quite a cycle of survivals
(Lubbock, Kovalevsky, Post have fully discussed
them), all pointing in the same direction.   Of course
it may be said that kinship is reckoned on the maternal

side "because the child remains more with its mother," or we may explain the fact that a man's children by several wives of different tribes belong to their mothers' clans in consequence of the savages' "ignorance of physiology;" but these are not arguments even approximately adequate to the seriousness of the questions involved—especially when it is known that the obligation of bearing the mother's name implies belonging to the mother's clan in all respects: that is, involves a right to all the belongings of the maternal clan, as well as the right of being protected by it, never to be assailed by any one of it, and the duty of revenging offences on its behalf.

Even if we were to admit for a moment the satis-factory nature of such explanations, we should soon find out that a separate explanation has to be given for each category of such facts—and they are very numerous. To mention but a few of them, there is: the division of clans into classes, at a time when there is no division as regards property or social condition; exogamy and all the consequent customs enumerated by Lubbock; the blood covenant and a series of similar customs intended to testify the unity of descent; the appearance of family gods subsequent to the existence of clan gods; the exchange of wives which exists not only with Eskimos in times of calamity, but is also widely spread among many other tribes of a quite different origin; the looseness of nuptial ties the lower we descend in civilization; the com-pound marriages—several men marrying one wife who belongs to them in turns; the abolition of the marriage restrictions during festivals, or on each fifth, sixth, etc., day; the cohabitation of families in "long houses;" the obligation of rearing the orphan falling, even at a late period, upon the maternal uncle; the considerable number of transitory forms showing the gradual passage from maternal descent to paternal descent; the limitation of the number of children by the clan—not by the family—and the abolition of this harsh clause in times of plenty; family restrictions coming after the clan restrictions; the sacrifice of the old relatives to the tribe; the tribal *lex talionis* and many other habits and customs which become a "family matter" only when we find the family, in the modern sense of the word, finally constituted; the nuptial and pre-nuptial ceremonies of which striking illustrations may be found in the work of Sir John Lubbock, and of several modern Russian explorers; the absence of marriage solemnities where the line of descent is matriarchal, and the appearance of such solemnities with tribes following the paternal line of descent—all these and many others[1] showing that, as Durkheim remarks, marriage proper "is only tolerated and prevented by antagonist forces;" the destruction at the death of the individual of what belonged to him personally; and finally, all the formidable array of survivals,[2] myths (Bachofen and his many followers), folk-lore, etc., all telling in the same direction.

Of course, all this does not prove that there was a period when woman was regarded as superior to man, or was the "head" of the clan; this is a quite distinct matter, and my personal opinion is that no such period has ever existed; nor does it prove that there was a time when no tribal restrictions to the union of sexes existed—this would have been absolutely contrary to all known evidence. But when all the facts lately brought to light are considered in their mutual dependency, it is impossible not to recognize that if isolated couples, with their children, have possibly existed even in the primitive clan, these incipient

---

[1] See *Marriage Customs in many Lands*, by H. N. Hutchinson, London, 1897.

[2] Many new and interesting forms of these have been collected by Wilhelm Rudeck, *Geschichte der öffentlichen Sittlichkeit in Deutschland*, analyzed by Durckheim in *Annuaire Sociologique*, ii. 312.

families were *tolerated exceptions only*, not the institution of the time.

———

## VIII.—DESTRUCTION OF PRIVATE PROPERTY ON THE GRAVE.

### (To p. 99.)

In a remarkable work, *The Religious Systems of China*, published in 1892-97 by J. M. de Groot at Leyden, we find the confirmation of this idea. There was in China (as elsewhere) a time when all personal belongings of a dead person were destroyed on his tomb—his mobiliary goods, his chattels, his slaves, and even friends and vassals, and of course his widow. It required a strong reaction against this custom on behalf of the moralists to put an end to it. With the gipsies in England the custom of destroying all chattels on the grave has survived up to the present day. All the personal property of the gipsy queen who died a few years ago was destroyed on her grave. Several newspapers mentioned it at that time.

———

## IX.—THE "UNDIVIDED FAMILY."

### (To p. 124.)

A number of valuable works on the South Slavonian *Zadruga*, or "compound family," compared to other forms of family organization, have been published since the above was written; namely, by Ernest Miler (*Jahrbuch der Internationaler Vereinung für vergleichende Rechtswissenschaft und Volkswirthschaftslehre*, 1897), and I. E. Geszow's *Zadruga in Bulgaria*, and

*Zadruga-Ownership and Work in Bulgaria* (both in Bulgarian). I must also mention the well-known study of Bogisic (*De la forme dite 'inokosna' de la famille rurale chez les Serbes et les Croates*, Paris, 1884), which has been omitted in the text.

---

## X.—THE ORIGIN OF THE GUILDS.

### (To p. 176.)

The origin of the guilds has been the subject of many controversies. There is not the slightest doubt that craft-guilds, or "colleges" of artisans, existed in ancient Rome. It appears, indeed, from a passage in Plutarch that Numa legislated about them. "He divided the people," we are told, "into trades . . . ordering them to have brotherhoods, festivals, and meetings, and indicating the worship they had to accomplish before the gods, according to the dignity of each trade." It is almost certain, however, that it was not the Roman king who invented, or instituted, the trade-colleges—they had already existed in ancient Greece; in all probability, he simply submitted them to royal legislation, just as Philippe le Bel, fifteen centuries later, submitted the trades of France, much to their detriment, to royal supervision and legislation. One of the successors of Numa, Servius Tullius, also is said to have issued some legislation concerning the colleges.[1]

Consequently, it was quite natural that historians should ask themselves whether the guilds which took

---

[1] A Servio Tullio populus romanus relatus in censum, digestus in classes, curiis atque collegiis distributus (E. Martin-Saint-Léon, *Histoire des corporations de métiers depuis leurs origines jusqu'à leur suppression en* 1791, etc., Paris, 1897).

such a development in the twelfth, and even the tenth
and the eleventh centuries, were not revivals of the
old Roman "colleges"—the more so as the latter, as
seen from the above quotation, quite corresponded to
the mediæval guild.[1] It is known, indeed, that corporations of the Roman type existed in Southern Gaul
down to the fifth century. Besides, an inscription found
during some excavations in Paris shows that a corporation of Lutetia *nautæ* existed under Tiberius; and
in the chart given to the Paris "water-merchants" in
1170, their rights are spoken of as existing *ab antiquo*
(same author, p. 51). There would have been, therefore, nothing extraordinary, had corporations been
maintained in early mediæval France after the barbarian
invasions.

However, even if as much must be granted, there
is no reason to maintain that the Dutch corporations,
the Norman guilds, the Russian *artéls*, the Georgian
*amkari*, and so on, necessarily have had also a Roman,
or even a Byzantine origin. Of course, the intercourse
between the Normans and the capital of the East-
Roman Empire was very active, and the Slavonians
(as has been proved by Russian historians, and especially by Rambaud) took a lively part in that intercourse.
So, the Normans and the Russians may have imported
the Roman organization of trade-corporations into
their respective lands. But when we see that the
*artél* was the very essence of the every-day life of all
the Russians, as early as the tenth century, and that
this *artél*, although no sort of legislation has ever
regulated its life till modern times, has the very same
features as the Roman college and the Western guild,
we are still more inclined to consider the eastern
guild as having an even more ancient origin than the
Roman college. Romans knew well, indeed, that

---

[1] The Roman *sodalitia*, so far as we may judge (same author, p. 9),
corresponded to the Kabyle *çofs*.

their *sodalitia* and *collegia* were "what the Greeks called *hetairiai*" (Martin-Saint-Léon, p. 2), and from what we know of the history of the East, we may conclude, with little probability of being mistaken, that the great nations of the East, as well as Egypt, also have had the same guild organization. The essential features of this organization remain the same wherever we may find them. It is a union of men carrying on the same profession or trade. This union, like the primitive clan, has its own gods and its own worship, always containing some mysteries, specific to each separate union; it considers all its members as *brothers and sisters*—possibly (at its beginnings) with all the consequences which such a relationship implied in the gens, or, at least, with ceremonies that indicated or symbolized the clan relations between brother and sister; and finally, all the obligations of mutual support which existed in the clan, exist in this union; namely, the exclusion of the very possibility of a murder within the brotherhood, the clan responsibility before justice, and the obligation, in case of a minor dispute, of bringing the matter before the judges, or rather the arbiters, of the guild brotherhood. The guild—one may say—is thus modelled upon the clan.

Consequently, the same remarks which are made in the text concerning the origin of the village community, apply, I am inclined to think, equally to the guild, the *artél*, and the craft- or neighbour-brotherhood. When the bonds which formerly connected men in their clans were loosened in consequence of migrations, the appearance of the paternal family, and a growing diversity of occupations—a new *territorial bond* was worked out by mankind in the shape of the village community; and another bond—an *occupation bond*—was worked out in an imaginary brotherhood—*the imaginary clan*, which was represented: between two men, or a few men, by the "mixture-of-blood brotherhood " (the Slavonian *pobratımstvo*), and between a greater number of men of different origin, *i. e.* originated from different clans, inhabiting the same village or town (or even different villages or towns)—the *phratry*, the *hetairiai*, the *amkari*, the *artél*, the guild.[1]

As to the idea and the form of such an organization, its elements were already indicated from the savage period downwards. We know indeed that in the clans of all savages there are separate secret organizations of warriors, of witches, of young men, etc.—craft mysteries, in which knowledge concerning hunting or warfare is transmitted ; in a word, " clubs," as Miklukho-Maclay described them. These " mysteries " were, in all probability, the prototypes of the future guilds.[2]

---

[1] It is striking to see how distinctly this very idea is expressed in the well-known passage of Plutarch concerning Numa's legislation of the trade-colleges :—"And through this," Plutarch wrote, "he was the first to banish from the city this spirit which led people to say : ' I am a Sabine,' or ' I am a Roman,' or ' I am a subject of Tatius,' and another : ' I am a subject of Romulus ' "—to exclude, in other words, the idea of different descent.

[2] The work of H. Schurtz, devoted to the " age-classes " and the secret men's unions during the barbarian stages of civilization (*Altersklassen und Männerverbände: eine Darstellung der Grundformen der Gesellschaft*, Berlin, 1902), which reaches me while I am reading the proofs of these pages, contains numbers of facts in support of the above hypothesis concerning the origin of guilds. The art of building a large communal house, so as not to offend the spirits of the fallen trees ; the art of forging metals, so as to conciliate the hostile spirits ; the secrets of hunting and of the ceremonies and mask-dances which render it successful ; the art of teaching savage arts to boys ; the secret ways of warding off the witchcraft of enemies and, consequently, the art of warfare ; the making of boats, of nets for fishing, of traps for animals, and of snares for birds, and finally the women's arts of weaving and dyeing—all these were in olden times as many " artifices " and " crafts," which required secrecy for being effective. Consequently, they were transmitted from the earliest times, in secret societies, or " mysteries," to those only who had undergone a painful initiation. H. Schurtz shows now that savage life is honeycombed with secret societies and " clubs " (of warriors, of hunters), which have as ancient an origin as the marriage " classes " in the clans, and contain already all the elements of the

With regard to the above-mentioned work by E. Martin-Saint-Léon, let me add that it contains very valuable information concerning the organization of the trades in Paris—as it appears from the *Livre des métiers* of Boileau—and a good summary of information relative to the Communes of different parts of France, with all bibliographical indications. It must, however, be remembered that Paris was a "Royal city" (like Moscow, or Westminster), and that consequently the free mediæval-city institutions have never attained there the development which they have attained in free cities. Far from representing "the picture of a typical corporation," the corporations of Paris, "born and developed under the direct tutorship of royalty," for this very same cause (which the author considers a cause of superiority, while it was a cause of inferiority—he himself fully shows in different parts of his work how the interference of the imperial power in Rome, and of the royal power in France, destroyed and paralyzed the life of the craft-guilds) could never attain the wonderful growth and influence upon all the life of the city which they did attain in North-Eastern France, at Lyons, Montpellier, Nîmes, etc., or in the free cities of Italy, Flanders, Germany, and so on.

---

XI.—THE MARKET AND THE MEDIÆVAL CITY.

(To p. 190.)

In a work on the mediæval city (*Markt und Stadt in ihrem rechtlichen Verhältnis*, Leipzig, 1896),

---

future guild : secrecy, independence from the family and sometimes the clan, common worship of special gods, common meals, jurisdiction within the society and brotherhood. The forge and the boat-house are, in fact, usual dependencies of the men's clubs ; and the "long houses" or "palavers" are built by special craftsmen who know how to conjure the spirits of the fallen trees.

Rietschel has developed the idea that the origin of the German mediæval communes must be sought in the *market*. The local market, placed under the protection of a bishop, a monastery or a prince, gathered round it a population of tradesmen and artisans, but no agricultural population. The sections into which the towns were usually divided, radiating from the market-place and peopled each with artisans of special trades, are a proof of that: they formed usually the Old Town, while the New Town used to be a rural village belonging to the prince or the king. The two were governed by different laws.

It is certainly true that the market has played an important part in the early development of all mediæval cities, contributing to increase the wealth of the citizens, and giving them ideas of independence ; but, as has been remarked by Carl Hegel—the well-known author of a very good general work on German mediæval cities (*Die Entstehung des deutschen Städtewesens*, Leipzig, 1898), the town-law is not a market-law, and Hegel's conclusion is (in further support to the views taken in this book) that the mediæval city has had a double origin. There were in it "two populations placed by the side of each other: one rural, and the other purely urban ; " the rural population, which formerly lived under the organization of the *Almende*, or village community, was incorporated in the city.

With regard to the Merchant Guilds, the work of Herman van den Linden (*Les Gildes marchandes dans les Pays-Bas au Moyen Age*, Gand, 1896, in *Recueil de travaux publiés par la Faculté de Philosophie et Lettres*) deserves a special mention. The author follows the gradual development of their political force and the authority which they gradually acquired upon the industrial population, especially on the drapers, and describes the league concluded by the artisans to oppose their growing power. The

idea, which is developed in this book, concerning the
appearance of the merchant guild at a later period
which mostly corresponded to a period of decline of
the city liberties, seems thus to find confirmation in
H. van den Linden's researches.

———

XII.—MUTUAL-AID ARRANGEMENTS IN THE VILLAGES
OF NETHERLANDS AT THE PRESENT DAY.

(To p. 250.)

The Report of the Agricultural Commission of
Netherlands contains many illustrations relative to
this subject, and my friend, M. Cornelissen, was kind
enough to pick out for me the corresponding passages from these bulky volumes (*Uitkomsten van het
Onderzoek naar den Toestand van den Landbouw in
Nederland*, 2 vols. 1890).

The habit of having one thrashing-machine, which
makes the round of many farms, hiring it in turn, is
very widely spread, as it is by this time in nearly
every other country. But one finds here and there a
commune which keeps one thrashing-machine for the
community (vol. I. xviii. p. 31).

The farmers who have not the necessary numbers
of horses for the plough borrow the horses from their
neighbours. The habit of keeping one communal
bull, or one communal stallion, is very common.

When the village has to raise the ground (in the
low districts) in order to build a communal school, or
for one of the peasants in order to build a new house,
a *bede* is usually convoked. The same is done for
those farmers who have to move. The *bede* is
altogether a widely-spread custom, and no one, rich
or poor, will fail to come with his horse and cart.

The renting in common, by several agricultural labourers, of a meadow, for keeping their cows, is found in several parts of the land; it is also frequent that the farmer, who has plough and horses, ploughs the land for his hired labourers (vol. I. xxii. p. 18, etc.).

As to the farmers' unions for buying seed, exporting vegetables to England and so on, they become universal. The same is seen in Belgium. In 1896, seven years after peasants' guilds had been started, first in the Flemish part of the country, and four years only after they were introduced in the Walloon portion of Belgium, there were already 207 such guilds, with a membership of 10,000 (*Annuaire de la Science Agronomique*, vol. I. (2), 1896, pp. 148 and 149).

# INDEX

AACHEN, 206
"Aba," common hunt, 141
Abbeville, 177, 207
Abyssinia, village community, 122
Adalbert, St., 167
Adlerz, Prof. Gottfried, on ants, 303, 304
Africa, animal population of, 39, 47; village community, 122; barbarian monarchies, 162; compensation laws of various stems, 133, 134; customary law, 148, 149; village community, 260
Agricultural co-operation in Netherlands, 327; in Belgium, 328. *See* also *Syndicats, Artéls.*
Agricultural implements, improved in village communities, 257
Aids, in guilds, 193
Aids: in Kabyle villages, 143; in Georgia, 143 *n.*; amongst French peasants, 243 *seq.*; in Caucasia, 244 *n.*; in Germany, 248
Aire, "friendship" of, 177
Alans, 136
Aleoutes, 91, 95 *seq.*; in stone age still, 96; peacefulness, 96; periodical distributions of accumulated wealth, 97; code of morality, 99
Alfurus, the, 149
Algeria, 144
Allthing, law recited at, 158
Alpine Clubs, 280
Altum, Dr. B., on destruction of the pine-moth, 71; of mice, 71, 72
Amalfi, 168
America, animal population of, 38
America, Northern, 32
Amiens, 177, 182, 183 *n.*, 194 *n.*; acting as arbiter, 207 *n.*
*Amitas*, 193

*Amkari*, 169, 170, 274, 322
*Amt*, 192
Amu river, 118 *n.*
Amur river, viii, ix, 48, 49, 130
Anabaptism, 225
"Anaya" custom, 145, 148
Ancher, Kofod, on old Danish guilds, 172
Anglo-Saxon law, 161
Annam, village community, 127
Antelopes, 47, 48
Anthropological Society of Paris, questions answered, 92, 93; on cannibalism, 105
Anticosti Island, 305
Ants, mutual support with, 12–16; feeding each other, 12; agriculture and horticulture of, 14; federations of their nests, 18; their play, 55; book of Pierre Huber on, 302; Mr. Sutherland's appreciation, 303; Prof. Adlerz on, 303; nations of, 304
Antwerp, 183 *n.*
Apes, sociability of, 50–52; family relations, 315
Aquatic birds, 33, 34; family habits of, 36 *n.*; on St. Lawrence river, 305
Arabs, invasion of, 165
Aral, lake, 118 *n.*
Arani, the, 149
Arbiter, city acting as, 207
Architecture, mediæval, 210 *seq.*; communal inspiration, 210; mechanical achievements, 211
Arctic America Eskimos, 84
Arctic archipelagoes, 33
Ardennes, re-allotting of land, 242
Ariège, village life in, 243 *seq.*; communal culture, 247
Armadillo, 310

329

Arnold, Dr. Wilhelm, 122 *n.*, 162 ; on German cities, 180

Art, mediæval and Greek, communal inspiration of, 210, 213

*Artél*, 169, 174 and *note*, 193 ; modern developments in Russia, 272–274, 322, 324

Arthur, King, legends of, 135

Aryans, early, 87, 119

Asara, on sociability in the Tree-creepers' family of birds, 312

Asia, Northern, 32

Assemblée Constituante, 231

Associations of animals : family, group, society, 53 ; in villages, 53

Athens, Acropolis, 212

Audubon, 5 ; on parrots, 30 ; packs of Labrador wolves, 40 ; Canada musk-rats, 44 ; his "Journals," 304 ; on aquatic birds on St. Lawrence river, 304 ; on eagles, 305

Augsburg, 167, 206

Augustin, St., 283

Aunt, maternal, sacrificing herself to follow dead child, 101 ; her duties in the tribe, 317

Australasia, Southern, 84

Australia, 29 ; droves of cattle, 59

Australians, 84, 91–93 ; *The Folklore, Manners, etc., of A. Aborigines*, 92 *n.* ; code of morality, 100 *n.*

Austria, destruction of village community, 235

Autumn, societies of birds, 36

Babeau, old village in France, 122 *n.* ; old towns, 184 ; village community, 231, 232

"Baby language," 317

Bachofen, on late origin of family, 79, 313–319

Bacon, Francis, 215

Bacon, Roger, 215

Baden, 247, 248

Bain, Eb., on merchant and craft guilds, 199 *n.*

Baker, S. W., hunting associations of lions, 40 ; societies of elephants, 50

Bakradze, Dm., on common culture, 127 ; on common ownership of serfs, 147

"Balai," or "barla," 94

Balkan peninsula, village community, 250

Bancroft, on common culture, 127

Baptists, 255

Barbarians, mutual aid among the, 115–152 ; migrations dissociating them, 120 ; village-community institutions worked out by, 120 ; justice rendered by village folk-mote, 131 ; *fred* and *wergeld*, 132, 133 ; amount of composition payment, 133 ; settling of peace, 134 ; mild punishments, 134 ; tribes now living under the institutions of, 138 *seq.* ; clearing forests, colonizing, 155

Barbarossa, 204

Barrow, 91

Barthold, German mediæval cities, 189 *n.*

Basel, 205 ; cathedral, 212

Bassano, 174 *n.*

Bassoutos, 148 *n.*

Bastian, Adolf, on blood-revenge and justice, 108 *n.*, 112 ; obligation to aid travellers, 145 ; Oceania islands, 150 *n.*

Batavians, 125 *n.*

Bates, W., on Darwinism, xiv ; "campos" of termites, 18 ; on Brazil vultures, 22 ; destruction of winged ants, 70 ; on swarms of butterflies, 301 ; scarcity of animal population in Brazil, 309 ; bird-societies, 312

Baudrillart, A., on rural populations of France, 246, 247

Baudrillart, H., on rural populations of France, 246

Bavaria, 249

Bears, sociable in Kamtchatka, 42 ; in Tibet, 306

Beaumont chart, 178

Beavers, colonies of, 39, 45

Becker, A., on sudden disappearance of *Sousliks*, 72

*Bede*, 327

"Bee," 143 *n.*

Bees, mutual aid with, 16–18 ; anti-social instincts among them, 17

Beetles, burying : mutual aid among, 10

Behring, his crew and polar foxes, 41

Belgium, forced sale of communal lands, 235 ; farmer's unions, 328
Bentham, 112
Berkshire, 237
Bern, 199 *n.*, 203 *n.*
Besançon, 201
Besseler, on formation of private land-ownership, 125
Bink, G. L., on New Guinea Papuas, 93
Bird-mountains, 34
Birds ; breeding associations, 32–35 ; autumn societies of, 36 ; migrations, 36–38
Blanchard, on Insect metamorphoses, 12
Blavignac, J. D., on labour in Fribourg, 194 *n.*
Bleck, W., on Bushmen, 89
Blood covenant, 318, 319, 324
Blood revenge—a conception of justice, 107 ; its survival amongst ourselves, 108 ; a tribal affair, 108 ; Ad. Bastian on, 108 *n.* ; "head-hunting," 109 ; with the barbarians, 132 *seq.*, 173
Boars, societies of, 50
Bock, Carl, on "head-hunting" among Dayaks, 109 ; grossly exaggerated, 109 *n.*
Bogisic, on joint family with the Serbs and the Croates, 321
Bogos, 148 *n.*
Bohemia, cities, 166, 210
Boileau, "Livre des métiers," 325
Bolivia, 60
Bologna, 203 *n.*, 205
Bonnemère, village institutions in France, 122, 232 *n.*
Borneo, 52
Botta and Leo : early accumulations of wealth, 157 ; Lombardian code, 161, 162, 180, 189 *n.*
"Bratskiye," 140
Braunschweig, 199 *n.*
Brazil, ants, 13 ; falcons, 22, 309 ; natives, 84 ; common culture, 127
Brehm, A., xi, 22, 23, 25, 26, 27, 28, 33, 36, 46 ; on fight of hamadryas against his caravan, 52, 56 ; on sociable life of monkeys, 80 *n.*
Brehon laws, 135
Bremen, 168, 213
Brentano, L., on trade-unions, 199 ; struggle within cities, 218 *n.*

Brescia, 205
Breslau, 210 ; bell-tower, 213
Brighton Aquarium, 11
Bristol miners, 269
Brittany, common culture, 127
Bruges, 168, 198
Buchenberger, A., on destruction of village community in Belgium, 235 ; on agricultural co-operation in Germany, 248, 249.
Bücher, K., addenda to Laveleye's *Primitive Property*, 122, 239 *n.* ; agricultural co-operation in Germany, 248
Büchner, Dr. Louis, xii, xviii ; on animal intelligence, 7 *n.* ; "Love," 7, 12, 41 ; on compassion among animals, 59
Budding of new communities, 129
Buffon on rabbits, 46
Bulgares, 254
*Buphagus.* *See* Sea-hen.
Burchell, 89
Burgdorf, 203 *n.*
*Bürgernutzen*, 239.
Burghers, struggle against feudalism, 200
Burgundy, 217
Burial, private property destroyed at, 320
*Burrichter*, 180
Buryates : joint families, 138 ; common meals, 138 ; confederations, 139 ; brotherly habits, 139 ; common hunts, 139
Bushmen, 84, 88, 90
Butterflies, swarms of, 301
Buxbaum, L., 37
Buzzards attacked by lapwings, 25
Byelaeff, Prof., Russian History, 162, 166 *n.*, 181, 189 *n.*

Cæsar, Julius, 126
Cæsarism, development of, 216 *seq.*, 224
Calonne, A. de, on communal purchases, 182 *n.*, 183, 194 *n.*
Cambrai, 200 *n.*
Canada, musk-rats, 44
Cannibalism : discussion at Paris Anthropological Society, 105 ; probably originated during Glacial period, 106 ; religious character of, in Fiji and Mexico, 107

Capponi, Gino, history of Florence, 198, 214 n.

Caprides, sociability of, 48

Capuchins, 51

Carnivores, sociability among, 40

Carpes, 136

Casalis, on common law of Bassoutos, 148

Caspian Sea, previous extension, 118 n.

Cassiques, mocking eagles, 26

Cathedrals, mediæval, 210

Caucasian Mountaineers, common culture, 127, 146, 147; growth of feudalism, 146; joint stock feudalism, 147; criminal law, 147; folk tribunals, 148; "aids" in villages, 244 n.

Celt-Iberians, 126

Celts, 87, 119

Central America, common culture, 129

Central Asia, herds of mammals, 39; dessication of, 118 n., 119

Centralization, growth of ideas of, 217 seq.

Centralization in France, 233

Ceylon, 50

Chakars, singing in concert, 56

Chambers' Encyclopædia, 167 n.

Charitable associations, 282.

Charities, 291

Charroi, 242

Checks, natural, to over multiplication, 70 seq.; mice, 307, 308; coypù, 308

Chernigov, village community in, 253

Cherusques, 136

Children, mutual support amongst, 285; purchase of, for factories in England, 290

Chinese, common hunts, 141

Chukchis, 91

Church, Christian, 125; and kings, 161; and Emperor in Italy, 204; favours Cæsarism, 217; studies of Roman law, 220; revolt against the Catholic Church, 224

Cibrario, L., mediæval economics in Italy, 183; on slavery and serfdom, 219

Civets, 40

Clan, its organization with primitive men, 78–88; opposed to

other clans, 112; dissociated by migrations, 119; App. VII, 313

Clan-marriage, 86; with Semites, Aryans, Australians, Red Indians, Eskimos, etc., 87; Appendix VII, 313

"Classes," marriage-, among savages, 316; age-classes and guilds, 325

Clements, Dmitri, on Lukchun antiquities, 119

Cliff swallows, 35

Clode, Ch. M., on Guild of Merchant Taylors, 175 n., 183 n.

"Clubs" of savages, 325

Cockroaches, one species driving another, 61

Code Napoléon, 198.

"Çof" of Kabyles, 145, 146, 171

Collegia, 169, 324, 323

Collins, Col., 89

Cologne, 167, 171 n.; neighbour guilds, 180; guilds, 198, 205, 206; cathedral, how it was built, 212

"Colonies Animales," 53

Colonization, by village communities, 130; by mediæval cities, 219

Colonna, 219

Colorado, 35

Combination Laws repealed, 266

Com-bourgeois, 202

Communal culture, modern, in Ariège, 247; in Westphalia, 248; in Kursk, 256

Communal hay-mowing, 128

Communal lands in France, 232 and note, 241, 242

Communal meals, 128

Commune of Laonnais, 207

Communes of France, 232

Compayne, 193

Compensation for murder, 133, 134; for stealing, 156

Competition in Nature, theory of, analyzed, 60–75; Darwin's arguments to prove it, 61; indirect argument in favour of it, 63 seq.; natural checks to it, 70; is it an element of progressive evolution, 70; adaptations to avoid it, 74, 309–312

Compiègne, 177

Conclusion, 293–300

Congresses, mediæval, of working-men, 196
Conrad, 205
Constance, 206
Constantinople, 217
Consular Reports (British), 252
Convention, French, edicts to destroy village community, 231; law against coalitions of workers, 265
*Convivii*, 175
Co-operation, among peasants, in Switzerland, 241; in France, 246 *seq.*; in Germany, 248; in Russia, 250–258; creameries in West Siberia, 258; in Britain, 271; 'on the continent, 272; in Russia, 272–274
Copernicus, 215
Cornelissen, on mutual support in Dutch villages, 327
Cornelius, on Münster insurrection, 225
Cornish, C. J., on animals at play, 307
Corporations of France, 195 *n.*
Couës, Dr. E., on Birds of Kerguelen Island, 25; cliff-swallows, and falcon, 35; birds of Dacota, 36, 305
Coulanges, Fustel de, 121, 313
Country life in England, 237 *n.*
Cources, village community of, 122
Cranes, sociability of, 27
Crema, 205
Crofters' Commission, 237
Cross, market-, 190
Cunow, Heinrich, village community in Peru, 127
Cyclists' Alliance, 279
"Cyvar," 127

Dacota, 36
Daghestan, feudal relations in, 146
Dahn, F., old Teutonic institutions, 122 *n.*; early accumulation of wealth, 157; old Teutonic law, 164
Dall, on Aleoutes, 97
Dalloz, on communal lands in France, 232
Dancing among birds, 55, 56
Danish co-operators in Siberia, 258
Danish guild, old, 171; Pappenheim on, 175 *n.*

Dante, 215
Dareste, 232 *n.*
Dargun, L., on altruism in economics, 284 *n.*
Darwin, Charles, on struggle for life, vii, 1; on same subject, in *Descent of Man*, x, 2; Bates on his ideas, xiv; Malthusian influence, 3; his followers, 4, 6, 9; hunting associations of kites, 21; fight of hamadryas, 52; dancing among birds, 55; features useful in struggle for life, 57; compassion among pelicans, 59; struggle for life and competition, analysis of this theory, 60–75; arguments of Darwin in favour of, 61; metaphoric sense of "extermination" more probable, 63; Malthus' "arithmetical argument," 68; over-population and natural checks to, 68–72; how animals avoid competition, 72–75; misuse made of his terminology, 78; Man originated from a sociable species, 79; on Man's sociable qualities as a factor of evolution, 110
Darwin, Dr. Erasmus, on moulting crabs, 12
Darwinism and Sociology, ix; Bates on Darwinism, xiv
Darwinists, vii; Russian, 9
Dasent, George, Burnt Njal saga, 135
Dayaks, their habits, 91; their conception of justice, 109; exaggerations of recent writers, 109, 110
Death-sentences amongst moderns, 107 *n.*
Decay of mediæval cities, causes of, 215 *seq.*
Deer. *See* Fallow Deer.
*Defensor* of the city, 188
"Degenerated" tribes, 83
Dellys, 144
Demidoff, A., 23
Dendrocolaptidæ, 312
Denmark shell-heaps, 80
Denton, Rev., on mediæval Scotland, 210 *n.*
Desiccation, a cause of migrations, 118; of post-pliocene lakes, 118 *n.*
Desmichels, 156
*Dessa*, 122

Destruction of animal life by natural agencies, vii
De Stuers, on Malayan village community, 150
Dhole dogs, packs of, 41
Diodorus, 126
Dixon, Ch., flights of birds for pleasure, 23; bird-mountains, 34; gatherings before migrating, 38; destruction of bird-life by cold, 72; on aquatic birds' associations, 305; on lapwings, 310
Djemmâa, 142, 147, 244, 259
Dock-labourers' strike, 269
Doniol, on village institutions in France, 122, 232 n.
D'Orbigny, 5; on kites, 21
Dordogne, palæolithic relics in, 80
Doren, A., on merchant guild, 191 n.
Dousse-alin, 48, 49
Dragon-flies, migrations, 302
Drought, effects of, 47
Drummond, H., xviii
Druzhestva, 169
Dunlins, 23
Durkheim, Prof., on human marriage, 314, 319

Ebrard, on ants, 12
Eckermann, Gespräche, xi
Eckert, 180 n.
Edward III., 219
Edward VI., confiscates estates of guilds, 264
Efimenko, Mme., on village community in Russia, 123 n.
Eghiazarov, S., on Georgian guilds, 170, 274 n.
Egypt, 325
Eichhorn, 165 n.
Elephants, societies of, 50
Elizabeth, Queen, statute of, for regulating wages, 265
Elphinstone, village community of the Afghans, 122
Emperor and Church in Italy, 204
Emprount, 243
Enclosure Acts, 234
England, village community in, 121; mediæval, 182 n.; destruction of village community, 233 seq.; present survivals of village community, 236 seq.
Ennen, Dr. Leonard, Cologne Ca-

thedral, 171; Cologne, 180 n., 212 n.
Ennett, J. T., mediæval art in small parishes, 213
Equality, institutions for maintaining it, 113
Equus Przewalski, 47, 67
Erskine, on self-sacrifice of old relatives, 104
Eskimos, their institutions, 84, 91, 94; their nearest congeners, their habits, 95 seq.
Esnafs or esnaifs, 169, 274
Espinas, on animal societies, xii, 7, 53
Europe, Northern, 32
Evolution, progressive: mutual aid its main factor, x, 7–9; is it fostered by competition, 73
Exchange artél, 273

Fabre, J. B., on insects, 12
Fagniez, on mediæval industry at Paris, 182 n., 196 n.
Falcon, prairie, 35
Falcon, red-throated, in bands, 22; in South Russian Steppes, 22
Falke, Joh., on mediæval conditions of labour, 194; on Hansa, 205
Falkenau, strikes, 268
Fallow deer, migrations, 8, 9, 48, 49.
Falsifications in agriculture, 246 n.
"Families, The," 191, 218
Family, paternal, amidst the clan, 113
Family, tribal origin of, 78–88, Appendix VII, 313–320
Farne Islands, 305
Federalism, principles of, 220
Federations of barbarian stems, 136; of cities, 204 seq. Also see Leagues
Fée, 6
Ferdinand I., 195
Ferrari, on Italian cities, 167, 168 n., 189 n.; on wars between them, 204
Feudalism, growing in Caucasia, 146; joint stock, 147; with Malayans, 149, 166
Feuds amongst savages, 94, 108, 109
Fielden, H. W., on musk-ox, 49
Fiji, religious cannibalism in, 195

Filial love with savages, 101 *n.*
Finns, village community of, 122
Finsch, O., on New Guinea, 93 *n.* ; on Hyperboreans, 100 *n.*
Fishing by pelicans, 23 ; co-operative in Russia, 273
Fison, L., and A. W. Howitt, on tribal origin of family, 85, 92
Flanders, 187
Flemish cities, 213
Florence, revolution of minor arts, 198 and *n.* ; fights against landlords, 202, 203 *n.* ; head of a league of cities, 205 ; league of villages in its *contado*, 206 ; flourishing state of country dependent upon it, 210 *n.* ; words of its Council, 213 ; its schools and hospitals, 214 *n.* ; its fifteenth-century revolution, 221
Folkmote, its attributions in villages, 121 ; judicial functions of, 131 ; extent of jurisdiction, 132 ; supreme in mediæval cities, 160 ; jurisdiction retained in feudal times, 164, 165 ; electing the *defensor*, 166, 167 ; in London, 167 *n.* ; its abolition, 226 ; Gomme on its functions, 237 *n.*
Food shared in common by savages, 112 ; by Hottentots, 112 *n.*
Forbes, James, on feeling of sympathy in monkeys, 51
Forel, Prof., on ants, 12, 13, 14, 15 ; on federations of ants' nests, 18
Forts, 165
Four-fields' system in village communities, 258
Foxes, hunting packs of, 41 ; polar, 41 ; gregarious, 307
France. *See* Village Community, Guilds, Mediæval Cities.
Franche-Comté, 201
Franconian period, 180
Franks, 125 *n.* ; common culture, 126
*Fred*, paid to village community, 132 ; origin, 133 ; in later periods, 158–161
Freemasonry, 282
Fribourg, 194 *n.*
Fritsch, on Bushmen, 89
Froebel Unions, 281
*Fruitières*, 247

Fuegians, part of savage belt, 84 ; in recent descriptions, 95
Fuego, Terra di, shell-heaps, 82
Fustel de Coulanges, on village community, 121, 313

Galicia, towns, 210
Galileo, 215
*Gau*, 123 *n.*
Gazelle, 48
*Geburschaften*, 180
Geddes, Prof. P., on Malthus argument, 68
Geelwink Bay, Papuas, 93
Geneva, 197
Genoa, 205, 213
*Gens*, gentile organization, 85. *See* Clan, Savages.
Georgians, 147
German Expedition on Eskimos, 96
Germans of Tacitus, 87
Gerona, 210
*Geselle*, 193–195
Geszow, I. E., on joint family in Bulgaria, 320
Ghent, 168, 246 *n.*
Gibelins, 204
Giddings, Prof. F. A., xviii
Gill, on New Hebrides savages, 101
Giraud Teulon, on tribal origin of family, 85
Gironnais, St., syndicate of, 247 *n.*
Giry, on Rouen commune, 201
Glaber, Raoul, 168
Glacial period, 81 ; time of probable origin of cannibalism, 106
Glarus, Alpine meadows, 240
Gleditsch, 10
"God's Peace," 167, 168 *n.*
Goethe, on Mutual Aid, xi
Gomme, G. L., on folkmote in London, 167 *n.* ; modern survivals of village community, 236, 237
Gorilla, a decaying species, 52
Gothic architecture, 178, 181, 210 *seq.*
Gramich, W., on mediæval Würzburg, 181, 182, 193
Grasshoppers, gregarious, 302
Great Inquest, 234
Greece, 165 *n.* ; antique cities of, 162, 169. 219 ; guilds of ancient Greece, 321

Greek art, 211, 213
Greeks of Homer, 87
Green, J. R., early accumulation of wealth, 157; on frith guilds, 175 *n.*; early London, 180, 189 *n.*; cities and country, 219
Green, Mrs., on mediæval cities in England, 162; on guilds, 175 *n.*; on communal purchases, 191 *n.*; labour and craft guilds, 199 *n.*
Greenland, ice age in, 84
*Greve*, 180
Grey, Adm., on Australians, 92; savage conception of justice, 112 *n.*
Groot, J. M. de, on religious systems in China, 320
Gross, Carl, *Play of Animals*, 54, 307
Gross, Ch., on guild merchant, 183; on communal purchases, 184, 185, 191; struggles between guilds, 199 *n.*
Guelfs, 204
Guilbert de Nogent, 178
Guilds, their universality, 169; their character on board ship, 170; for building, 171; Danish *skraa*, 171; obligations of guild brothers, 172; of serfs, beggars, teachers, etc., 173, 174; common meal, 175; merchant, 175 *n.*; frith guilds, *id.*; federation of, in the city, 176; its sovereignty, 181; sale of products and buying of necessaries, 181-183; merchant guilds, 184-6; organization of work, 191-194; hours of labour, 195; their own militia, 197; union of, symbolized in cathedrals, 212; donations, 212; spoliated by State, 226; estates confiscated by Henry VIII. and Edward VI., 263, 264; State legislation instead of self-jurisdiction, 264; wages, 265; guilds and trade unions, 266 *seq.*; origin of, Appendix X, 321 *seq.*; in old Rome, 321; with the Normans and the Slavonians, 322; in old Greece, 323; in the East, 323; modelled upon the clan, 323; in old France, 324; relation to "age classes" and secret societies of early barbarians, 325

Guizot, on early accumulation of wealth, 157
Gurney, G. H., on house-sparrow, 24
Gutteridge, Joseph, on artisan life, 287
Gymnasts' societies, 279

Hamburg, 199 *n.*
Hanoteau, on Kabyles, 141-145
Hansa, ship guild, 170, 189 *n.*; labour congresses in Hanseatic towns, 196 and *n.*; league, 205; Flemish, North German, 208
Hares, 45; sociable, 305
Harvest supper, 128
Hawk chased by sparrows, 26
Haygarth, on cattle in Australia, 59
Hearne, on musk-ox, 49
Heath, Richard, on Anabaptism, 225, 226
Hedin, Sven, on Lob-nor, 119 *n.*
Hegel, Carl, history of German mediæval cities, 189; their origin, 326
*Heimschaften*, 180
Heinrich V., 201
Henry VII., 234
Henry VIII., enclosures, 234; ruins the guilds, 263
Heribert, St., 167
*Hetairiai*, 323, 324
Himalaya natives, 84
Hippopotamus, societies of, 50
*Hirdmen*, 156
Historical documents, chiefly relating struggles, 116
History, begins several times anew with the tribe, 117
Hobbes, xv; war of each against all, 77; his followers, 78; his main error, 78
Hodder, Edwin, *Life of Seventh Earl of Shaftesbury*, 288, 290
Hohenzollern, 247, 248
Holm, Capt., on Greenland Eskimos, 96
Horses, 46; half wild in Asia, 46; effects of droughts upon, 47; wild in Tibet, 47; origin of, 66; after a drought, 73
Hottentots, 90, 91, 228
Houzeau, on animal sociability, 6; prairie-wolves, 41; sociability diminishes in decaying species, 53

Howitt, A. W., Australians, 85, 92
Huber, Pierre, on ants, 12, 13, 14, 15; their play, 55; Mr. Sutherland's appreciation of, 303
Hudson, W. H., on viscachas, 46; on pigs, 50; on music and dancing in Nature, 54, 55, 56, 302; want of animal population in South America, 309; adaptation to avoid competition, 309-312
Hugues, Archbishop, 201
Humber district, birds in, 23
Humboldt, Alexander, on tee-tees, 51
Huns, 136
Hunting associations : of male and female, 19; among eagles, 20; of kites, 21; of pelicans, 23; lions, 40; in dogs' tribe, 40; of wolves, 40; prairie-wolves, 41; foxes, 41; hyenas, 41
Hunting in common, 141
Hussite wars, 219
Hutchinson, H. N., on marriage customs, 319
Hüter, E., on foxes, 41
Huxley, on struggle for life, xiv, 4, 5; origin of society, 54; on Hobbesian war, 78

Ice age, extension of ice cap, 81
Iceland, *Allthing*, 158
Ihering, Dr., on importance of free mutual support, 284 *n.*
Ile de France, 217
Inama-Sternegg, on formation of private property in land, 125, 157
India, 26, 29; village community in, 122, 123; guilds in ancient India, 169
Indians of Vancouver, 100 *n.*
Individualism preached in modern society, 228
Infanticide with the savages, 101
Innes, Cosmo, on mediæval Scotland 210 *n.*
Innocent III., 220
Insectivores, associations, 42, Appendix IV, 306
Intellectual development due to sociability, 27
International law, 137
Inter-tribal relations, 113
Inventions, mediæval, 214

Ipswich merchant guild, 185
Ireland, village community in, 121
Iron, cost of, in early mediæval times, 156
Irrigation, co-operative, in France, 246
Isolation of species, 65
Italian art, 174
Italian cities, 178; struggles against nobles, 202; slavery maintained, 203
Italian language, 215
Ivanisheff, Prof., on village community in Russia, 123 *n.*
*Izvestia* of Russian Geographica Society, 95

Jackals, hunting associations of, 41
Jackdaws chasing kites, 25
Jacobsen on Bering Strait Eskimos, 98 *n.*
Jacqueries, 219
*Jagdschutzverein*, 280
Janssen, history of Germany, 122 *n.*, 162, 170 *n.*, 189 *n.*, 193, 194 *n.*, 195; 225
Jerdon, Dr., on ants, 14 *n.*; jackdaws and kites, 26
Jobbé-Duval, village community in Annam, 127
Joint family as a phasis of civilization, 123; with the Ossetes, 146; with the South Slavonians, 320, 321
Joint household. *See* Joint Family.
"Joint team" in Wales, 127; in Caucasia, 147 *n.*
Judge in mediæval times, 175 *n.*
Justice, sense of, developed by sociability, 59

Kabardia, 134 *n.*
Kabyles, village community with, 122 *seq.*; their institutions, 141-149; return to tribal law, 142; *djemmâa*, 142; work in common, 143; rich and poor, 143; aid in travels, 144; feeding destitutes, 144; *anaya* custom, 145; the *çof*, 145, 160
*Kada*, common-hunt, 141
Kafir laws, 148 *n.*
Kaimani Bay, Papuas, 93 *n.*
Kallsen, Dr. Otto, on German mediæval cities, 167, 189 *n.*, 190

*n.*; on mortmain, 201; cities the makers of national unity, 206 *n.*

Kalmucks, customary law, 131

Kamilaroi-speaking Australians, 85, 86

Kamtchatka bears, 42

Karl the Great, 164

Kaufmann, early significance of "King," 161, 162

Kautsky, K., on sixteenth-century communism, 225

Kavelin, 123 *n.*

*Kegelbrüder*, 279

Keller, on Anabaptism, 225

Kerguelen Island, 25

Kessler, Prof., x; lecture on the law of Mutual Aid, 6–8; our imperfect knowledge of mammals, 19; Societies of youngsters, 307

*Kharouba*, 142

Khevsoures, return to tribal law, 146; women stopping quarrels, 147

Khingan, Great, Little, 48, 49

Khoudadoff, N., on common culture, 127; on Khevsoure common law, 146

*Kihlakunta*, 122

Kilkenny ordinance, 184

King, double origin of authority of the, 157, 158; duke equal to, 160; early meaning of the *kong*, 161; Canute, 161; compensation for a slain king, 161

Kingsley, Mary, on the Fans, 109 *n.*

Kinship systems, 85

Kirk, T. W., on house-sparrows, 26

Kites, sociability, 21; attacking eagles, 25; chasing the hawk, 25

Klaus, on village community in Russia, 123 *n.*

Kluckohn, on God's Peace, 168

Knights of Labour, 268

Knowles, James, xiv, xix

*Knyaz*, 157, 159

Kohl, horses against wolves, 41

Kolben, P., on Hottentots, 90, 91.

Koloshes, 95

Königswarter, on primitive justice, 131; on compensation, on *fred*, 133

*Konung, kong*, 161

Koskinen, early institutions of Fins, 122 *n.*

Kostomaroff, Prof. N., early Russian history, 162; origin of autocracy, 166; twelfth-century Rationalism, 169; free cities, 189 *n.*

*Kota*, 122, 150

Kovalevsky, Prof. Maxim, on tribal origin of family, 79, 85; on primitive law, 85; on origin of family, 87; on village community in Britain, 121 *n.*; in Russia, 123 *n.*; evolution of family and property, 123 *n.*, 125 *n.*; Ossetes' hay-stacks, 129; compensation laws, 134; origins of feudalism, 146; on cities of Bohemia, 166; on Russian feudalism, 167 *n.*; tribal marriage, 313, 315

Kozloff, P., fight with monkeys in Tibet, 52

Kudinsk Steppe. *See* Buryates

Kulischer, on primitive trade, 190 *n.*

Kursk, communal culture, 256

Kuttenberg ordinance, 195

*La Borne*, 245

Labour, conditions of, in free cities, 193–196; mediæval congresses of, 196

Labourers, obstacles to their combinations, 263; wages settled by State, 265; Combination Laws repealed in 1825, 266; Robert Owen's Trades' Union, 266; prosecutions, 266, 267; modern unions, 267; strikes, 267–269; part taken in political agitation, 270; in socialist work, 270, 271; co-operation, 271 *seq.*

Lake-dwellings, 83

Lakes, bird-nesting on their shores, 32

Lamarckians, 65

Lambert, Rev. J. M., on guild life, 169 *n.*

Lamprecht, on Franconian law and economics, 156; on mediæval economics in Germany, 210 *n.*

Lanessan, J. L., lecture on Mutual Aid, xii, 7

Langobard institutions, 122 *n.*

Laon, commune of, 206; cathedral, 213

Laonnais, federation of villages, 206

La Plata, 309
Lapwings, attack of a buzzard, 25 ; their dances, 56 ; adaptations to varied food, 310
Laudes, on village community in Annam, 127
Laveleye, "Primitive Property," 122, 239 n., 250 n.
Law, customary, kept in certain families, 157–159 ; recited at All-things, 158
Law, Sir Hugh, on Dayaks, 109 n.
Lawyers, their influence, 219 ; of Bologna, 220 n.
Leagues, of towns, 204 ; of villages, 206
Lebret, on mediæval Venice, 180
Lendenfeld, R., on cacadoos, 29
Leo and Botta, 161, 162, 180, 189 n.
Lesholztag, 248
Letourneau, on common culture, 127
Letourneux, on Kabyles, 141–145
Le Vaillant, 5, 22
Lezghines, having joint feudal rights, 146, 147
Lichtenstein, journeys in South Africa, 89
Lifeboat Association, 275
Limulus, 11
Lincecum, Dr., on harvesting ants, 14 n.
Linden, Herman van den, on merchant guilds in Netherlands, 326
Linlithgow, 183 n.
Linnæus, on aphides and ants, 14
Lipari Islands, 126
Little Russians, village community with the, 253
Lives, village community of, 122
Loans in mediæval cities, 220
Lombardian law, 161
Lombardian League, 205
Lombardy, struggle against nobles, 202 ; canals, 213
London, communal purchases in, 182 n., 183–186
Long houses (balai), 94 ; of Eskimos, 96 ; 316
Lorris, commune, 178
Louis le Gros, 168 n.
Louis XIV., 230
Love and sociability, xiii
Lövsögmathr, 158
Lozère, 245

Lubbock, Sir John, on ants, bees, and wasps, 12 ; palæolithic men, 80, 81 ; Denmark shell-heaps, 82 ; neolithic men—not degenerated specimens of mankind, 83 ; on tribal origin of family, 79, 85 ; on Hottentots, 90 ; tribal marriage, 313–320
Lübeck, 199 n.
Lucca, 203 n., 205
Luchaire, A., on French mediæval cities and guilds, 162, 168, 177, 189 n., 201 ; on village leagues, 206, 207
Lukchun depression, antiquities of, 119
Lumholtz, on North Queensland natives, 92
Luro, village community in Annam, 127
Lutchitzky, Prof., on village community, 122 n., 203 n. ; on slavery in Florence, 219 n.
Lutetia, 324
Luxembourg, Jardin du, sparrows, 24
Lyons, unsuccessful revolution of minor crafts, 199 n. ; duration of struggles for emancipation, 200 n.

MacCook, on ants, 14 n. ; nations of ants, 18
Maclean, on common law of Kafirs, 148
MacLennan, J. F., on tribal origin of family, 79 ; Studies in Ancient History, 85, 313–315
Madrid, 216
Maeterlinck, on bees, 304
Maine, Sir Henry, on primitive institutions, 79 ; on village community in Britain, 122 ; in India, 123 n., 130 ; on common law, 132 ; origin of international law, 137 ; village community, 157 ; on communal lands, 237
Mainz, 205, 207 n.
Malayans, common culture, 129
Mammals, prevalence of sociable species, 38
Manchuria, vii
Manitoba, 130
Marin, on mediæval Venice, 180
Market in mediæval city, Appendix XI, 326 ; its protection, 189 seq.

Markoff, E., on Shakhseven common law, 134 *n.*

Marmots, 43, 44

Marriage-institutions of savages, 85 *seq.* ; institutions of Eskimos, 96 ; exchange of wives, 96 ; "communal," 313-320 ; compound, 318 ; restrictions abolished during festivals, 318 ; solemnities, 319

Marshall, on communal lands, 237

Martial, L. F., on Cape Horn aborigines, 95

Master and Servant Act, 267

Maurer, on village community, 122 *n.*, 157, 313 ; common culture, 126 ; communal jurisdiction, 132 ; supremacy of folkmote, 165 ; on evolution of village community into city, 165 *n.*; mediæval cities, 189 *n.*

Maures, invasion, 217

Maynoff, on Mordovian common law, 135, 136

Mediæval cities: uprise in the tenth to twelfth centuries, 163 ; its unanimity, 163 ; co-jurations, 163 ; double origin, 164 ; folkmote and *defensor*, 166 ; "God's Peace," 167 ; foundations of commercial and international law, 168 ; fine monuments, 168 ; the guilds, 169 ; their origin, 170 ; their functions, 171 ; their diversity, 173 ; secondary importance of yearly festival, 174, 175 ; federation of parishes and guilds in the city, 176 ; extension of the revolt all over Europe, 178 ; self-jurisdiction, 179 ; sovereignty, 179 ; labour, position of, 181 ; communal buying of necessaries of life, 182 ; for the guilds, 185 ; variety in, 187 ; the market, 189 ; growth of the merchant oligarchy, 190 ; conceptions about honesty in work, 191 ; master and companion, 193 ; wages of the latter, 193 ; hours of labour, 194 ; eight hours' day, 193 ; guild and town militia, 197 ; "minor arts," 198 ; battles fought, 199 ; wars against feudal barons, 200; the surrounding peasants, 202 ; leagues of cities, 204, 205 ; unions of villages, 206 ; commercial treaties, 207 ; results achieved in, 209 ; prosperity, 209 ; French, German, Italian, Russian (literature), 189 *n.* ; as arbiters, 207 ; architecture, 209 ; city buildings, 212 ; growth of arts and industries, 213 ; progress in science, 214 ; causes of decay, 215 ; ideas of sanctity of kings spread by Church and lawyers, 217 ; city oligarchy, 217 ; city and village, 218 ; principles of centralization, 220 ; influence of Church and Roman law, 220 ; example of Florence, 221

Mediterranean, 37

Medley, Mr., 91

Meitzen, on Swiss communes, 239 *n.*

Mennonites, village community, 255

*Mercati personali,* 192

Merchant guild, 183, 185, 191 *n.* ; H. van den Linden on, 327

Merghen, 48

Merovingian France, 137

Mexico, religious cannibalism in, 105 ; common culture, 129

Miaskowski, on struggle within communes, 218 *n.*, 239

Mice, destruction by changes of weather, 71

Michel Angelo, 213

Michelet, 156, 162

Middendorff, A. Th., 100 *n.*

Middle Russia, village-community movement, 254

Migrations: fallow deer on the Amur, viii ; of birds, 32-38 ; of nations, causes of, 118

Miklukho-Maclay, on Papuas, 94, 95 ; savages sharing food, 112 ; savage "classes," 316 ; clubs, 324

Milan, 168, 204 *n.*

Miler, Ernest, on South Slavonian joint family, 320

Milgaard shell-heaps, 82

Miller, Prof. Orest, on common law of Caucasian mountaineers, 134

Miners, mediæval, 195 ; Radstock, 269 ; Bristol, Yorkshire, 269 ; Rhonda Valley, 276 ; support of orphans, 288

*Minne,* 169

*Mir*, 126

Missour, 305

Moeller, Alfred, on ants' gardens, 14 *n.*

Moffat, 89 ; on self-sacrifice of old relatives with the savages, 104

Moggridge, J. T., on harvesting ants and trapdoor spiders, 12, 14 *n.*

Molucca crab, endeavours to lift a comrade, 11

Mongolia, 46, 119

Mongols, 5 ; village community, 122, 138 ; aid to traveller obligatory, 145, 160 ; invasion, 217

Monkeys, sociability of, 50–52 ; fight of hamadryas against Brehm, 52 ; against the Kozloff expedition in Tibet, 52 ; family habits of, 315

Montana, 36

Montaugé, Théron de, 231

Montbéliard, 201

Montrozier, on common culture, 127

Moodie, 91

Moors, English, 72

Moral feeling developed in animals by sociability, 58, 59

Morality, Aleoute code of, 99

Moravia, communities, 225

Morbihan, 127

Mordovians, common law, 135, 136 ; aids, 143

Morgan, Lewis H., on tribal origin of family, 79 ; *Ancient Society*, 85 ; on "Hawaian" group-system, 88 *n.*, 123 *n.* ; tribal marriage, 313–316

Morocco, 144

Mortmain, 201

*Moscou, Bulletin des Naturalistes de*, 72

Moscow, 167 *n.*, 216, 217, 257, 304

*Motacilla alba.* See Wagtails

Mother's clan, child belonging to, 318

Mothers, mutual support amongst, 285

Mount Tendre, 18

Mugan Steppe. 134 *n.*

Müller and Temminch, on Dayaks, 110 *n.*

Münster. 225

Münzinger, on common law of Bogos, 148

Musk-ox, 49

Musk-rats, 44

Mutual Aid : Kessler on, x ; law of, x ; Goethe on, xi ; works on, xii ; M. A. and love, xiii ; as a law of Nature, xiv ;—institutions, xv ; struggle against, xvi ; M. A. and individualism, xvii ; "M. A.," articles on, xviii ; lecture by Kessler, 6 ; lecture by Lanessan, 7 ; Büchner on, 7 ; among animals, 1–75 ; among savages, 76–114 ; among barbarians, 115–152 ; in the mediæval city, 153–222 ; amongst ourselves, 223–292 ; tendency in history, 116, 117 ; tendency developed in the mediæval city, 154 ; destruction of institutions by State, 226–229

" Mysteries " and guilds, 325

" Mystery," 192

Napoleon III., 233

Nassau, 248

Nasse, on the village community in Britain, 121 *n.* ; on communal lands, 234, 237, 313

*Nature*, quoted, 100 *n.*

*Nautœ*, guild of, 322

*Naviglio Grande*, 213 *n.*

Navy in free cities, 209

Nazaroff, on common hunts, 141

Neath. 183 *n.*

*Necrophorus*, 10

*Negaria*, 149

Negroes, common culture with, 127

Neolithic man's relics, 81, 82

Nesting associations of birds, 32–35, 304–306

Netherlands, anabaptism in, 226 ; agrarian inquest, 327 ; on mutual support in villages, 327 ; merchant guilds, 328

Nets'-king, 161

New Caledonia, common culture in, 127

New England, 130

New Guinea, 93

New Hebrides savages, 101

Newton, Prof. A., on thrushes 61 *n.*

New Zealand, 26 ; multiplication of pigs and rabbits, 68

Nitzsch, 162, 206 *n.*
Nobles and cities in Italy, 202
Nordenskjöld, A. E., on bird-mountains, 33
Nordmann, on falcons, 22
Normans, invasions of, 159, 165
Northamptonshire, 237
Notre Dame de Paris, 212
Novgorod, communal dépôts, 184; "Sovereign N." carries on commerce, 185; *povolniki*, 192; leagues, 206
Numa, 321, 324
Nuremberg, 210, 214 *n.*; learning and technical skill, 215
Nys, Prof. E., on military executions, 108 *n.*; on Old Irish law, 135; on origin of international law, 137

O'Brien, on Swiss villages, 240 *n.*
Ochenkowski, on mediæval England, 182; communal lands, 234, 264 *n.*
Old relatives, self-sacrifice of, 103; Moffat and Erskine on it, 104
"Old Transformist." *See* Tchernyshevsky
Orang-utan, a decaying species, 52
Orkhon, inscriptions, 119
Ornithological Society, 280
Orsini, The, 219
Ory, village community in Annam, 127
Ossetes, hay-stacks free in spring, 129; compensation laws of, 133; common law of, 146
Ostyaks, 91; mild character, 100 *n.*
Oucagas, The, 149
*Oulous* of Buryates, 122, 138, 139, 140
Outlaws, 131
Over-population, animal, not proved, 68; natural checks to, 68-72, Appendix V, 307. *Also see* Checks
Overstolzes, The, 219
Ovides family, sociability of, 48
Owen, Robert, Trades' Union, 266
Oxfordshire, 237

Pacific Islands, 94, 95
Padua, 174 *n.*, 205
Pa frey, on village communities in New England, 130
Painters, guilds of, 174

Pappenheim on Danish guilds, 175 *n.*
Papuas, 84, 91; description by G. Bink, 93, 94; by M. Maclay, 94, 95
Parental love with savages, 101 *n.*
Parentship relations among savages, 317
Paris, 167 *n.*; mediæval conditions of labour, 195 *n.*; guilds, 198; Notre Dame, 212; a royal city, 216; early guilds, 322; mediæval guilds, 324
Paris Exhibition, bees at, 17
"Parricide," supposed, among savages, 103
Parrots, sociability of, 27-31; with jays and crows, 29; vigilance, 29; high intelligence, 30; mutual attachment, 30
Patagonia, 84
Pavloff (Pawlow), Marie, on origin of modern horse, 67
Peasant War, 219, 225; massacres to stop it, 225 *n.*, 226
Pelicans, fishing associations, 23
People, constructive genius of the, 162
Periodical distributions of wealth, 97; of land, 98; remittance of debts, 98
Perrens, history of Florence, 168, 198
Perrier, Ed., on animal colonies, 53
Perty, Maximilian, 7, 24; on compassion among animals, 59
Peru, 60; village community, 127
Pfeiffer, Ida, on Dayaks, 109 *n.*, 110
Phear, Sir John, village in India, 123 *n.*
Philippe le Bel, 321
Phillip, Count of Flanders, 177
Phillips-Wolley, Clive, on big game shooting, 47
Phylloxera, 246
Piacenza, 205
Piepers, M. C., on mass-flights of butterflies, 301
Pine-moth, 71
Pisa, 168, 204 *n.*
Pistoia, 205
*Pittäyä*, 122
Plata, La, W. H. Hudson on, 54, 56

Play of animals, 307
Plimsoll, Samuel, on life of poor, 218 ; on altruism with poor and rich, 289
Plovers, ringed, 23
Plutarch on guilds, 321, 324
Polyakoff, Ivan, on struggle for life, 9 ; on gulls, 35, 47
*Polynesian Reminiscences*, 107
Polynesians, 87
Poor, The, mutual support among, 284 *seq.*
Poor and rich, 181
Population, animal, want of, 309
Porter, list of Enclosure Acts, 235
Posnikoff, Prof. A., 123 *n.*
Post, A., on tribal origin of family, 79, 85 ; on clan-marriage, 87 *n.* ; on exchange of wives, 96 *n.* ; common culture, 127 ; compensation laws of Africa, 134 ; common law of African stems, 148, 149 ; development of family rights, 149 ; on Sumatra, 150, 260 ; origin of family, 313–315
Post-glacial period, 84
Post-pliocene lakes, 118 *n.*
Powell, on the village community in Sumatra, 150
Prague, 167
Prairie-dogs, societies of, 43 ; keep sentries in Zoological Gardens, 307
Primitive men, supposed war between, 76 *seq.* ; their tribes, 78
Prisoner, escaped, self-sacrifice of, 278 *n.*
Pritchard, W. T., on Polynesian cannibalism, 107
Private property destroyed on grave in China, 320
Private property in land, 125 *n.*
Prjevalsky on sociable bears in Tibet, 306
Prussia, destruction of village community, 235, 249
Pskov, city walls, 160 ; commune of, 178 ; communal dépôts, 184 ; "Sovereign P." carries on commerce, 185 ; leagues, 206
Purchases by the guild. *See* Guilds
*Purra*, 259
Pyrenees, 23

Quades, 136
Quagga, 47

Rabbits, 46
Radstock miners, 269
Rails, their dances, 56
Rambaud, history of Russia, 167 *n.* ; on early relations between Normans and Slavonians, 322
Ranke, Leopold, on Roman law, 216 *n.*
Rationalism, twelfth century's, 169
Ratisbon, 167
Rats, mutual support, 44 ; brown and black, 62
Ravenna, 168
Reclus, Élie, on savages' reluctance towards infanticide, 102
Reclus, Elisée, on Hottentots, 90 ; on Dayaks, 110 *n.*
Redemption of land, 251 *n.*
Red Indians, 94 ; common hunts, 141
Reform, character of its beginnings, 216, 224
Rein, on village community with Finns, 122 *n.*
Renaissance, twelfth century's, 169
Republic, Third, in France, 233
Rheims, 213
Rhine, league of cities on, 205
Rhinoceros, societies of, 56
Rhonda Valley miners, 276
Rietschel, on market in mediæval city, 326
Rink, Dr. H., on Eskimos, 96, 97, 98
Riparian law, 156 *n.*
Roads built by village communities, 129
Robert, King, 201
Rocquain, F., on twelfth century Renaissance, 169
Rogers, Thorold, mediæval conditions of labour, 194, 195
Romanes, Georges, 7, 12 ; agriculture of ants, 14 *n.* ; sociable jackals, 41 ; sympathy in monkeys, 51
Roman law, its growth, 125 ; changes the sense attached to the King, 161 ; renewal of study of, 216 ; Christian Church accepting its principles, 220
Roman "municipia," 165 *n.*

Rome, Imperial, 125 *n.*; mediæval, 165; struggles against nobles, 202

Roncaglia, congress of, 220 *n.*

Ross, Denman, 121

Rossus, *Historia*, 234

Rostock, 199 *n.*

Rothari, code of, 161

Rouen, 201

Rousseau, J. J., Huxley's appreciation of, 5; on origin of society, 54, 77; idealizes savages, 111

Royal cities, 167 *n.*, 216, 324

Rudeck, Wilhelm, on marriage customs in Germany, 319

Rumohr, on proletariat in colonies of Toscana, 203 *n.*

Russia: eleventh century, 137; annals about calling Norman princes, 159, 165 *n.*; independent cities, 166; feudal period, 166 *n.*; history of, 167 *n.*; criminal law, 173; making of, by *artéls*, 174; village co-operation in, 250 *seq.*

Russian Geological Survey, 81

Russian peasants, saying of old people, 103

Sacrifices made by workers, 265–271

Sagas: on blood revenge, 133; *Story of Burnt Njal*, 135

Saint-Léon, Dr. E. Martin, history of trade unions in France, 195 *n.*; on Roman guilds, 321; on Paris guilds, 324

St. Ouen, 213

Sakhsevens, 134 *n.*

Sales by the guild. *See* Guilds

Salève mountain, 18

Salic law, 156 *n.*

Samara, 255

Samoyedes, kindness of, 91; mild character, 100 *n.*

Sanderlings, 23

Sarmates, 136

"Savage-belt," 84

Savages, xv; described as the gentlest people, 91; idealized by Rousseau, 111; identify themselves with the clan, 112

Savannahs, 34, 36

Savonarola, Gieronimo, 221

Saxon barbarian codes, 134

*Scabini*, 170, 207

Scandinavians, 119; village community of, 122

*Schaar*, 197

Schmoller, on Strassburg crafts, 199 *n.*

*Schöffen*, 180

*Scholæ* of warriors, 155–157, 159–161

Schönberg, mediæval conditions of labour, 194; craft guilds, 196

Schrenck, Leopold, 100 *n.*

Schultz, Dr. Alwin, mediæval conditions of labour, 195 *n.*

Schurtz, H., on age classes and secret societies of savages, 324

Science in free cities, 209, 214, 215

Scot, Michael, 215

Scotland, sociable weasels in, 40; village community in, 121; common culture, 126, 187; roads, 210, 217

Sea-hen (*Buphagus*) chasing gulls, 24

Secret societies among savages, 88 *n.*, 325 *seq.*

"Sections" in mediæval city, 179

Seebohm, F., on village community in Britain, 121, 122, 157; on common culture, 126; on Enclosure Acts, 234, 313

Seebohm, H., on migrations, 23; bird-mountains, 34; gatherings of birds before nest-building, 38

Self-jurisdiction, 190

Self-sacrifice, traditions among fishermen, miners, 277; of an escaped prisoner, 278 *n.*

Sémichon, L., on God's Peace, 168

Semites, primitive, 87

Senlis, 177

Serfs, their guilds, 173; revolts of, 173

Sergievitch, Prof., on folkmote and prince in Russia, 166 *n.*

Servius Tullius, 321

Seyfferlitz, 33

Shaftesbury, seventh Earl of, on flower-girls, 288 *n.*; on purchase and slaughter of children, 290

Sheffield, 72

Shell-heaps, 82

Shooting amongst moderns, 107 *n.*

Siberia, animal life in, vii; birds of, 23; animal population of, 38, 46; lakes, 118 *n.*

Sicambers, 136
Sienna, 203 *n.*
*Signoria*, 221
Silesia, village community, 235, 249
Singing in concert, of birds, 56
Sioux, 91
Sismondi, on Italian republics, 189 *n.* ; wars between cities, 204 ; agriculture in Tuscany, 210 ; Lombardy canals, their subsequent decay, 213 ; growth of royal authority, 216 *n.*
*Skraa* of Danish guild, 171
Slavery in Italian cities, 203, 219
Slavonians, primitive, 87, 119 ; village community of, 122, 123, 131 *n.*, 159 ; cities, 220
Smith, Adam, on State intervention in corporations, 197 *n.*
Smith, Miss Toulmin, on woman in guilds, 172 *n.* ; on guilds, 196 *n.*
Smith, Mr. Toulmin, English guilds, 172 *n.* ; Cambridge guilds, 175 *n.*, 196 *n.*, 199 *n.* ; confiscation of guilds' property, 263, 264
Sociability, greater in regions uninhabited by man, 20 ; with all animals before the appearance of man, 52 ; cultivated for love of society, 54 ; "joy of life," 54 ; distinctive feature of animal world, 55 ; expressed in dancing and singing, 56 ; best weapon in struggle for life, 57 ; develops moral instincts, 58 ; also sense of justice, 59 ; and sympathy, 60
Socialism, sacrifices for, 270
Societies, opposed by State, 227 ; growing now for all possible purposes, 279
Society, pre-human origin of, 54
*Sodalitia*, 323
Sohm, on Teutonic village community, 122 *n.*
Soissons, 177, 206
Sokolovsky, 123 *n.*
Soudan, village community, 122
*Souslik* of South Russia, 43 ; sudden disappearance of, 72
Spain, 36
Sparrows warning each other, 24 ; Mr. Gurney on, 24 ; chasing a hawk, 26
Speier, 205

Spencer, Herbert, on struggle for life, xv ; on animal colonies, 53 ; influence of surroundings, 65
Sproat, Gilbert, on Vancouver Indians, 98 *n.*, 100 *n.*
Squirrels, 42
Stansbury, Capt., on compassion among pelicans, 59, 60
Starcke, Prof. C. N., on primitive family, 313, 314
Starkenberg, province, 247
State, interference in corporations, 196 *seq.* ; growth of, in sixteenth century, 216 ; aided by Church, 217 ; its ideals within the cities, 218 ; its victory over the cities, 225 ; spoliation of guilds, 226 ; absorption of all their functions, 227 ; destruction of mutual-aid institutions, 228 ; interference in guilds, 264 *seq.* ; its ideals favoured, 278
Steffen, Gustaf, on mediæval conditions of labour in England, 194 *n.*
Steller on polar foxes, 51 ; on Kamchatka bears, 42
Steppes, Russian and Siberian, lakes of, 32, 47
Stieda, W., on Hansa towns, 196 *n.*, 207 *n.*
Stobbe, on "movable" property, 124
Stoltze, on Dayaks, 110
Strassburg, 199 *n.*, 205
Strikes prosecuted, 265 ; right to, slightly won in England, 267
Struggle for life, its proper sense, v ; checks to multiplication, vii ; viii ; "a law of Nature," ix ; Kessler on, x ; its philosophical importance, 1 ; metaphorical sense, 2 ; Darwin on, 2 ; Darwinists on, 4 ; Huxley on, 4, 5 ; in Nature, 5 ; Kessler on, 6 ; who are the fittest in it? 57 ; and competition, theory of, analyzed, 60–75, 307
Struggles, the part they play in history, 115 ; subject of historical documents, 116
Suabian League, 206
Sueves, 126, 136
*Suka*, 150
Sumatra, 52

Sungari river, 48
Surrey, 237
Sutherland, A., on moral instinct, xviii; appreciation of Huber's work, 303
Swallows, one species displacing another, 61
Swiss Confederation, 197, 207
Switzerland, sociable weasels, 40; lake-dwellings, 83; roads, 129, 218; village communities selling lands, 235 *n.*, 238
Syevertsoff, N., on Mutual Aid, 9; on hunting associations of white-tailed eagles, 20; nesting associations of birds, 33, 312
Sykes, Col., 14 *n.*
Sylvestre, village community in Annam, 127
Sympathies, " stratified " with the rich, 289
Sympathy, xii, xiii
*Syndicats agricoles*, 246

Tachart, 91
Taine, 231
*Taisha*, 139
Tartar villages, 147
Taurida, province, village community in, 253
Taylors, guild of the Merchant 175 *n.*
Tchany, Lake, desiccation of, 119 *n.*
Tchernyshevsky, N. G., essay on Darwinism, 74
Tchuktchis, 91; infanticide prevented, 103
Tennant, Sir E., on Ceylon, 41
Territorial union, grows up instead of bonds of common descent, 119 *seq.*; gods, 120
Terssac, M., 247 *n.*
Tessino, 213 *n.*
Teutons, The, village community, 122; common culture, 126, 131 *n.*
*Thaddart*, 122, 142
Thierry, Augustin, early sense of word "king," 161; free cities, 162, 168, 177, 188 *n.*
Thlinkets, The, 95
Threshing machines kept in common, 244
Thrushes, one species displacing another, 61
Thun, 203 *n.*

Thurso, commune of, 183; communal purchases, 184
Tibet, 47
*Tofa*, 122
Tolstoi, Lev Nikolaevich, hay-making in a Russian village, 256
Tortona, 205
Toucans, mocking-eagles, 26
Toussenel, 6
Town halls, mediæval, 210
Trade unions. *See* Labourers, Strikes
Transbaikalia, ix
Transcaspian kites, 22
Tree-creepers' family, 311, 312
Trevisa, 174 *n.*, 205
Tribal marriage, 316 *seq.*
Tribal organization of primitive men, 78–88, Appendix VII
Tribal stage, proved by an immense array of facts, 314 *seq.*
Tschudi, animal life in the Alps, 40
Tuetey, on municipalities, 201
Tungus, hunter, 48; on European morality, 105
Tunguses, 91
Tupi, The, 149
Turgot, measures against folkmotes, 121
Turkestan, East, 119
Turks, invasion, 217
Tuscany, 203 *n.*; league of, 205; agriculture, 210 *n.*
Tver, 217
Tylor, Edward H., on tribal origin of family, 79; on degeneration theory, 83

*Udyelnyi period* in Russia, 166 *n.*
Ugrians, invasion of, 165
Ulm, 206
Ulrik, St., 167
Uncle, maternal, 318
Uncle Toby's Society, 280
Under-population, 69, Appendix VII, 313
*Universitas*, 126
Universities, Italian, 215
Unterwalden, 40
Ural Altayans, 119
Ural Cossacks, 273
Uri, 197
*Urmans* of West Siberia, 84
Urubú vultures, 22

Usuri river, viii, 141
Utah, 59
Uthelred,.St., 167

Vandals, federations, 136
*Vanellus.* See Lapwing.
Variety of adaptations in one bird's family, 311, 312
Vaud, canton, 238, 239
Veniaminoff, missionary, later Metropolitan of Moscow, on the Aleoutes, 99 *seq.*; their code of morality, 99; infanticide among Tchukchis prevented by, 103
Venice, St. Marc of, 168; art, 174 *n.*, 198; distribution of provisions, 183; league, 205
*Verein für Verbrestung gemeinnützlicher Kenntnisse*, 281
Verona, 174 *n.*, 204, 205
Versailles, 197
Vicenza, 205
Vicunas, 60
Village and town, 203
Village community, worked out to resist disintegration, 120; its universal extension, 121, 122; explorers of, 121 *n.*; its different names, 122; not a servile growth but anterior to, 122; bibliography of, 122 *n.*; relation to joint family phasis, 123; common possession of land, 124; clearing of woods, 125; work in common, 126; common cultivation, 128; roads built, 129; forts, 129; budding of new villages, 130; judicial functions of folkmote, 131; of feudal lord, 132; the *fred*, 132; extent of jurisdiction, 132; "composition," 133; its amount, 133; moral principles, 134; confederations of, 136; military protection, 137; with the Buryates, 138–141; Kabyles, 141–146; mountaineers of Caucasia, 146–148; in Africa, 148; with the Brazil Tupis, 149; the Arani, 149; the Oucagas, 149; the Malayans, 149; the Alfurus, 149; the Wyandots, 150; in Sumatra, 150; universality of, 150–152; achievements, 157; independence retained in early mediæval times, 164; federation of village com-

munities in the city, 166–168; efforts of rich and State to get rid of, 229; destruction of, in France, 230–233; in England, 233–235; in Germany, Austria, Prussia, Belgium, 235; persevering till now, 236; laws and institutions derived from, in Britain, 236; in Switzerland, 238–240; in France, 241–247; in Germany, 247–249; village community in Russia, 250–259; in Turkey, Caucasia, 259; in Asia and Africa, 259, 260; recent spontaneous growth in Russia, 252 *seq.*
Village life in France, 241 *seq.*
Villages, leagues of, 206
Vinogradov, Prof., on village community in England, 121, 157; on pillage of communal lands, 234, 313
Viollet, P., on old institutions, 122
Viscacha, 45
Vitalis, 168 *n.*
Vogt, reception of the, 164; functions, 170, 180
Votkinsk iron-works, 274
Vultures, sociable, 22
V. V., on peasant community, 250 *seq.*
*Vyeche, Weich* (folkmote), 166 *n.*, 179

Wages, State regulation by, in England, 264, 265
Wagner, A., 235 *n.*
Wagner, Moritz, on isolation, 65
Wagtails chasing sparrow-hawk, 25; also fishing-hawk, 25
Waitz, 89, 91, 101 *n.*, 104; common culture, 127; Oucagas, 149; Malayans, 150
Wales, village community in, 121
Wallace, A. R., on struggle for life, 1; on orang-utans, 51; features useful in struggle for life, 57; struggle for life and competition, theory of, analyzed, 60–75; arguments of Wallace in favour of, 62, 63; metaphoric sense of "extermination" more probable, 63; migration factor, 65–67; over-population and natural checks to, 68–72; how animals

avoid competition, 72–75; on thrushes, 61 *n.*

Walt, Johan van der, 89

Walter, on village community in Wales, 122; common culture, 127

Warriors, bands of, 155–157

Warwickshire, 237

Waterford, 183 *n.*, 184, 185

Wauters, A., Belgian mediæval cities, 189 *n.*

Weasels, 40

Weather, effect on insects, on birds, 69–72

Webb, Sidney and Beatrice, *History of Trade-Unionism*, 266, 267, 268

Weddell, H. A., mutual protection among vicunas, 60

*Weichbild,* 190

Welsh, The, common culture, 126; "triads," 135

*Wergeld,* 158

Westermarck, Prof. Edward, on history of human marriage, 313 *seq.*

Westminster, 324

Westphalia, 207 *n.*; communal culture, 248

Westphalian League, 205

Whewell, on mediæval inventions, 214

White, *Natural History of Selborne,* 36 *n.*

Whitechapel, mutual support in slums, 286, 287

Wied, Prince, on eagles mocked by toucans, 26

Wilman, R., on Westphalian federations, 207

Wilmot Street, 287

Wiltshire, 237

Winchester, 167

Winckell, Dietrich de, on hares, 45; *Handbook,* 46

Wises, The, 219

Wives, exchange of, among Eskimos, 96; in Australia, 96 *n.*; 318

Wolfgang, St., 167

Woman, inferior position in clan, 319

Women, mockeries in case of small faults, with the Eskimos, 96; in the tribe, 112; educational institutions for, in Russia, 281 *n.*

Wood, J. C., on compassion among animals, 59

Woodhewers' family, 311, 312

Workers. *See* Labourers

Worms, 207 *n.*

*Wormser Zorn,* 205

Wunderer, J. D., guild on board ship, 170

Württemberg, co-operation in, 247 *n.*

Würzburg, 181, 182, 193

Xanten, labourers of, 194 *n.*

Yadrintseff, desiccation of Siberian lakes, 118 *n.*

Yenisei, 37

Yorkshire, 237; miners, strike of, 269, 289

Young, Arthur, on French agriculture, 231

Yukon river, Aleoutes, 97

*Zadruga,* 123, 320, 321

Zakataly district, 147

Zarudnyi, N., on sociability of kites, 22; of hares, 306

Zebras, 46, 47

*Zemstvos,* house-to-house inquiry, 250

*Zoologische Garten, Der,* 37

Zöpfl, on *Weichbild,* 190 *n.*

Zürich, 199 *n.*

Zwickau, 225

*Richard Clay & Sons, Limited, London and Bungay.*

Made in the USA
Lexington, KY
02 May 2010